Multiple Model Approaches
to
Modelling and Control

Multiple Model Approaches
to
Modelling and Control

EDITED BY

RODERICK MURRAY-SMITH

AND

TOR ARNE JOHANSEN

Taylor & Francis
Publishers since 1798

Published by Taylor & Francis
2 Park Square, Milton Park, Abingdon, Oxon, OX14 4RN
270 Madison Ave, New York NY 10016

Transferred to Digital Printing 2010

British Library Cataloguing in Publication Data
A catalogue record for this book is available from the British Library.

ISBN 0-7484-0595-X

Library of Congress Cataloging in Publication data are available

Cover design by Amanda Barragry
Typeset in Times 10/12pt by Digital by Design

Publisher's Note
The publisher has gone to great lengths to ensure the quality of this
reprint but points out that some imperfections in the original may be apparent.

Contents

Series Introduction

Control systems has a long and distinguished tradition stretching back to nineteenth-century dynamics and stability theory. Its establishment as a major engineering discipline in the 1950s arose, essentially, from Second World War driven work on frequency response methods by, amongst others, Nyquist, Bode and Wiener. The intervening 40 years has seen quite unparalleled developments in the underlying theory with applications ranging from the ubiquitous PID controller widely encountered in the process industries through to high-performance/fidelity controllers typical of aerospace applications. This development has been increasingly underpinned by the rapid developments in the, essentially enabling, technology of computing software and hardware.

This view of mathematically model-based systems and control as a mature discipline masks relatively new and rapid developments in the general area of robust control. Here intense research effort is being directed to the development of high-performance controllers which (at least) are robust to specified classes of plant uncertainty. One measure of this effort is the fact that, after a relatively short period of work, 'near world' tests of classes of robust controllers have been undertaken in the aerospace industry. Again, this work is supported by computing hardware and software developments, such as the toolboxes available within numerous commercially marketed controller design/simulation packages.

Recently, there has been increasing interest in the use of so-called intelligent control techniques such as fuzzy logic and neural networks. Basically, these rely on learning (in a prescribed manner) the input–output behaviour of the plant to be controlled. Already, it is clear that there is little to be gained by applying these techniques to cases where mature mathematical model-based approaches yield high-performance control. Instead, their role is (in general terms) almost certainly going to lie in areas where the processes encountered are ill-defined, complex, nonlinear, time-varying and stochastic. A detailed evaluation of their (relative) potential awaits the appearance of a rigorous supporting base (underlying theory and implementation architectures for example) the essential elements of which are beginning to appear in learned journals and conferences.

Elements of control and systems theory/engineering are increasingly finding use outside traditional numerical processing environments. One such general area in which there is increasing interest is intelligent command and control systems which are central, for example, to innovative manufacturing and the management of advanced transportation systems. Another is discrete event systems which mix numeric and logic decision making.

It was in response to these exciting new developments that the present book series of Systems and Control was conceived. It publishes high-quality research texts and reference works in the diverse areas which systems and control now includes. In addition to basic theory, experimental and/or application studies are welcome, as are expository texts where theory, verification and applications come together to provide a unifying coverage of a particular topic or topics.

The book series itself arose out of the seminal text: the 1992 centenary first English translation of Lyapunov's memoir *The General Problem of the Stability of Motion* by A. T. Fuller, and was followed by the 1994 publication of *Advances in Intelligent Control* by C. J. Harris. Since then a number of titles have been published and many more are planned. A full list is given below.

Advances in Intelligent Control, edited by C. J. Harris

Intelligent Control in Biomedicine, edited by D. A. Linkens

Advances in Flight Control, edited by M. B. Tischler

Forthcoming

Sliding Mode Control: Theory and Applications, by C. Edwards and S. K. Spurgeon

Neural Network Control of Robot Manipulators and Nonlinear Systems, by F. L. Lewis, S. Jagannathan and A. Yesildirek

Generalized Predictive Control with Applications to Medicine, by M. Mahfouf and D. A. Linkens

Control of Linear Multivariable Systems with Saturating Actuators, by Z. Lin, A. Saberi and P. Sannuti

A Unified Algebraic Approach to Linear Control Design, by R. E. Skelton, T. Iwasaki and K. Grigoriadis

Sliding Mode Control in Electro-Mechanical Systems, by V. I. Utkin, J. Guldner and J. Shi

From Process Data to PID Controller Design, by L. Wang and W. R. Cluett

E. ROGERS
J. O'REILLY

Foreword

Modelling and control of nonlinear dynamical systems is one of the most challenging areas of system theory. A great deal of recent research activity has focused on approaches such as neural networks and fuzzy logic. Much of this work on identifying nonlinear input–output models was, however, based on architectures and approaches which provided little insight into the underlying data generating process (the so called transparency problem). These methods were also not ideally suited for adaptive control applications where tracking of non-stationary plant behaviour is necessary.

For 'intelligent' modelling and control methods to make a practical contribution to real problems they should fulfil a number of requirements. They should be able to work online, with provable convergence (in learning) and provide a framework for analysing system stability. It is important that these methods are parsimonious, in that they should have good regularisation characteristics which lead to models which have low complexity and are therefore easier to interpret. Moreover, it must be feasible to implement them in real systems, and it should be easy to incorporate both mechanistic and rule-based or symbolic *a priori* knowledge into the learning schema.

In this regard basic research into local approaches to modelling and control, including classical control approaches, statistical methods, neuro- and fuzzy architectures (e.g. B-splines and Radial Basis Function networks) has shown substantial progress in recent years. In this book, these issues are addressed by authors who have made a substantial contribution towards resolving the problems. Many chapters are concerned with the use of local models and controllers which are parameterised by operating point conditions as a means of exploiting classical linear control methods for resolving global nonlinear control and modelling problems. This approach has been previously successfully exploited as an effective means of providing smooth bumpless control. The book provides a very effective and comprehensive coverage of current research in this area and provides in the opening chapter a welcome architectural viewpoint that integrates these apparently disparate approaches. The different methods include probabilistic interpretations, such as Jordan's mixture of models, fuzzy interpretations of local model methods, and Kuipers' and Åström's heterogeneous models.

Specialised topics in identification are included, e.g. the application of construction algorithms to model structure identification is analysed, active learning is used, identification methods are interpreted, and there is a specialist chapter on the normalising effects of basis functions. These effects have been observed before in neurofuzzy systems (the partition of unity requirement), but their significance to local modelling is particularly valuable. A number of chapters are concerned with control, addressing local model-based controllers such as the Takagi–Sugeno fuzzy controllers (equivalent to operating point-dependent local linear controllers), Multiple Model Adaptive Control methods, as well as local approaches in combination with more classical methods such as H_∞, and Laguerre polynomials. The stability of these methods is also investigated. Several chapters include applications which provide the reader with considerable insight into the practical implementability of these ideas.

The most valuable contribution of the book is that it brings together basically similar approaches which have appeared in different fields. Local techniques for intelligent modelling and control are shown to be a widely used approach with many practical advantages for use in analysis and design of complex systems. This provides an important contribution for researchers involved in nonlinear modelling and control, as the methods are easy to implement, have a number of useful properties, are relatively transparent, and have the ability to incorporate *a priori* knowledge with measured data from a real process.

C. J. HARRIS
Southampton University

Preface

This book presents a number of approaches which produce complex models or controllers by piecing together a number of simpler subsystems. This divide-and-conquer strategy is a long-standing and general way to cope with complexity in engineering systems, nature and human problem solving.

More complex plants, advances in information technology, and tightened economical and environmental constraints in recent years have lead to practising engineers being faced with modelling and control problems of increasing complexity. When confronted with such problems, there is a strong intuitive appeal in building systems which operate robustly over a wide range of operating conditions by decomposing them into a number of simpler linear modelling or control problems, even for nonlinear modelling or control problems. This appeal has been a factor in the development of increasingly popular 'local' and multiple-model approaches to coping with strongly nonlinear and time-varying systems.

Such local approaches are directly based on the divide-and-conquer strategy, in the sense that the core of the representation of the model or controller is a partitioning of the system's full range of operation into multiple smaller operating regimes each of which is associated with a locally valid model or controller. This can often give a simplified and transparent nonlinear model or control representation. In addition, the local approach has computational advantages, it lends itself to adaptation and learning algorithms, and allows direct incorporation of high-level and qualitative plant knowledge into the model. These advantages have proven to be very appealing for industrial applications, and the practical, intuitively appealing nature of the framework is demonstrated in chapters describing applications of local methods to problems in the process industries, biomedical applications and autonomous systems. The successful application of the ideas to demanding problems is already encouraging, but creative development of the basic framework is needed to better allow the integration of human knowledge with automated learning.

The underlying question is 'How should we partition the system – what is 'local'?'. This book presents alternative ways of bringing submodels together, which lead to varying levels of performance and insight. Some are further developed for autonomous learning of parameters from data, while others have focused on the ease with which prior knowledge can be incorporated. It is interesting to note that researchers in Control Theory, Neural

Networks, Statistics, Artificial Intelligence and Fuzzy Logic have more or less independently developed very similar modelling methods, calling them *Local Model Networks*,
*Operating Regime based Models, Multiple Model Estimation and Adaptive Control, Gain
Scheduled Controllers, Heterogeneous Control, Mixtures of Experts, Piecewise Models,
Local Regression* techniques, or *Tagaki–Sugeno Fuzzy Models*, among other names. Each
of these approaches has different merits, varying in the ease of introduction of existing
knowledge, as well as the ease of model interpretation. This book attempts to outline much
of the common ground between the various approaches, encouraging the transfer of ideas.

Recent progress in algorithms and analysis is presented, with constructive algorithms
for automated model development and control design, as well as techniques for stability
analysis, model interpretation and model validation.

OVERVIEW OF THE BOOK

Part I – Basic principles

The editors introduce the basic ideas of multiple model approaches in Chapter 1, where the
existing paradigms for the application of multiple model and operating regime approaches
to nonlinear modelling, identification and control are explored. The chapter also provides
a survey and overview of the state of the art in terms of procedures, algorithms and tools
for visualisation and interpretation.

Part II – Modelling

Part II of the book deals predominantly with modelling methods and applications, including
methods for estimation and experiment design.

In Chapter 2, Babuška and Verbruggen describe the Takagi–Sugeno fuzzy model, used as
an interpolating scheduler for a set of multiple linear models which are valid locally around
certain operating conditions. The antecedent of the fuzzy rule provides the local region and
the interpolating mechanism, while the consequent is the locally valid model. The difficult
task of learning the antecedents and consequents from data is reviewed and a constructive approach incorporating fuzzy clustering is developed. The identification methods are
demonstrated using experimental data from problems in biotechnology and medicine.

Gollee, Hunt, Donaldson and Jarvis show in Chapter 3 that nonlinear models based on
multiple local ARX models are able to capture the nonlinear effects apparent in experiments
with electrically stimulated muscles, and provide high accuracy over a wide range of input
signals. This chapter is an interesting example of local methods being applied to a problem which has been studied intensively with a variety of complex mathematical modelling
techniques, and comparing well. The identification methods used are those from Chapter 7.

The 'functional state' approach described by Halme, Visala and Zhang in Chapter 4 uses
a discrete representation to describe the current 'functional state' of the system. This concept is close to the idea of an operating regime, and each functional state has an associated
local model. A finite state automaton is used to describe the possible transitions between
operating regions of dynamical processes, and multi-layer perceptron neural networks with
Laguerre filters are trained to recognise transitions between states. The concept is illustrated
by experiments with a two-tank problem, and a fermentation process.

In Chapter 5 Meilă and Jordan review the *Mixture of Experts* model structure and extend
it to a *Markov Mixture of Experts*, where a Markov graph is used to define the transitions

between multiple models in the system. This is similar in many ways to the functional state approach described in Chapter 4, but this time placed in a probabilistic framework where the transitions are described by a Markov model. The method is applied to fine motion control in robotics.

In Chapter 6 by Cohn, Ghahramani and Jordan, experiment design with the *mixture of Gaussians* model representation is studied. Local representations make it easier to produce local confidence limits, which can be used as the basis for an active learning algorithm, where the optimal search for new data can be guided by the model structure. Robotics simulations are used to illustrate the ideas. This chapter also, as with Chapter 5, gives useful insight into the probabilistic interpretation of multiple model approaches. The Expectation Maximisation algorithm is used for learning – this is also studied in Chapter 7. A further interesting aspect of the representation used in this chapter is that multiple local models are used to represent the joint input–output *density* of the data, which does not distinguish between inputs and outputs, unlike the other approaches which explicitly represent input–output or input–state mappings.

In Chapter 7, Murray-Smith and Johansen show that the commonly used global least squares method for parameter identification in multiple local models can be very sensitive and lead to ill-conditioning. They propose a cheaper locally weighted least squares identification method as a solution. The interactions between model structure and parameter identification methods are discussed – this theme reappears in several other chapters. The smoothing analysis used to illustrate the effects discussed, is also a general technique, which can be applied to other frameworks.

Chapter 8, by Shorten and Murray-Smith examines some of the side-effects of basis function normalisation – a common technique used in the weighting functions in local model and control structures to ensure that the operating range is completely covered. Normalisation also appears naturally in fuzzy and probabilistic representation of the weighting functions. Normalisation has a number of side-effects which alter the global properties of the model or controller, with respect to robustness and interpretability. These become especially important when automatic learning algorithms are used to adapt the basis functions. As well as the graphical and intuitive explanations of the side-effects, the chapter also describes some mathematical tools which can help gain a deeper understanding of the trade-offs involved.

Part III – Control

Part III of the book is dedicated to applications of multiple model methods for nonlinear control.

In Chapter 9, Kuipers and Åström present methods for developing a nonlinear controller by combining multiple heterogeneous local control laws appropriate to different operating regions. Operating regions are described using fuzzy set membership, as in Chapter 2, but the local controllers can be classical control laws with their own internal states. Qualitative simulation is suggested as a method for validation of the global behaviour of the heterogeneous controls. Some aspects of the control law can, even in the case of incomplete knowledge, be represented as a qualitative differential equation, and qualitative simulation can be used to predict the possible behaviours of the system. The methods are demonstrated on a water level controller and a highly nonlinear chemical reactor.

In Chapter 10, Sbarbaro uses operating regime based models with multiple *local Laguerre models* for identification and control. Local Laguerre models potentially have advantages over the more common ARX local model as they can cope more easily with uncertainty in time delays and model orders. The method is compared in a simulation of a chemical reactor using a model predictive control algorithm.

Chapter 11 by Schott and Bequette describes multiple model adaptive control (MMAC), which is a classical model-based control strategy. Multiple models are used and a probabilistic weighting chooses which model or combination of models best represents the current plant input/output behaviour. The authors review MMAC theory, including model bank estimation and control, and describe applications to biomedical control problems.

In the work presented by Banerjee, Arkun, Pearson and Ogunnaike in Chapter 12, the composition of multiple linear state-space models is described as a parameter-varying model. The parameters of the global model are the local model weights which are estimated on-line using a Bayesian approach similar to Chapter 11. A globally stable controller composed of multiple local linear controllers is then designed for the linear parameter varying model using H_∞ design based on Linear Matrix Inequalities. The theory is applied to a simulated chemical reactor.

Zhao, Gorez and Wertz present in Chapter 13 a method for identification and structured analysis and design of Takagi–Sugeno fuzzy models and controllers. The Takagi–Sugeno fuzzy model is based on multiple local linear state-space models that are weighted using fuzzy membership functions. An identification method based on fuzzy clustering (see also Chapter 2) is presented and experimental results from application on a glass furnace are included. The control design method guarantees stability and robustness properties. The methods are based on modern tools such as Linear Matrix Inequalities, being closely related to Chapter 12. Simulation examples are used to illustrate the methods.

We hope that this book will bring to a wider audience the progress being made in both practical and theoretical use of the multiple models philosophy, and that the workers in the field will be able to gain a deeper understanding of the relations between the different existing approaches, and tools.

RODERICK MURRAY-SMITH
Berlin
TOR ARNE JOHANSEN
Trondheim

Contributors

Karl Johan Åström
Lund Institute of Technology, Department of Automatic Control, Box 118, S-221 00 Lund, Sweden
E-mail: kja@control.lth.se
WWW address: http://control.lth.se:80/~kja/

Yaman Arkun
Georgia Institute of Technology, School of Chemical Engineering, Atlanta, GA 30332-0100, USA
E-mail: yaman.arkun@che.gatech.edu
WWW address: http://www.chemse.gatech.edu/people/ya.html

Robert Babuška
Delft University of Technology, Control Laboratory, Department of Electrical Engineering, P.O. Box 5031, 2600 GA Delft, The Netherlands
E-mail: R.Babuska@et.tudelft.nl
WWW address: http://dutera.et.tudelft.nl/people/babuska/babuska.html

Atanu Banerjee
Georgia Institute of Technology, School of Chemical Engineering, Atlanta, GA 30332-0100, USA
E-mail: atanu@cezanne.chemse.gatech.edu
WWW address: http://www.chemse.gatech.edu/grads/ab/atanu.html

B. Wayne Bequette
Rensselaer Polytechnic Institute, Howard P. Isermann Department of Chemical Engineering, Troy, NY 12180-3590, USA
E-mail: bequeb@rpi.edu
WWW address: http://www.rpi.edu/~bequeb

David Cohn
Harlequin Inc., 1010 El Camino Real, Menlo Park, CA 94025, USA
E-mail: cohn@harlequin.com
WWW address: http://www.ai.mit.edu/people/cohn/cohn.html

Nick Donaldson
University College London, Department of Medical Physics and Bioengineering, London, England
E-mail: nickd@medphys.ucl.ac.uk

Zoubin Ghahramani
University of Toronto, Dept. of Computer Science, 6 Kings College Road, Toronto, Ontario M5S 1A4, Canada
E-mail: zoubin@cs.toronto.edu
WWW address: http://www.cs.utoronto.ca/~zoubin/

Henrik Gollee
University of Glasgow, Centre for Systems and Control, Glasgow, Scotland
E-mail: henrik@eng.gla.ac.uk
WWW address: http://www.mech.gla.ac.uk/~henrik/

R. Gorez
Université Catholique de Louvain Centre for Systems Engineering and Applied Mechanics (CESAME), Bât. EULER, Avenue Georges Lemaître, 4, B-1348 Louvain-la-Neuve, Belgium
E-mail: gorez@auto.ucl.ac.be

Aarne Halme
Helsinki University of Technology, Laboratory of Automation Technology, SF-02150, Finland
E-mail: aarne.halme@hut.fi
WWW address: http://www.automation.hut.fi/

Ken Hunt
Daimler-Benz Research, Berlin, Alt-Moabit 96a, D-10559 Berlin, Germany
E-mail: hunt@dbresearch-berlin.de

J. Jarvis
University of Liverpool, Department of Human Anatomy and Cell Biology, Liverpool, England
E-mail: jcj@liverpool.ac.uk

Tor Arne Johansen
SINTEF Electronics and Cybernetics, N-7034 Trondheim, Norway
E-mail: Tor.Arne.Johansen@ecy.sintef.no
WWW address: http://www.itk.ntnu.no/SINTEF/ansatte/Johansen_Tor.Arne/

Michael Jordan
Dept. of Brain and Cognitive Sciences, Massachusetts Institute of Technology,
Cambridge, MA 02139, USA
E-mail: jordan@psyche.mit.edu
WWW address: http://www.ai.mit.edu/projects/cbcl/web-pis/jordan/homepage.html

Benjamin Kuipers
University of Texas at Austin, Department of Computer Sciences, Austin TX 78712, USA
E-mail: kuipers@cs.utexas.edu
WWW address: http://www.cs.utexas.edu/users/kuipers/

Marina Meilă
Dept. of Elec. Eng. and Computer Sci., Massachusetts Institute of Technology,
Cambridge, MA 02139, USA,
and *Department of Automatic Control and Computers, Politehnica University of*
Bucharest, Romania.
E-mail: mmp@psyche.mit.edu
WWW address: http://www.ai.mit.edu/people/mmp/mmp.html

Roderick Murray-Smith
Daimler-Benz Research, Berlin, Alt-Moabit 96a, D-10559 Berlin, Germany,
from 1997, Department of Mathematical Modelling, Technical University of Denmark,
Building 321, DK-2800 Lyngby, Denmark
E-mail: murray@dbresearch-berlin.de

B. Ogunnaike
E. I. DuPont deNemours & Co., Inc. Wilmington, DE 19880-0101, USA
E-mail: ogunnaike@esspth.dnet.dupont.com

R. Pearson
E. I. DuPont deNemours & Co., Inc. Wilmington, DE 19880-0101, USA
E-mail: pearsork@esspth.dnet.dupont.com

Daniel Sbarbaro-Hofer
Universidad de Concepcion, Dept. of Electrical Engineering, Casilla 53-C, Concepcion,
Chile
E-mail: dsbarbar@vangogh.die.udec.cl
WWW address: http://www.die.udec.cl/~dsbarbar/

Kevin Schott
Rensselaer Polytechnic Institute, Howard P. Isermann Department of Chemical
Engineering, Troy, NY 12180-3590, USA
E-mail: schotk@rpi.edu

Robert Shorten
Daimler-Benz Research, Berlin, Alt-Moabit 96a, D-10559 Berlin, Germany
E-mail: shorten@dbresearch-berlin.de

H. B. Verbruggen
Delft University of Technology, Control Laboratory, Department of Electrical Engineering, P.O. Box 5031, 2600 GA Delft, The Netherlands
E-mail: H.B.Verbruggen@et.tudelft.nl
WWW address: http://lcewww.et.tudelft.nl/~vbruggen

Arto Visala
Helsinki University of Technology, Laboratory of Automation Technology, SF-02150, Finland
E-mail: arto.visala@hut.fi
WWW address: http://www.automation.hut.fi/

V. Wertz
Université Catholique de Louvain Centre for Systems Engineering and Applied Mechanics (CESAME), Bât. EULER, Avenue Georges Lemaître, 4, B-1348 Louvain-la-Neuve, Belgium
E-mail: wertz@auto.ucl.ac.be

Xia-Chang Zhang
Helsinki University of Technology, Laboratory of Automation Technology, SF-02150, Finland
E-mail: xiachang.zhang@hut.fi
WWW address: http://www.automation.hut.fi/

J. Zhao
Université Catholique de Louvain Centre for Systems Engineering and Applied Mechanics (CESAME), Bât. EULER, Avenue Georges Lemaître, 4, B-1348 Louvain-la-Neuve, Belgium
E-mail: zhao@auto.ucl.ac.be

PART ONE

Basic Principles

CHAPTER ONE

The Operating Regime Approach to Nonlinear Modelling and Control

TOR ARNE JOHANSEN and RODERICK MURRAY-SMITH

Multiple model approaches have appeared more or less independently in several branches of science and engineering, disguised in different terminology. The purpose of this chapter is to introduce the basic ideas of multiple model approaches, focusing on their similarities rather than their differences. Existing paradigms for the application of multiple model and operating regime approaches to nonlinear modelling, identification and control are explored. A survey and overview of the state of the art in terms of procedures, algorithms and tools for visualisation and interpretation are also provided.

1.1 INTRODUCTION

Technological development is steadily increasing the complexity of process plants, vehicles and other engineered systems, and economical and environmental constraints are raising the awareness of the need for practical approaches which can be used to aid engineers to better understand and perform such complex modelling and control tasks.

These highly complex systems are characterised by a large number of components which are strongly coupled and have a wide operating range, among other factors. However, despite the increasing mathematical sophistication of research in modelling and control over the last decades, and the greater use of powerful computing facilities, many of the advanced methods have not been regularly applied to real problems under normal working conditions. This is often because of the theoretical sophistication required to understand the methods. Hence, if we want to deal with complex high-dimensional, coupled, nonlinear and non-stationary systems, tomorrow's automatic control systems must be more autonomous, robust, intelligent and user-friendly. However, dealing with complexity is obviously an inherently difficult problem, by the very definition of the word. The principle of incompatibility (Zadeh 1973) tries to make the implications of complexity more precise: *As the complexity of a system increases, our ability to make precise and yet significant statements about its behaviour diminishes until a threshold is reached beyond which precision and significance (or relevance) become almost mutually exclusive characteristics.* A consequence of this principle is obviously that models and analysis of complex systems will be less precise than for simple systems. A further consequence is that one should perhaps look for other model representations and tools that can make use of less precise system knowledge than the traditional approaches which worked well for low and medium complexity modelling and control problems in the past. This is indeed the trend in the area of intelligent control where fuzzy logic, qualitative modelling, neural networks, expert systems and probabilistic reasoning are being explored, e.g. (Åström and McAvoy 1992, Antsaklis *et al.* 1991, Åström *et al.* 1986).

In everyday life, as well as in solving engineering problems, the standard approach to complex problem solving is the divide-and-conquer strategy: *A complex problem is somehow partitioned into a number of simpler subproblems that can be solved independently, and whose individual solutions yield the solution of the original complex problem.* The key to successful problem solving with this approach is to find suitable axes along which the problem can be partitioned. In this work we will focus on one approach to the decomposition of modelling and control problems that has recently attracted significant attention, namely operating regime decomposition.

The core of the operating regime approach is to make use of a partitioning of the operating range of the system in order to solve modelling and control problems. The operating regime approach thus leads to multiple model or multiple controller approaches, where different local models or controllers are applied under different operating conditions, see Figure 1.1. The supervisor (or scheduler) will coordinate the local models or controllers. This coordination may include selection of a single one, or combining the actions or parameters of a number of local models or controllers.

The operating regimes can often be characterised by different sets of phenomena or behaviours of the system. The rationale behind this approach is basically that the development of local models (or controllers) is simpler because the interactions between the relevant phenomena in each operating regime are simpler locally than globally. For instance, if the system phenomena or behaviour change smoothly with the operating point, then a linear model (or controller) will always be sufficient locally by making each operating regime sufficiently small, even though the system may contain complex nonlinearities when viewed globally. Other motivations for operating regime approaches include:

- The model/controller structure is easy to understand and interpret, both qualitatively and quantitatively, and the approach has its roots in traditional engineering methods.
- Various types of knowledge can be incorporated and integrated within the framework, including qualitative knowledge, empiricism, measured data and available models. Operating regime based modelling is closely related to grey-box modelling.
- Reduced computational complexity compared to other nonlinear methods.

Chapter overview

In the remainder, we will discuss this approach in some detail. In section 1.2 the fundamentals are introduced and numerous simple examples are given. Next, in section 1.3, a broader perspective is taken, and the operating regime approach is seen in relation to other approaches. Details on the available algorithms, procedures and tools for operating regime based modelling and control can be found in sections 1.4 and 1.5, respectively. Further details on analysis, validation and interpretation can be found in section 1.6. Finally, some concluding remarks are given in section 1.7.

Input

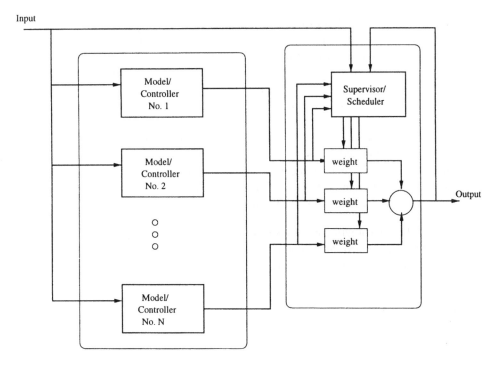

Figure 1.1 The multiple model/controller approach. This is a general visualisation of the methods described in this book, where multiple sub-models or -controllers are organised in some way by a scheduling or gating block. Each method varies in the style of scheduling done, the variables used for the scheduling, whether parameters or states are scheduled and in the structures of the individual local models or controllers.

1.2 OPERATING REGIME APPROACHES

1.2.1 The basics of the framework

Any model or controller will have a limited range of operating conditions in which it is sufficiently accurate or performs sufficiently well in order to serve its purpose. This range may be restricted by several factors, such as validity of linearisation, modelling assumptions, stability properties, or experimental conditions. A model or controller that is useful in a region less than the full range of operating conditions is called a local model or controller, as opposed to a global model or controller which is useful over the full range of operating conditions. Of course, the ultimate goal is a global model or controller. However, we will argue that in some cases it may be beneficial to achieve this goal by developing a number of local models or controllers.

The basis of the framework is a decomposition of the system's full range of operation into a number of possibly overlapping operating regimes, as illustrated in Figure 1.2. In each operating regime, a simple local model or controller is applied. These local models or controllers are then combined in some way to yield a global model or controller. Hence, model or controller development within this framework typically consists of the following tasks:

■ Decompose the system's full range of operation into operating regimes. This task includes a definition of the full operating range, as well as the identification of variables that can be used to characterise the operating regimes.

■ Select simple local model or controller structures within each operating regime. These structures will often be determined by the relevant system knowledge that is available under different operating conditions, as well as the intended purpose of the model or controller.

■ The local model or controller structures are usually parameterised by certain variables that must be determined.

■ A method for combining the local models or controllers into a global one must be applied. Numerous approaches exist, and can be characterised according to deterministic vs. stochastic assumptions, soft or hard partitions etc.

For practical problems it will not always be easy to find a natural sequence in which these tasks should be approached. Several iterations of the same tasks are usually needed before a satisfactory model or controller is found.

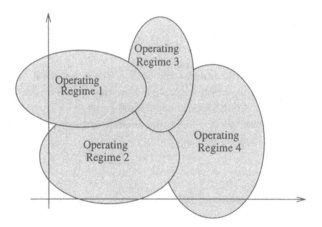

Figure 1.2   The operating range of a complex system is decomposed into a number of operating regimes. These can then be used to represent the system using associated simple subsystems. Much of this book is concerned with how to determine these regions, how to bring subsystems together, and how to interpret the resulting systems.

1.2.2   Some introductory examples

Before we proceed, this section will let us give some simple examples of how operating regime based approaches can be used to approach some nonlinear modelling and control design problems. The more fundamental aspects of partitioning approaches will then be discussed in section 1.2.3.

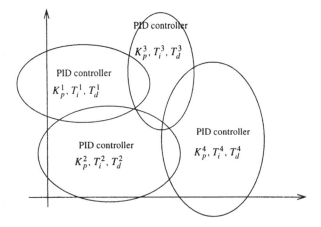

Figure 1.3 A nonlinear PID-like controller is designed by patching together local linear PID controllers.

A nonlinear PID-like controller

Suppose we want to design a nonlinear PID-like controller of the form

$$u(t) = K_p(z(t))e(t) + \frac{1}{T_i(z(t))} \int_0^t e(t)dt + T_d(z(t))\frac{d}{dt}e(t),$$

where the gain K_p, integral time T_i and derivative time T_d are functions of the system's operating point $z(t)$. In standard PID controllers these functions are all constant, and there exist numerous design procedures which guarantee that various stability, performance and robustness specifications are met, e.g. (Åström and Hägglund 1988). Of course, with the nonlinear PID-like controller above, the design problem is much more difficult. However, the operating regime based approach offers an engineering-friendly solution to this design problem. One can for instance design a number of standard linear PID controllers to meet the desired stability, performance and robustness criteria locally when the system is operating in neighbourhoods of some selected operating points $z_1, z_2, ..., z_{n_\mathcal{M}}$:

$$u(t) = K_p^i e(t) + \frac{1}{T_i^i} \int_0^t e(t)dt + T_d^i \frac{d}{dt}e(t), \qquad \text{when } z(t) \text{ is close to } z_i,$$

for $i = 1, 2, ..., n_\mathcal{M}$, see Figure 1.3. Designing such local PID-controllers is often simpler than approaching the nonlinear PID-control design problem directly, even if a nonlinear dynamic model of the system exists. In addition, we need an algorithm for combining or switching between the local PID-controllers. For instance, a weighting of the local PID-parameters as a function of the operating point $\rho_i(z(t))$:

$$u(t) = \sum_{i=1}^{n_\mathcal{M}} \left(K_p^i e(t) + \frac{1}{T_i^i} \int_0^t e(t)dt + T_d^i \frac{d}{dt}e(t) \right) \rho_i(z(t)),$$

i.e.

$$K_p(z(t)) = \sum_{i=1}^{n_\mathcal{M}} K_p^i \rho_i(z(t)),$$

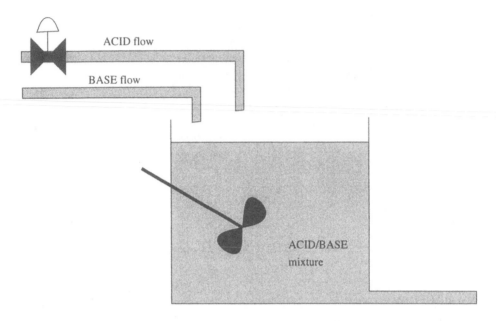

Figure 1.4 A pH neutralisation tank.

$$\frac{1}{T_i(z(t))} = \sum_{i=1}^{n_{\mathcal{M}}} \frac{1}{T_i^i} \rho_i(z(t)),$$

$$T_d(z(t)) = \sum_{i=1}^{n_{\mathcal{M}}} T_d^i \rho_i(z(t)).$$

It should be noted that even though all the locally designed PID controllers satisfy the design criteria locally, extensive analysis or simulation must usually be applied to validate and verify the global properties of the control system.

A nonlinear input–output model of a neutralisation tank

Consider, as a second example, a pH neutralisation tank where acid and base flows into a stirred tank through pipes with valves, see Figure 1.4. The mixture exits through another pipe. Suppose we select a nominal tank level, pH and flow-rates corresponding to an equilibrium point for the tank. Exciting the system about this equilibrium point and collecting experimental data, we can identify a linear transfer function model

$$\text{pH}(s) = \frac{K e^{-\tau s}}{1 + T s} q(s),$$

where $q(t)$ is the flow-rate through the valve, $\text{pH}(t)$ is the pH value in the tank, and s is the complex variable in the Laplace transform. Suppose we repeat this experiment for a number of different nominal tank levels, pH and flow-rates corresponding to different equilibrium points. Then we would get different identified values of K, τ and T, because the gain would depend on the pH value, and the time-delay would depend on the flow-rate, and the time-constant would depend on both the flow-rate and the tank level. Hence, we have different linear transfer function models which are reasonable descriptions of the system only within some small operating regimes. Again, these local models can be combined using some weighting function into a globally accurate model.

Modelling the longitudinal dynamics of a car

The speed (v) of a car can be described by a dynamic model that takes into account variables such as gear position (g) and throttle angle (α):

$$v(t) \quad = \quad f(v(t-1), \alpha(t-1), g(t-1)).$$

The dynamics of a linearised model

$$\Delta v(t) \quad = \quad a_1 \Delta v(t-1) + b_1 \Delta \alpha(t-1) \tag{1.1}$$

will obviously depend on the gear position, but also on the throttle angle (due to e.g. the nonlinear engine characteristics) and the speed (due to e.g. rolling and drag forces). A model could be based on the operating regime decomposition in Figure 1.5 and local linear models of the form (1.1). A complete example can be found in (Hunt *et al.* 1996*a*).

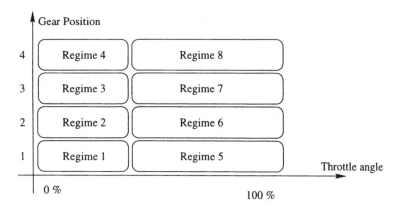

Figure 1.5 Operating regimes for a vehicle.

A nonlinear semi-mechanistic model

Consider a chemical reactor where a set of chemical reactions take place. A mechanistic model typically consists of mass and energy balances. The structure of the balance equations can typically be developed quite easily. The main problem of modelling such systems is to specify how the reaction-rates, mass-transfer coefficients, enthalpies and other model variables depend on the system's state. This typically requires a good understanding of the reaction kinetics, thermodynamics, fluid dynamics, mass- and heat-transfer phenomena present in the reactor. If such knowledge is unavailable or only qualitative knowledge about these phenomena is present, the use of operating regime based approaches can be used to formulate a semi-mechanistic model of the reactor. To illustrate this, let us consider a specific reactor, namely a semi-batch fermentation reactor where the fermentation of glucose to gluconic acid takes place, Figure 1.6.

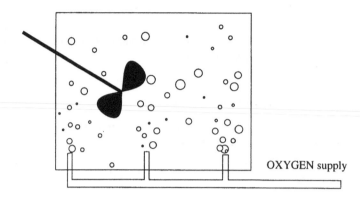

Figure 1.6   A fermentation reactor.

The main overall reaction mechanism can be described by

$$\text{Cells} + \text{Glucose} + O_2 \quad \rightarrow \quad \text{More Cells}$$
$$\text{Glucose} + O_2 \quad \overset{\text{Cells}}{\rightarrow} \quad \text{Gluconolactone}$$
$$\text{Gluconolactone} + H_2O \quad \rightarrow \quad \text{Gluconic Acid}$$

The first reaction is the reproduction of cells, consuming the substrate glucose and oxygen. The second reaction is the production of gluconolactone, again consuming glucose and oxygen. This reaction is enzyme-catalyzed by the cells, while the final product, gluconic acid, is formed by the last reaction.

Let x be the state vector that contains the compositions of cells, glucose, gluconolactone, gluconic acid, and oxygen. A mass balance can be written in the form

$$\dot{x} \;=\; r(x, T, pH) + q(x), \tag{1.2}$$

where $r(x, T, pH)$ is a vector of reaction rates, $q(x)$ is a vector of flow-rates that describes the uptake of oxygen, and T is temperature. An examination of the reaction mechanism (Johansen and Foss 1993b) reveals that the operation of the reactor can naturally be decomposed into four operating regimes:

1   **Initial regime** when it is the number of cells that limits the chemical reactions. Only the first reaction will be significant.
2   **Growth regime** when all reactions proceed at a high rate, limited only by the oxygen supply. All reactions will be significant.
3   **Growth termination regime**, characterised by shortage of glucose which limits the reaction rates.
4   **Termination regime** when all glucose-consuming reactions have terminated, and it is only the production of gluconic acid that is significant.

These operating regimes can for instance be characterised in terms of glucose and oxygen concentrations, see Figure 1.7. Within each operating regime the dynamics of the reactor can be described by linear differential equations with high accuracy, if the dependence on temperature and pH are neglected. Otherwise, one must decompose the regimes further on the basis of these variables, or apply nonlinear local models which take these dependences into account.

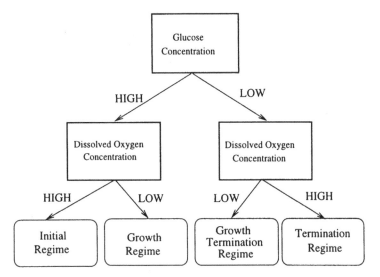

Figure 1.7 Operating regimes for batch fermentation reactor.

Similarly, the flow-rate term q in (1.2) can also be decomposed. The uptake of oxygen will depend on the oxygen concentration, stirring and other fluid dynamics phenomena that in turn may depend on the concentration of cells, since it affects the viscosity and other key properties.

Process operator heuristics

Consider a polymerisation reactor where the operator determines the feed (monomer) flow-rate (Sugeno and Yasukawa 1993). The operator's control procedures can be represented as a set of heuristic rules such as

```
IF (monomer concentration is high)
   AND (monomer concentration is increasing)
   AND (monomer feed-rate is low)
     THEN monomer flow-rate set-point is small.
```

The premise of this rule clearly determines a set of operating conditions that can be viewed as an operating regime. In (Sugeno and Yasukawa 1993), six control rules are identified on the basis of the observed behaviour of the operator.

1.2.3 Fundamental properties

The local versus global dilemma

No matter what underlying model or controller representation one chooses, e.g. state-space or input–output, lumped or distributed, discrete-time or continuous time, the nonlinear model or controller development will typically involve the specification of one or more nonlinear functions. Examples of such functions can be the dependence of the gain on certain process variables, or relations that can be linked to physical phenomena like chemical reactions. The core of the operating regime based approach is to decompose the domain

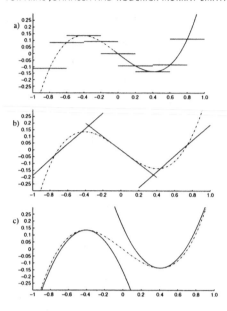

Figure 1.8 Function approximation using local models. The dashed-dotted curve is the function to be approximated, while the solid lines are local approximations. Part a) applies local constant functions, part b) applies local linear functions, while part c) applies local quadratic functions.

of functions into operating regimes and the design of simpler functions that are adequate approximations to the relationships we want to model within their respective operating regimes, see Figure 1.8. Obviously, there is a trade-off between the number and size of the operating regimes on the one hand, and the complexities of the local models on the other. For instance, at one extreme one can have only one large operating regime that covers the full range of operation. The corresponding local model must typically be complex, since it is actually the global model. In general, a decomposition into a few 'large' operating regimes will require more complex local models than a decomposition into numerous 'small' operating regimes, see Figure 1.8b) and c). On the other extreme, one can have a very fine partitioning into operating regimes based on a large number of characterising variables. In this case, the functions we are representing can be approximated by constant values locally, see Figure 1.8a).

Provided the functional relationship we want to approximate is smooth, it should be intuitively clear that approximations based on operating regimes and local models can be made arbitrarily accurate either by making the decomposition into operating regimes sufficiently fine, or by making the local models sufficiently complex. This is studied in detail by (Johansen and Foss 1993a), see also (Johansen 1994c) for some slight improvement. Explicit bounds on the approximation error as a function of the regularity (smoothness) of the underlying function, the granularity of the decomposition, and the complexity of the local models are given. Moreover, it is shown that any continuous function can be approximated to arbitrary uniform accuracy this way using polynomial local models of arbitrarily low order (like linear or constant local approximation) on a compact domain. Similar bounds and approximation results are given by (Omohundro 1987, Kosko 1994, Zeng and Singh 1994). As mentioned above, these results are quite intuitive and can almost be seen without any mathematical proof.

The curse of dimensionality

An inherent problem with all function approximation approaches that are based on partitioning of the function's domain is the curse of dimensionality: *With an increasing number of variables on which the function depends, the number of partitions required in a uniform partitioning will increase exponentially.* Consequently, a uniform partitioning is undesirable and unrealistic for anything else than low complexity problems (Friedman 1991). Fortunately, a uniform partitioning is usually not necessary. The reasons for this are diverse:

- First, the required accuracy of the model or controller may be significantly higher in some operating regions than in others. For instance, in some regions it may be necessary to have an accurate model or controller in order to optimise the control performance, while in other regions there are other criteria which are important. Fulfilment of these criteria need not always require an accurate model or controller.
- Second, in a dynamical system a large fraction of the state-space will typically be infeasible, in the sense that the system can never be in these states because they are not compatible with the physics of the system.
- Thirdly, if the local models or controllers are sufficiently complex, they will typically describe the system adequately along certain axes, while inadequately along other axes. It can be seen that it is sufficient to further decompose into operating regimes only along those axes which are inadequately described by the local model or controller (Johansen and Foss 1993a). This is a major advantage over simpler partitioning approaches that apply very simple local models (such as radial basis function and wavelet series expansions).
- Finally, the complexity of the system is typically not uniform. Hence, sometimes a simple local model or controller will be sufficient in a large operating regime, while in other cases a more complex local model or controller may be necessary in a smaller operating regime.

Development of a simple non-uniform decomposition of the operating range is often difficult. Prior knowledge about the system, or careful examination of large amounts of empirical data is the key to achieving this goal. Hence, the curse of dimensionality can often be 'warded off', at least to some extent.

Reducing the dimension of the scheduling variable

Consider the development of a model of the form

$$\dot{x} = f(x, u).$$

It is easy to see that such a model can always be written

$$\dot{x} = a(x, u) + A(x, u)x + B(x, u)u, \tag{1.3}$$

which is a form that emphasises the close relationship to linear models of the form

$$\dot{x} = a_i + A_i x + B_i u \tag{1.4}$$

and quasi-linear models of the form

$$\dot{x} = a(z) + A(z)x + B(z)u. \tag{1.5}$$

By comparing (1.3) and (1.5) we observe that the variable z clearly should depend on x and u. Suppose we decompose the operating range into a number of operating regimes where the system is described by linear models of the form (1.4). From (1.3) it is evident that the linearised model's parameters depend on the operating point. However, by a possible change of variables it is clear that under some conditions, one can find a characterising vector z satisfying

$$a(x, u) = a(z), \quad A(x, u) = A(z), \quad B(x, u) = B(z),$$

which contains fewer elements than the total number of elements in x and u (Johansen and Foss 1993a). In particular, if only an approximate model is sought, then this reduction in dimension can be significant, and the curse of dimensionality can be reduced directly. The use of prior knowledge about the system to reduce the effects of dimensionality is a recurrent theme throughout this work. How can we define modular or hierarchical structures which produce the right level of local complexity needed to model or control the real system?

Example of dimension reduction

As an example of the above considerations, consider the fermentation reactor described in section 1.2.2. A mass balance is

$$
\begin{aligned}
\dot{x}_1 &= \mu(x_4, x_5)x_1, \\
\dot{x}_2 &= v(x_4)x_1 - k_1 x_2, \\
\dot{x}_3 &= k_2 x_2, \\
\dot{x}_4 &= -k_3 \mu(x_4, x_5)x_1 - k_4 v(x_4)x_1, \\
\dot{x}_5 &= k_5(x_5^* - x_5) - k_6 v(x_4)x_1 - k_7 \mu(x_4, x_5)x_1,
\end{aligned}
$$

where x_1 is the cell concentration, x_2 is gluconolactone concentration, x_3 is gluconic acid concentration, x_4 is glucose concentration and x_5 is dissolved oxygen concentration. The functions (parameters) μ, v, and $k_1, ..., k_7$ depend on temperature and pH in addition to the states as written above. This fact follows directly from an examination of the reaction mechanism and basic principles of chemical reaction kinetics (Bailey and Ollis 1986). We clearly see that with local linear state-space models, then the operating regimes can be characterised by $z = (x_4, x_5)$ in addition to pH and temperature. Hence, the number of variables needed to characterise the operating regimes is reduced from seven to four – a significant improvement.

Such knowledge-based reduction of model complexity is fundamental to the approaches proposed in this book, and will appear in a number of guises, from the analysis of statistical distributions, hierarchical decomposition, graphical networks of behaviours and on to fuzzy logic rule bases.

1.2.4 Combining local models and controllers

Having partitioned the problem and developed a number of operating regimes and local models or controllers within each operating regime, the natural question is how to recombine the submodels – i.e. when and how to "switch" between the local models or controllers. In the discussion above we have not paid any attention to possible problems related to overlap between the operating regimes: for some operating conditions there may exist several local models that are partially relevant. Also, we have not argued whether there should be a sudden switching between the local models, or if there should be a smooth transition. The purpose of this section is to address these questions.

Hard partitions and discrete logic

The basic idea of discrete partitions is that at each operating point, exactly one local model or controller is chosen as a deterministic function of the operating point. This is often referred to as mode switching (Hilhorst *et al.* 1991, Hilhorst 1992, Söderman *et al.* 1993), or piecewise models and controllers (Opoitsev 1970, Dorofeyuk *et al.* 1970, Kasavin 1972, Rajbman *et al.* 1981, Bellman 1961, Haber *et al.* 1982, Omohundro 1987, Billings and Voon 1987, Farmer and Sidorowich 1987).

Related to the fermentation example decomposition in Figure 1.7, this means that the partition based on glucose concentration splits the set of operating points into two parts, depending on whether the glucose concentration is below or above the threshold that defines the boundary between low and high glucose concentration, see Figure 1.9.

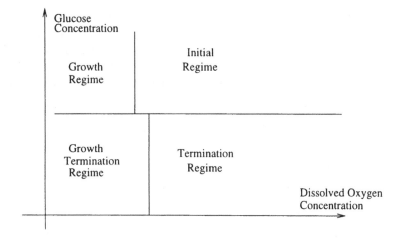

Figure 1.9 Rough decomposition of the operation of a fermentation reactor on the basis of two variables.

Convenient frameworks and representations that can be applied to describe such model or controller behaviour include decision trees (Breiman *et al.* 1984, Strömberg *et al.* 1991*b*, Sanger 1991*b*), discrete logic, expert systems, and hybrid systems (Barton and Pantelides 1994, Pettit and Wellstead 1993, Bencze and Franklin 1995, Simonyi *et al.* 1989, Konstantinov and Yoshida 1991, Konstantinov and Yoshida 1989) as well as variable structure systems (Utkin 1977, Badr *et al.* 1991). The representation in Figure 1.7 is essentially a decision tree. A discrete logic or expert system equivalent would be a number of logical statements of the form

```
IF glucose concentration IS HIGH,                              (1.6)
   AND dissolved oxygen concentration is low,
      THEN the system is operating in the Growth Regime.
```

In practice, such operating regime based models and controllers can be implemented using anything from simple programmable logic systems to sophisticated expert systems. A supervisory system based on discrete logic that lies on top of the traditional control system to handle various situations such as startup, shutdown, exceptions, safety and product changes, is the standard approach to dealing with wide operating range plants in industry.

Assume the modelling problem is a static function approximation problem, where local approximations $f_1, f_2, ..., f_{n_{\mathcal{M}}}$ are known for each operating regime. The operating regimes must form a complete partition of the operating range, and they must not overlap. The global approximation is then the piecewise approximation

$$f(u) = \sum_{i=1}^{n_{\mathcal{M}}} f_i(u)\mu_i(u),$$

where μ_i is the characteristic function for the set of points that defines the operating regime with index i.

Finite state automata

Switching between the operating regimes can also be described using a finite state automaton. A finite state automaton consists of a finite number of discrete states, each corresponding to an operating regime, or a functional state (Halme 1989, Branicky 1994, Zhang *et al.* 1994), see also Chapters 4 and 5. Transition between discrete states is described by a discrete state transition function. Suppose that within each discrete state the dynamics are described by difference equations. This leads to the hybrid model

$$q(t+1) = \mathcal{A}(q(t), x(t), u(t)), \tag{1.7}$$

$$x(t+1) = f_{q(t)}(x(t), u(t)), \tag{1.8}$$

where x is the continuous state and q is the discrete state. The deterministic function $\mathcal{A}$ represents the discrete state transition function.

Soft partitions and fuzzy logic

In some cases it may not be natural to have a sudden change between operating regimes. This may for example be the case when the operating regimes are characterised by different behaviours or mechanisms which change gradually as the operating point moves between different operating regimes. Most physical phenomena have this property. In such cases one can describe the operating regimes as overlapping sets and implement a smooth deterministic transition between them.

Consider again the decomposition described in Figure 1.9. We would like to represent that the boundary between the operating regimes characterised by low and high glucose concentration is soft. Hence, when the glucose concentration is near the overlapping boundary between low and high glucose concentration, then a blend of both local models or controllers is applied.

Frameworks that can be applied to describe such soft boundaries between operating regimes include fuzzy sets and fuzzy logic (Zadeh 1973, Takagi and Sugeno 1985) and interpolation methods (Stokbro *et al.* 1990b, Jones and co-workers 1991, Johansen and Foss 1993a). Fuzzy sets are characterised by gradual membership and are a very natural way of describing an operating regime. In fuzzy logic, the logical statement (1.6) is interpreted in terms of fuzzy sets definitions of the terms HIGH and LOW, see Figure 1.10. The theory of fuzzy logic also defines natural ways of making inference on the basis of such a rule-base. The resulting inference mechanism (Zadeh 1973, Takagi and Sugeno 1985) can be viewed as an interpolation algorithm that gives more or less weight on the local models or controllers in the different operating regimes, depending on the operating point.

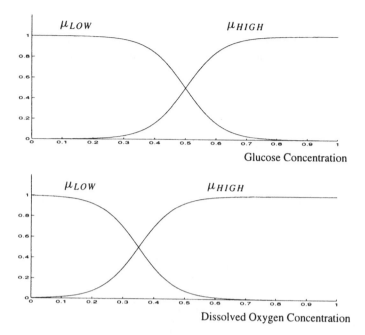

μ_{LOW} μ_{HIGH}

Glucose Concentration

μ_{LOW} μ_{HIGH}

Dissolved Oxygen Concentration

Figure 1.10 Fuzzy membership functions for representation of the terms HIGH and LOW.

Consider again the function approximation problem described above. The global approximation is now an interpolation of the local approximations:

$$f(u) = \sum_{i=1}^{n_{\mathcal{M}}} f_i(u)\rho_i(u),$$

where the smooth weighting functions $\rho_1, \rho_2, ..., \rho_{n_{\mathcal{M}}}$ provide soft transitions between the operating regimes. Typically, the weighting functions satisfy

$$\sum_{i=1}^{n_{\mathcal{M}}} \rho_i(u) = 1$$

for all u. In the case of a fuzzy logic inference (Takagi and Sugeno 1985), then

$$\rho_i(u) = \frac{\mu_i(u)}{\sum_{j=1}^{n_{\mathcal{M}}} \mu_j(u)}, \qquad (1.9)$$

which clearly satisfies the equation above. The function μ_i is now the membership function for the fuzzy set that represents the operating regime with index i. The inference mechanism (1.9) is studied in (Johansen 1995).

Probabilistic approaches to partitioning

Both with the hard and soft partitioning approaches mentioned above, the characterisation of operating regimes was deterministic. An alternative is to use statistical methods to infer which operating regime is most appropriate at each time instant.[1]

The basis for mixture models (Titterington *et al.* 1985, McLachlan and Basford 1988) is that each local model or controller has an associated probability density that indicates how correct or appropriate it is (Jacobs *et al.* 1991, Jordan and Jacobs 1994, Petridis and Kehagias 1996). The basis functions can be regarded as the components of a mixture density model. For example, the function approximation problem has the following solution

$$f(u) \;=\; \sum_{i=1}^{n_{\mathcal{M}}} f_i(u) P(i|u), \tag{1.10}$$

where $P(i|u)$ is the posterior probability for the local model or controller with index i being the correct one, given the data. Depending on the probabilistic assumptions, the posterior $P(i|u)$ can be computed in a number of different ways. For example, using just priors $\rho_1, \rho_2, ..., \rho_{n_{\mathcal{M}}}$

$$P(i|u) \;=\; \frac{\rho_i(u)}{\sum_{j=1}^{n_{\mathcal{M}}} \rho_j(u)} \tag{1.11}$$

(cf. (1.9)), or priors modified by the data according to Bayes' law

$$P(i|u) \;=\; \frac{p(u|i)\rho_i(u)}{\sum_{j=1}^{n_{\mathcal{M}}} p(u|j)\rho_j(u)},$$

where $p(u|i)$ is the probability density function for the input u given that local model or controller with index i is the correct one. For dynamic modelling, an approach based on statistical pattern recognition and decision theory is described in (Skeppstedt *et al.* 1992). Moreover, there is a long tradition within the control community with multiple model estimation based on Kalman filter banks and Markov models, e.g. (Lainiotis 1976*a*, Athans *et al.* 1977, Greene and Willsky 1980, Lund *et al.* 1991, Lund *et al.* 1992, Blom and Bar-Shalom 1988) and Chapter 11. The Markov model is a probabilistic relative of the finite state automaton. Transition between the discrete states is described in terms of probabilities or probability densities. Suppose that a discrete time process is modelled by a Markov chain with discrete state $q(t)$ taking values in a finite set $\{1, ..., n_{\mathcal{M}}\}$. At each time step, the probability of transition between j and i is

$$p_{ij} = P[q(t) = i|q(t-1) = j]$$

with

$$\sum_{i=1}^{n_{\mathcal{M}}} p_{ij} = 1, \quad \text{for all } j = 1, ..., n_{\mathcal{M}}. \tag{1.12}$$

In general, the transition probabilities may be densities depending on the system state, input and some parameters. In the static function approximation case, the density may depend only on the input u and the parameters W:

$$p_{ij} \;=\; p_{ij}(u(t), W).$$

---

[1]The relationship between probabilistic and fuzzy methods is interesting, as in some ways, fuzzy logic is deemed '*orthogonal to probability theory as it focuses on ambiguities in describing events, rather than uncertainty of occurrence or non-occurrence*' (Pearl 1988), but it is possible to describe fuzzy logic in terms of uncertainty of which label to use. In the multiple model case a fuzzy representation of the model validity functions would amount to uncertainty about which local model was the 'correct' one for a given non-fuzzy operating point.

In the Markov Mixture of Experts model (MME) described in Chapter 5 the same input u is also processed in parallel by a number of static modules called *experts*. Each expert is a local model which outputs a value $f(u(t), \theta_i)$, $i = 1, 2, \ldots, n_{\mathcal{M}}$.

The output y of the system is chosen from among the outputs of the experts, using the discrete state of the Markov chain as an indicator variable. Thus, if the discrete state is $q(t)$, then the output is $f_{q(t)}(u(t), \theta_i)$, cf. (1.8). Only the input u and the output y are usually measured; the discrete state q of the Markov chain is hidden to the exterior observer.

The model is schematically represented in Figure 5.2 on page 149. Taking the functions p_{ij} and f to be constant functions (independent of the input), it is easy to see that the MME model contains the Hidden Markov Model (HMM) as a special case; in fact, the former can be viewed as a time variant, continuous output distribution HMM. It is also a generalisation of the mixture of experts architecture (Jordan and Jacobs 1994), to which it reduces when the columns of the transition probability matrix are all equal.

Heterogeneous local models and controllers

As discussed above, there will be a transition between local models or controllers when the system moves between operating regimes. When the local models or controllers are homogeneous, in the sense that they have the same state-space, then information can be transferred between the local models and controllers quite easily. This is highly desirable, since it gives smooth behaviour and undesirable transients are avoided. A very simple example of such information transfer is "bumpless transfer" mechanisms which are applied when switching between controllers in process control systems. This means that the local controller is reinitialised to avoid transients when re-entering an operating regime.

When the local models or controllers have different structures, only in special cases is it possible to find one-to-one information transformation mappings between the different operating regimes. One solution is to let the local models or controllers have local states and simply accept possible transients (Kuipers and Åström 1994, Gawthrop 1995), see also Chapter 9. Alternatively, one can use state-estimators or observers to reduce the transients (Johansen and Foss 1993*b*, Gawthrop 1995, Gawthrop 1996, Shorten 1996).

1.2.5 Hierarchical modelling and control

We have discussed the use of locality in operating regime based models and controllers for the limitation of complexity, but how can we decide on the suitable level of locality for any given part of the operating range? For complex systems, the operating conditions of the system must often be described by a large number of variables.[2] How can we reduce the effect of high dimensionality by only concentrating on the operating conditions and variables of interest?

A common technique for the control of complexity and high-dimensionality, found in nature, society and technical systems, is *hierarchy*, and although touched on in almost every area of science, there is still little systematic analysis of the concepts involved.[3] Hierarchical methods offer, due to their structure, the ability to more effectively hide local complexity through a multi-resolution representation of the system.

---

[2] By a large number of variables we understand a number that is too large to allow simple visualisation of the operating regimes, i.e. larger than two or three.

[3] See (Mesarovic *et al.* 1970) for a thorough technical discussion of hierarchical systems, and (Simon 1976) for an interesting analysis of hierarchical structure in human organisations. Modern research in mathematical organisation theory can be found in the journal *Computational and Mathematical Organization Theory*.

The idea is first to roughly decompose the operating range according to the most important characterising variable. This gives rise to a number of operating regimes. Now, each of these operating regimes can be further decomposed according to another characterising variable, and so forth, see Figure 1.11. Of course, characteristic variables can reappear at different levels in the hierarchy, and the same questions of partitioning style are relevant, as discussed in the previous section.

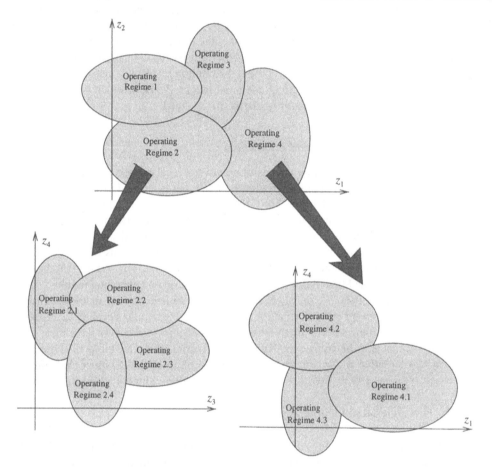

Figure 1.11   Hierarchical decomposition into operating regimes. On the top level, the four operating regimes are characterised using the variables z_1 and z_2. The second and fourth operating regimes are further decomposed using the variables z_3 and z_4, and z_1 and z_4, respectively.

Examples of hierarchical decomposition

As an example of hierarchical operating regime based modelling, consider the fermentation reactor discussed above, see (Foss *et al.* 1995) for more details. We have concluded that the operating range should be decomposed into four regimes on the basis of two variables such as dissolved oxygen concentration and glucose concentration. However, to also take into account the strongly nonlinear dependence of the reaction kinetics on temperature and

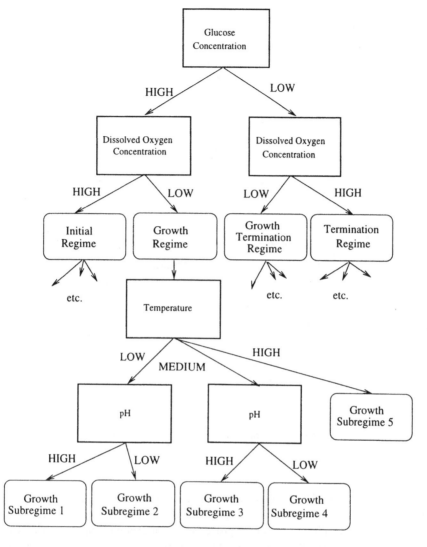

Figure 1.12 Hierarchical decomposition of the batch fermentation reactor into operating regimes.

pH, one can decompose further each of the four operating regimes into five new operating regimes on the basis of temperature and pH. This gives a total of twenty operating regimes, where each operating regime is characterised by four variables. It is clear that this is quite a complex decomposition, but as Figure 1.12 shows, it can be easily visualised and interpreted due to its hierarchical nature.

Likewise consider a hierarchical control system, Figure 1.13. On a high level in the hierarchy, the operating condition of the system can be characterised as startup, production or shutdown. The production regime can be further decomposed into normal operation, product change, and operating regimes corresponding to various failures, maintenance, and significant disturbances. Furthermore, the normal operation regime can be broken down

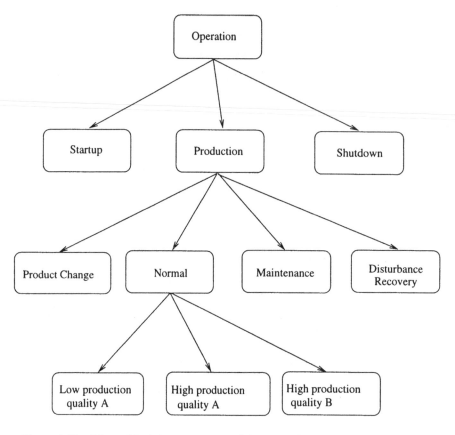

Figure 1.13   Hierarchical decomposition of the control system for a process plant.

into regimes characterised by various production levels and qualities, or raw material types. Within each operating regime there are typically different control objectives that are handled by different controllers or operating procedures. For example, during recovery after a major disturbance it is usually of primary interest to bring the plant safely back into normal operation as quickly as possible, while during normal production it is usually of primary interest to fine-tune the performance to optimise quality and production rate within the constraints given by the equipment, raw material and other factors. The number of operating regimes can be quite large, and we see that a hierarchical structure is very useful for efficiently visualising the model structure and handling the complexity. We review the literature of learning algorithms for hierarchical structures in section 1.4.5 on page 42.

1.3   OPERATING REGIME APPROACHES IN PERSPECTIVE

We will now attempt to present the operating regime approach in a broader perspective, focusing on relationships, advantages and disadvantages compared to other approaches to modelling and control of complex systems.

As discussed above, the advantage of the operating regime approach is that the modelling problem can be solved by developing simple local models that are assumed to be valid only when the system operates within their respective operating regimes. Likewise, the control problem can be solved by developing simple local controllers for each operating regime. The local models (or controllers) can then be combined by different means in order to give a global model (or controller) for the system, as described in earlier sections.

The drawback of this approach is that the number of operating regimes may increase rapidly as the complexity of the system increases. Also, it may be difficult to determine the operating regimes, and how they should be characterised, which is the key to successful modelling and control within this framework. One advantage is that qualitative, vague and imprecise system knowledge can more easily be incorporated into the model or controller than with the other decomposition approaches mentioned above.

1.3.1 Decomposition of complex modelling and control problems

In everyday life, as well as in solving engineering problems, the standard approach to complex problem solving is the divide-and-conquer strategy: *A complex problem is somehow partitioned into a number of simpler subproblems that can be solved independently, and whose individual solutions yield the solution of the original complex problem.* The key to successful problem solving with this approach is to find suitable axes along which the problem can be partitioned.

In the following, consider the problem of developing a dynamic model or a controller of a complex system. This problem can be decomposed along at least five axes:

1 decomposition into physical components,
2 decomposition based on phenomena,
3 decomposition in terms of mathematical series expansions,
4 decomposition into goals, and
5 decomposition into operating regimes.

Decomposition into physical components

The first approach we consider is to decompose the complex system into system components or unit processes that correspond to physical components like heat exchangers, chemical reactors, buffer tanks, compressors, valves etc.

The modelling problem can now be solved by assuming that simple sub-models can be built for each component. These can be combined into a total model by establishing relations between the components in terms of mass, energy, momentum and information flow, see Figure 1.14(a). It is clearly simpler to develop sub-models for each component, and such sub-models can be recycled for similar components.

The control problem can be solved by designing decentralised controllers for each unit process, see Figure 1.14(b). This is a great simplification since a number of SISO (Single Input Single Output) controllers can be designed, rather than a single MIMO (Multiple Input Multiple Output) controller. Decentralised controllers may work well together when the interactions between the units are not too strong.

The main difficulty with this approach is to identify a set of components that are loosely coupled. It does not make much sense to decompose a system into components that are strongly coupled, because it does not untie the complexity. A major problem may be that some of the components may still be quite complex because they cannot naturally be broken down into loosely coupled sub-components.

Consider the operating range of a system which consists of a number of interconnected

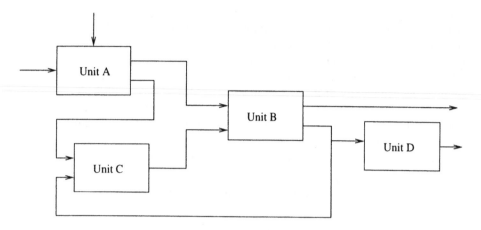

(a) A complex system is decomposed into physical components that are related by flows.

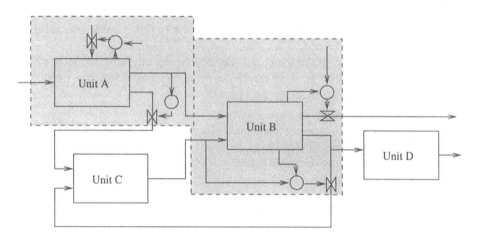

(b) A complex system is decomposed into physical unit processes that are controlled by decentralised controllers.

Figure 1.14   Complex systems can be decomposed in several different ways.

components. Clearly, when the system becomes very complex, it may be difficult to characterise the operation of this complex system. It makes more sense to consider the operating ranges of each of the components individually, which is the natural way of combining operating regime based and component based decompositions.

Decomposition into phenomena

A second approach is to decompose into the set of relevant phenomena that governs the system behaviour, see Figure 1.15. For instance, in a chemical reactor, relevant phenomena may be various chemical reactions, heat transfer and thermodynamical phenomena like heat conduction, compression and phase transition, or mass transfer phenomena like diffusion and convection. Models of each phenomenon can be developed and combined by taking into account their interactions. This approach is the basis for mathematical modelling using first principles.

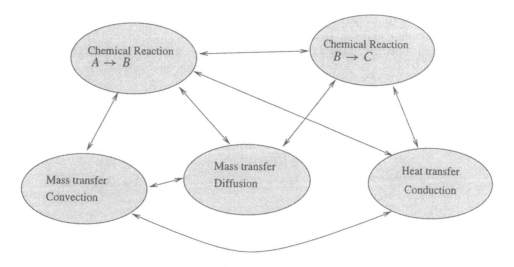

Figure 1.15 A complex system is decomposed into interacting phenomena.

The key to successful model development with this approach is to be able to identify the relevant phenomena, and describe their interactions correctly. Unfortunately, this may be difficult for a complex system that has strongly interacting phenomena. These interactions will appear in the balance equations as terms that contain a large number of variables. As the dimensionality and complexity of the system increase, the number and complexity of these terms increases rapidly and may make this approach unmanageable.

As we have illustrated above, operating regimes may at least sometimes be characterised by phenomena. The local models within operating regimes characterised by phenomena may be first principles models based on the assumed set of phenomena that is relevant within each operating regime. Likewise, local controllers with different structures may be designed to control the most important phenomena within each operating regime. Hence, we have established a very close similarity between decomposition on the basis of phenomena and operating regimes. As an example, consider the four operating regimes of the batch fermentation reactor discussed above. The first operating regime was characterised by the phenomena growth of cells and supply of oxygen, while none of the other possible phenomena were relevant. A mass-balance for this operating regime can therefore be simplified, since only one of the three chemical reactions need be taken into account. Likewise, there are different sets of chemical reactions (phenomena) that are relevant in the other operating regimes as well.

Decomposition based on mathematical series expansion

A third decomposition approach is the basis for black-box modelling and control. General classes of complex operators f can be mathematically built up from simpler basis-operators

f_1, f_2, ..., $f_{n_{\mathcal{M}}}$ in a series expansion

$$y \;=\; f(u) \;=\; \sum_{i=1}^{n_{\mathcal{M}}} f_i(u)$$

see also Figure 1.16. Clearly, such operators can be useful as input/output models or controllers for a complex system. In order to decouple the development of each basis-operator, the set of basis-operators is often chosen to be orthogonal, or at least close to orthogonal.

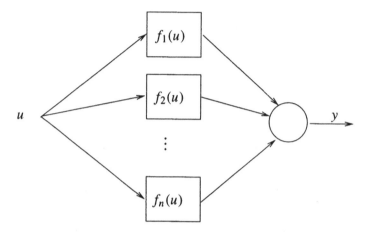

Figure 1.16   A complex system is decomposed mathematically by basis-operators that together describe the input/output behaviour.

For representing models, the basis-operators f_1, f_2, ..., $f_{n_{\mathcal{M}}}$ are typically chosen on the basis of empirical data and prior knowledge about certain system characteristics. For representing controllers, the basis-operators are either chosen on the basis of some system model and control design procedure, or they are chosen adaptively.

The key to successful modelling or control with this approach is to select the set of basis-operators carefully, and make sure that enough information is available to fit the parameters of the basis-operators.

The advantage of decomposition based on mathematical series expansions and black-box modelling is that it is a generic approach in the sense that the same set of basis-operators are suitable for defining the model or control structure for a wide range of systems. Only certain model or controller parameters must be tuned in order to adapt the model or controller to the particular application. However, this approach has serious disadvantages that may make it unsuitable for modelling highly complex systems: a sufficient amount of empirical data for every operating condition is required in order to "guarantee" (statistically) that the model is accurate. Various economic, safety and operational constraints, may make this assumption unrealistic. Moreover, system knowledge, over and above empirical data, may be difficult to incorporate directly into the series expansion.

Orthogonal basis-functions like Laguerre polynomials have been applied both for decomposing linear and nonlinear models (Wahlberg 1991, Wiener 1958) in series expansions. Non-orthogonal polynomial operators are used in the well known Volterra expansion of nonlinear system, e.g. (Rugh 1981). More recently, neural networks similar to series expansions and wavelet basis-functions have been suggested for nonlinear system modelling, e.g. (Sjöberg *et al.* 1994, Benaim 1994, Cannon and Slotine 1995) and nonlinear control, e.g. (Hunt *et al.* 1992, Narendra and Parthasarathy 1990).

Quite interestingly, there exists a very close relationship between operating regime based models and certain series expansions. This relationship is particularly evident for series expansions based on basis-functions which are local, in the sense that they are zero or close to zero over most of their input domain. Hence, the local basis-functions define a subset of the input domain that can be viewed as an operating regime. Such basis-functions act in exactly the same way as weighting functions, interpolation functions and fuzzy membership functions as described in section 1.2.4:

$$f(u) = \sum_{i=1}^{n_\mathcal{M}} \theta_i \rho_i(u),$$

where $\theta_1, \theta_2, ..., \theta_{n_\mathcal{M}}$ are real parameters and $\rho_1, \rho_2, ..., \rho_{n_\mathcal{M}}$ is the set of basis-functions.[4]

Furthermore, within the framework of series expansions, it has been suggested that the constant parameter θ_i should be replaced by a simple parameterised function of u, e.g. linear (Stokbro *et al.* 1990*b*, Jones and co-workers 1991)

$$f(u) = \sum_{i=1}^{n_\mathcal{M}} f_i(u, \theta_i) \rho_i(u),$$

which is clearly functionally equivalent to an operating regime approach. Relations between fuzzy models, certain series expansions, and operating regime based models are further discussed in (Jang and Sun 1993*a*, Foss and Johansen 1993, Hunt *et al.* 1995, Hunt *et al.* 1996*b*).

It is widely accepted that series-expansions based on orthogonal basis-functions have some useful properties compared to non-orthogonal basis function sets. It is clear that if the operating regimes are non-overlapping, the equivalent series expansion will be orthogonal since the weighting functions will be orthogonal

$$\int_u \rho_i(u) \rho_j(u) du = 0$$

for $i \neq j$. If the operating regimes do not overlap significantly, we can informally state that the basis-functions are almost orthogonal, in the sense that the inner product above is close to zero, see Figure 1.17. Hence, operating regime based models and controllers have much the same advantages as orthogonal series expansions: the local model or controller within each operating regime can be modified without significantly affecting the model accuracy or controller performance outside that operarating regime. Hence, single local models or controllers can be developed, added, analysed and modified more or less independently.

---

[4]Numerous basis-function sets satisfy the constraints

$$\sum_{i=1}^{n_\mathcal{M}} \rho_i(u) = 1$$

for all u, i.e. at any point in the input domain, the sum of all weighting functions should be 1. These include B-splines (Lane *et al.* 1992) and normalised basis-functions (Moody and Darken 1989):

$$\rho_i(u) = \frac{\Phi_i(u)}{\sum_{j=1}^{n_\mathcal{M}} \Phi_j(u)},$$

where $\Phi(u)$ is the general *unnormalised* basis-function. Normalisation can be important for such basis-functions, often making the model less sensitive to the parameterisation of the basis-functions, but it also has a number of side-effects which are discussed in detail in Chapter 8.

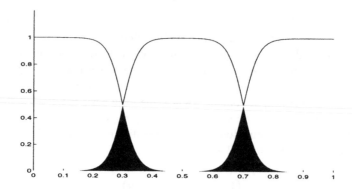

Figure 1.17   Basis-functions (or weighting functions) without too much overlap are close to orthogonal since the inner product (related to the shaded area) is small.

Decomposition based on goals

Most engineering control systems can be viewed as attempting to satisfy certain goals.[5] Complex systems are typically designed to satisfy multiple goals. Sometimes, several goals should be fulfilled simultaneously, but often the system must give priority to different goals under different conditions. Consequently, the behaviour and internal mechanisms of the system change with the changing priorities. Other approaches to coping with conflict when combining behaviours are discussed in (Yavnai 1993).

This attitude to problem decomposition is shown in the *subsumption architectures* applied to robotics problems in (Brooks 1986). These are layered architectures where each level is capable of control, independent of the higher levels, but where the higher levels are capable of increasingly complex behaviours. When a higher level behaviour is switched on it *subsumes* the lower level one, altering its internal states, while utilising its functionality. Different goal-oriented models are suggested in (Lind 1988, Lind 1991) as a basis for fault detection and diagnosis in process plants and by (Rasmussen 1986) in models of human problem solving.

Goal-oriented models and controllers are characterised by changes in goal under different conditions. These conditions may in many cases be naturally viewed as operating regimes, and it may be possible to describe the system's behaviour when attempting to satisfy different goals using local models or controllers.

For example, during normal operation, the goal of a control system is typically to control quality or production rate. On the other hand, during faulty operation, startup or shutdown, the goal of the control system is typically to bring the system quickly and safely to normal operation. In this case, the relations between the goal and the operating regimes are easily seen. Hence, we argue that the operating regime approach and goal-oriented modelling and control have very much in common and can benefit from each other.

---

[5]Whether or not *natural* systems should be viewed as teleological, or goal-directed, is a long-disputed topic, but in some cases such an analogy can, at a given level of abstraction, still be useful for the analysis of systems exhibiting a range of complex behaviours.

Which decomposition approach to apply?

Of course, all the above decomposition approaches are useful for solving practical problems. However, they have distinct characteristics that may clearly favour one over the other
for certain systems and applications, and they have overlapping characteristics such that the
choice between two of them may be more or less arbitrary in some cases.

It should be noted that the different approaches can (and must!) be combined in a number
of natural ways. In particular, both component-based decomposition and operating regime-
based decompositions must necessarily be combined with either series-expansion based or
phenomena-based models or controllers, since the local models or controllers, or component sub-models at some point must be represented in terms of equations based on series-
expansions or phenomena. In other words, goal-oriented, operating regime and component
approaches are mainly useful as an approach for breaking down the problem by introducing
a high level representation that brings the modelling or control problem to a higher and more
user-friendly level than equations. However, as discussed above, at some point one cannot
decompose further, but must deal with the sub-model or sub-controller in terms of low level
representations based on equations and numerical parameters.

1.4 SOLVING MODELLING PROBLEMS WITH OPERATING REGIME METHODS

The use of operating regime approaches is aimed at improving techniques for the design
of models of nonlinear dynamic systems with the aid of computationally intensive data-
driven techniques. This involves integrating knowledge about the system with data from
experiments. This allows the developer to produce more transparent model structures and
identify their parameters, as well as being able to validate the accuracy of the final model.
Modelling from data and knowledge is often viewed as an art form, mixing 'expert' insight
with the information in observed data, while using *ad hoc* simplifications to make the problem solvable. The typical sources of information could be:

- Experimental data, such as responses for perturbations about a number of operating
  points.
- A possibly incomplete nonlinear model or controller that may be too simple or too complex (computationally or conceptually) to be directly applicable.
- Qualitative knowledge, i.e. phenomena, behaviours, or operators' and engineers'
  heuristics.

A model will only be able to represent certain aspects of a system, so obviously, in order to
decide which aspects the model should capture, we need to know the purpose of the model.
Hence, modelling without a particular purpose in mind does not make sense. Typical applications of models include system analysis and optimisation, system design, control, automatic supervision and fault detection, prediction etc. The operating regime based approach
is not restricted to a particular purpose, but is a generic framework. So once the phase of
formulation of modelling goal and purpose of the model is complete we can view the typical
abstract modelling cycle to be as shown in Figure 1.18. This identifies a number of further
tasks that are essential in modelling:

- Experiment design and data acquisition.
- Raw data processing and analysis.
- Analysis of *a priori* knowledge, physical laws and available models.
- Structure and parameter identification – machine modelling.

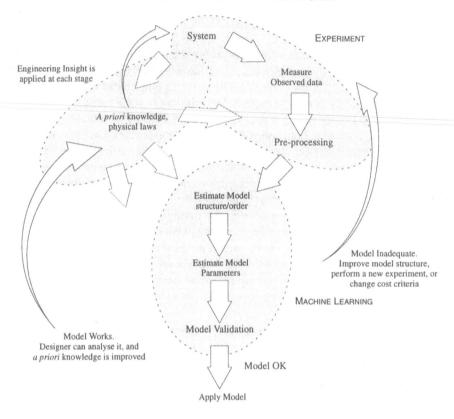

Figure 1.18 The Engineering Cycle for developing a model. Note the interaction of *a priori engineering insight, experiment design, data acquisition* and *pre-processing,* followed by *machine modelling* and *validation.*

- Model reduction and simplification.
- Model validation, analysis and interpretation.

The purpose of this section is to discuss the advantages of approaching these tasks with operating regime methods.

1.4.1 Experiment design

Experimental data is used for identifying the parameters and structure of the model, and for model validation. The general process of experimentation is shown in Figure 1.19.

It is obviously important that the data set has captured the important aspects of the system that we want to represent in the model. In particular, the data should cover the relevant areas of the operating space, but it is also important to consider the relative importance of the various areas of the operating space. The basic considerations being:

- Information about the relative significance of individual data samples is important. This could be in the form of probability distribution functions, showing the relative frequency of particular situations, noise levels, or the definition of regions which are particularly important.

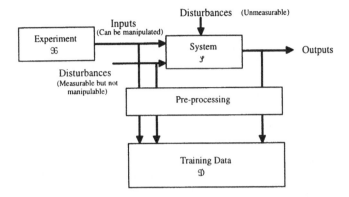

Figure 1.19 Acquiring and preparing experimental data.

■ In many processes it is important to have very accurate models in relatively stable areas, rather than less accurate models throughout the operating range, as the stable areas are often where most of the operating time is spent.

■ It is not only important to have identification data covering the operating range, but to have larger amounts of data where the system is most complex or its behaviour or control is most critical.

■ Certain operating conditions of the system may be associated with danger – how should these be treated in the model and in the acquisition of experimental data?

From this we can see that constructing a representative set of experimental data may often be more difficult, and require more consideration than the subsequent task of identifying the model! *Experiment Design* is the science of optimal data collection (Fedorov 1972, Goodwin and Payne 1977), and has been recently applied by workers in machine learning within the *active learning* framework. See Chapter 6 for a review of recent progress in this area.

When local operating regions are used, the experiment design phase is often simplified, because properties of the data set can often be directly linked to particular operating regimes and local models. For instance, lack of data from a certain operating condition can explain why the related local model's parameters are not accurately identified. If operating regimes are further decomposed, there is also a clearer link to the requirements of future experiments.

1.4.2 Pre-structuring operating regime based models

Operating regime based models can be viewed as general structures which are well suited for use in modelling dynamic systems, and especially well suited for use with learning algorithms. Determining the structure of the model is one of the most challenging tasks in model development. For operating regime models we describe structure identification as the task of determining a decomposition into operating regimes as well as determining the structure of the local models. Powerful algorithms for identifying a model structure from experimental data are emerging, and will be discussed in section 1.4.3, but as already mentioned, the analysis and use of *a priori* knowledge is essential in the search for a good model structure.

Analysis of a priori *information in structuring the model*

Empirical modelling from observed data is supposed to reduce the need to understand the detailed physical relationships within the system under investigation, but as can be seen in Figure 1.18 on page 30, a major feature of the cycle is the important role of *a priori* knowledge at each stage of the modelling process. *A priori* information is initial knowledge about the system, or problem in question. This includes aspects such as the goals of the problem, the characteristics of the process, its parameters, the effect of the environment (expected noise, disturbances), and the robustness requirements for different situations. Learning or adaptive systems are sometimes used because of the insufficiency of the *a priori* information, however, in practical applications it is usually necessary to constrain the learning task by introducing *a priori* knowledge into the model.

Depending on the available knowledge for a given problem, the approach for practical operating regime based modelling or controller development will differ considerably. We discuss two main classes of approach: *data-driven* and *model-driven*, but there will always be an element of both in any real problem. *A priori* knowledge can be used both *implicitly* and *explicitly*. It is used *implicitly* when framing the problem, in the act of creating a representative set of data, deciding which identification algorithms and structures are best suited to the problem and which inputs and pre-processing algorithms are likely to make the identification task easiest. The *explicit* use of *a priori* knowledge involves the direct integration of sub-models or rule-bases into the model. Model structure, dynamic order and sampling rate are dependent on *a priori* knowledge. Knowledge about physical constraints can be used to restrict the parameter set. Existing models or controllers (e.g. human operators) can also be used, where valid. Operating regime based models are characterised by their operating regimes, local model structures and local model parameters.

One of the major advantages of operating regime based modelling and control is that qualitative knowledge in various forms can be utilised directly when developing the model or controller structure. This includes the number of local models, the type and complexity of local model used, constraints on their parameters, and in particular, the decomposition of the operating range into operating regimes. This leads to more interpretable models which can be more reliably identified from a limited amount of experimental data.

Local models – number and type

The most general form of information is the expected dynamic order of the system, and the form of model to be identified (e.g. simple linear ARX models, state-space models etc). If more knowledge is available, the local models could be physically oriented models, possibly with only a subset of their variables to be identified, thus allowing the engineer to easily create *grey-box* models.[6]

Certain areas of the state-space may well be already accurately modelled using existing physically motivated models. These may be directly included as local models, where the associated weighting functions can be estimated by comparing the model outputs with the observed data. Another reason for including physical models is that there will not be sufficient data to identify the model from data throughout the state-space. This is especially true in areas outside normal desired operation, where the model may have to be very robust, and well understood, but where there are few data. These situations can be covered by fixing

---

[6] A generalised form would allow the designer to specify a pool of feasible local models, which could be locally tested for suitability in the various operating regimes defined by the basis functions. See for example (Johansen 1994c).

a priori models in the given areas, and applying learning techniques only where the data is available and reliable. Examples of this approach include (Kramer *et al.* 1992, Thompson and Kramer 1994, Foss and Johansen 1993, Su *et al.* 1992).[7]

A further option, for cases where a complex nonlinear model already exists and is valid for certain operating regimes, is to pre-set fixed local models to be locally accurate linearisations of this nonlinear function at operating points throughout these regimes. This is useful in cases where good models already exist, but they are too computationally expensive (or financially expensive in the case of proprietary software!) to be used in interactive control law development, or on-line in real-time systems. The use of a consistent framework also allows the comparison of parameters identified from data with the linearisations taken from existing models. This can then be added to the developer's toolkit as a validation method.[8]

For a linearisation to have any meaning in the classical sense, however, it must be around an equilibrium point of the dynamic system. Care must therefore be taken when choosing the scheduling inputs. This is discussed in section 1.6.

The problem of choosing the local model dynamic order parameters is certainly not a trivial problem. The general problem of order selection is widely discussed in the system identification literature, and several statistical criteria and general guidelines exist, e.g. (Box and Jenkins 1970, Ljung 1987). In the operating regime based modelling framework, there may be a need for a varying dynamic order of the different local models in different areas of the state space. Sbarbaro has suggested representing continuous-time local linear models with Laguerre networks to reduce the sensitivity of the model with respect to the choice of dynamic order (see Chapter 10).

A further possibility is to include physical constraints on the relationship between multiple parameters. Examples of *a priori* knowledge of physical constraints being used to improve generalisation in models can found in (Kramer *et al.* 1992, Thompson and Kramer 1994), and in (Röscheisen *et al.* 1992) who demonstrate the use of *a priori* models in identifying an RBF model of a rolling mill. A general method for constraining the parameters of physically motivated models is to select a prior distribution for parameters based on knowledge of, or beliefs about, the physical system. A theoretically elegant approach is to define a prior probability distribution for each parameter, where the developer must choose a distribution for the parameter. This helps prevent spurious combinations of variables, given sparse and noisy data, but is also a time-consuming method and may often be difficult to apply due to the lack of prior knowledge. Supporters of this approach, such as MacKay (1991), also show how priors can be parameterised with hyperparameters which can be derived from the training data during learning. These concepts are important and it seems likely that many ideas from Bayesian statistics (Bernardo and Smith 1994) are likely to become more widely used in data-driven modelling work in the near future.

Using transformed inputs to characterise the operating regimes

It is often possible to characterise the operating regimes using variables transformed in such a way that they still 'capture the nonlinearity', while having a reduced dimension (see

---

[7]In (Psichogios and Ungar 1992, Aoyama and Venkatasubramanian 1993, Brockett 1993), combinations of neural network structures and mechanistic model structures are suggested. Prior knowledge has been used for structuring neural nets (Mavrovouniotis and Chang 1992), initialisation of neural network parameters and interpretation of the resulting model through linearisations (Scott and Ray 1993).

[8]Note that the linearised models produced by such an approach will often have different parameters to those created by fitting a linear model from data using a least squares approach.

also the discussion of design of gain scheduling systems in section 1.5.1). This can range from the use of a subspace of the state and input space, for example some sort of principal components transformation, such that the operating regimes are characterised using as few variables as possible, to some form of knowledge based transformation.

For example, when modelling dynamic systems, we can often produce a dramatic reduction in the operating point vector by assuming that the operating point can be approximated using a lower order dynamic state (i.e. fewer delays in an ARX representation), while the local models use the full state vector.

Using prior knowledge to place the operating regimes

The partition of the problem is achieved by weighting functions associated with each local model which define its operating range, and the style of transition made from one behaviour/model to another. Locality of representation provides advantages for learning efficiency, and transparency. It is, however, very difficult to automatically find the 'correct' level of locality for a given subspace of an arbitrary problem, significantly more difficult than optimising the parameters of the local models. The resulting weighting functions have a major effect on the performance and interpretability of the trained model or controller. It is therefore important to constrain wherever possible the freedom of the weighting functions with prior knowledge.

The probabilistic interpretation of the weighting functions used in (Jordan and Jacobs 1994) (see also Chapter 6) views the weighting function ρ_i as the conditional prior probability of the local model f_i being the correct model. We therefore see the initialisation of the weighting functions as providing an initial set of priors for the likelihoods of the individual local models being correct (it is also possible to provide meta-priors for these weighting function priors, which constrain their parameters during learning (Waterhouse *et al.* 1996, McMichael 1995). As mentioned earlier, the probabilistic interpretation allows us to incorporate techniques from the areas of Belief networks, and uncertain reasoning (Pearl 1988, Shafer and Pearl 1990) to provide more sophisticated representation.

Using fuzzy rules for the model structure

It is possible to use other knowledge-based methods to position the weighting functions, or initialise their parameters. Because of the strong links between fuzzy membership functions and basis functions,[9] the *a priori* knowledge of how best to decompose the problem could be expressed as linguistic rules with accompanying membership functions. A key point here is that each rule will typically only be conditioned on a subspace of the state space, reducing the dimensionality problems of the learning algorithm.

Vague human intuition about the relative complexity of the operating range can often be expressed in IF .. THEN rules of the form

$$R_i \; : \quad \text{IF } z_1 \text{ is } Z_{i1} \text{ AND } \ldots \quad \text{AND } z_d \text{ is } Z_{id} \text{ THEN } y = \hat{f}_i(u)$$

which can either be used to provide initial parameters, or constraints on possible model structures. The use for initialisation of a model structure is fairly clear, the premise forming the basis function and the consequent forming the local model or controller, but their use as constraints is maybe less so: the membership functions which determine the distance

---

[9]See section 1.4.4 for a review of the relevant literature, and section 1.5.3 for a review of fuzzy control. Chapters 2 and 13 also deal with fuzzy controllers.

or similarity of a variable z_j from the fuzzy set Z_{ij} can still be optimised using learning algorithms, but the variables they are conditioned on must remain limited to the subset defined in the rule. This not only constrains the solution to one which should be better (assuming useful constraints were chosen), but to one which is also more interpretable, because of the 'conjunctive' or axis-orthogonal partition of the operating range (see Figure 1.21, and also the discussion in section 2.3.3 on page 85).

Using graphical structures to structure models

Graphical network models are being increasingly applied to represent human knowledge of causal links or of independence or irrelevance relationships between variables (e.g. (Whittaker 1990, Buntine 1994)). When used to structure Bayesian networks (Jensen 1996) they can be viewed as alternatives to the fuzzy rule-based approach to basis function representation just described.

Graphical structures can also be used to describe possible transitions between model states. This can be a very useful way of constraining operating regime models and controllers with prior knowledge. Many applications can usefully be abstracted in terms of a transition graph, where each node of the graph corresponds to a given submodel or behaviour. Figure 1.20 shows a general graph of behaviours. The actual transition functions can be purely random, state or input dependent, and can also be learned from data, but the very act of eliminating some transition possibilities will simplify the learning of the other transition functions. The transition functions can use the same sort of transformations described in the previous section. These methods are used in Chapters 4 and 5 to constrain the possible transitions between operating modes. This approach has the advantage that many engineers are already used to thinking in terms of finite state automata, as well as using them for documentation. This should make it easier to include useful prior knowledge, and to maintain it in future projects.

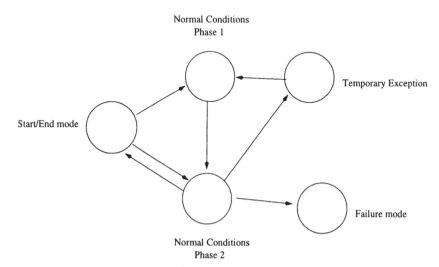

Figure 1.20 Example of a general graph method for constraining behaviour transitions. Only transitions marked by arrows are permitted.

1.4.3   Learning operating regimes from data

As many problems are not well enough understood for the model structure to be fully specified in advance, it will often be necessary to adapt the structure for a given problem, based on information in the data. The optimisation of the model structure is, however, a difficult non-convex optimisation problem, and is probably one of the most important areas of research for data-driven modelling approaches.

The goal of the structure identification procedure is to provide a problem-adaptive learning scheme which automatically relates the density of operating regimes and their size to the local complexity and importance of the system being modelled. The desirable features of a structure identification algorithm are:

- *Convergence* – as the number of data points increases the algorithm should produce models which approximate the real process more accurately.
- *Parsimony* – the model structure produced by the algorithm should be the simplest possible which can represent the process to the required accuracy.
- *Robustness* – the model structures produced should be as robust as possible with regards to noisy data or missing data.
- *Interpretability* – the model structure produced should ideally be as interpretable as possible, given the available data, local models and basis functions.

The model set is typically large, and parameterised by elements of both function spaces, discrete spaces, and Euclidean spaces. *Model structure identification* deals with search and optimisation in function spaces and discrete spaces. It is often an iterative process, where the model structure from the previous step in the iteration is used as a basis for the selection of a more promising model structure, either refined or simplified. Often, the structure identification problem is a combinatorial one, so good heuristics are important to minimise computational effort. Structure identification algorithms for operating regime based models must necessarily rely on some particular representation of the operating regimes, see Figure 1.21, and the options discussed in section 1.2.4 provide a range of options.

Robustness is an important aspect, as nonlinear models can obviously be very powerful, possibly representing the identification data very accurately by using a large number of parameters, but usually then leading to a high variance. The choice of model structure plays a major role in the *bias-variance trade-off* (see (Geman *et al.* 1992) for details about the trade-off), and this should be reflected in the cost functions.

Algorithms for structure identification from data should take into account the complexity of the system, the representational ability of the local models and their associated parameter estimation algorithms and the local availability of data. As shown in Figure 1.22, this can often lead to a number of confusing interacting effects.

Model reduction and simplification

Usually, one wants a model structure which is as simple as possible, but still able to capture the main aspects of the system that are relevant for the given application. A simple structure is desired because it is usually more easy to interpret and understand. Moreover, experience has shown that simpler models are typically more robust and reliable and the task of parameter identification and validation may be simpler.

Certain aspects of model reduction are simplified with operating regime based models. For example, a simplified model can be generated by merging two or more neighbouring operating regimes into a larger one, provided the local models within the regimes can be merged in a natural way (see Figure 2.6 on page 87). Another example is the reduction of single local models. For example, a high-order local model may be needed in some operating regimes, while a simpler low-order model may be sufficient in other operating regimes.

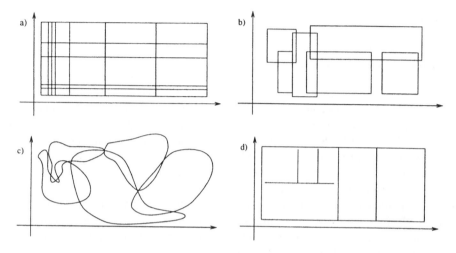

Figure 1.21 Examples of different operating regime representations. In a) we represent operating regimes with a simple regular grid, in b) we represent the operating regimes as rectangles, in c) the operating regime boundaries are parameterised by complicated functions, and d) shows a hierarchical partition of the operating range (as would be produced by a decision tree).

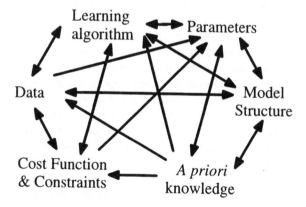

Figure 1.22 The structure learning process involves a number of complex interactions. In practice, many problems are successfully solved only by the intensive use of knowledge of the system structure. Automated approaches can help in this process, but should not be relied on alone.

1.4.4   Literature of local methods in learning and modelling

The earliest model we have found directly based on local models, is from the field of mathematical biology, and dates back to Kolmogoroff (1936). The model is an alternative to the classical Lotka–Volterra model, and based on a decomposition of the system into different regimes (called zones) with qualitatively different behaviour, see (Johansen 1994c) for a detailed discussion.

Piecewise models and splines

Piecewise linear models clearly fall into this framework. Early contributions describing modelling procedures based on piecewise linear models and an optimisation formulation of the regime decomposition problem are found in (Bellman 1961, Opoitsev 1970, Dorofeyuk et al. 1970, Kasavin 1972, Rajbman et al. 1981). Various other procedures are described in (Breiman and Meisel 1976, Cyrot-Normand and Mien 1980, Haber et al. 1982, Omohundro 1987, Farmer and Sidorowich 1987, Strömberg et al. 1991a, Hilhorst 1992, Söderman et al. 1993, Štecha and Havlena 1994).

Pao's *Functional Link Network* (Pao 1992), and Billing's closely related *Extended Model Set* ideas (Billings and Chen 1989), use links with a fixed non-linear function built in to expand the input vector, resulting in the production of extra 'higher-order' inputs. These are linearly weighted by the network's parameters.

(Skeppstedt et al. 1992) describes the use of local dynamic models for modelling and control purposes, but with hard transfers from one model regime to the next. (Pottmann et al. 1993) describes a multi-model approach where although the local models overlap there is still a sharp transition from one model to the next.[10]

A particularly interesting piecewise modelling approach from the biochemical engineering literature is described in Chapter 4, where the application of *a priori* knowledge for the decomposition into regimes (called functional states) is highlighted. The Markov Mixtures of Experts architecture described in Chapter 5, is another example of such switching methods, where the transition functions are identified from data.

The use of more complex local models than linear, for example a piecewise combination of several local neural network models (Sørheim 1990) and local polynomial models (Pottmann et al. 1993) are also described in the literature. Clearly, piecewise polynomials are closely related to the spline-based modelling approach, e.g. (Wahba 1990), with the main difference lying in the way splines handle multi-dimensionality. While piecewise polynomials are multi-dimensional by nature, splines are not, and multi-dimensionality is typically introduced by the means of tensor products (this is often also true of fuzzy logic when a conjunction of univariate fuzzy sets is used). The *Gaussian Bar* nets (Hartman and Keeler 1991, Kurcova 1992) are also closely related. These are nets with 'semilocal' basis functions formed by taking one-dimensional local basis functions and forming their tensor product to approximate a multi-variable function, as with tensor product spline models such as the ASMOD system (Kavli 1993). The advantage of these basis functions is that they can better cope with some classes of high dimensional problems, as they do not need to partition 'uninteresting' dimensions. A disadvantage is that the representation does not cope well with processes where the nonlinearity varies with several variables.

---

[10]To minimise model switching a heuristic criterion is introduced which only allows switching after three consecutive steps in the direction of the new model.

Basis function nets

Basis Function Networks and their equivalents have been used for function approximation and modelling in various forms for many years. The original Radial Basis Function methods came from *interpolation theory* and are described in (Powell 1987), where a basis function is associated with each data point, as in (Specht 1991). *Potential Functions* (Aizermann *et al.* 1964), *Kernels* (Wahba 1992) and *Spline Models* (Wahba 1990) are all similar structures. Albus' CMAC ideas have a great deal of overlap with basis function nets with uniformly distributed local basis functions (Albus 1975, Lane *et al.* 1991, Brown and Harris 1994).

Basis function nets have also received attention from the neural network community, starting with the early papers (Moody and Darken 1989, Broomhead and Lowe 1988). (Hlaváčková and Neruda 1993) gives a brief review of the use of RBF nets. (Poggio and Girosi 1990), (Girosi *et al.* 1993) and (Mason and Parks 1992) describe the networks within the mathematical framework of *Regularisation Theory* for function approximation. (Hutchinson 1994) describes the use of RBF nets for financial time series modelling. (Park and Sandberg 1991) and (Park and Sandberg 1993) proved the universal approximation abilities of RBF nets.

The well known *polynomial methods* such as Volterra models and Group Method of Data Handling (GMDH) (Ivakhnenko 1970, Ivakhnenko 1971) can also be viewed as Basis Function systems, as there is again a single layer of nonlinear functions, and the parameter optimisation is a linear process. The problem here is to find the suitable model structure – i.e. which set of basis functions can approximate the system adequately. Some off-line structure identification algorithms are described in (Chen and Billings 1994, Ivakhnenko 1971). Holden describes a general framework for basis function nets, calling them *Phi-nets* in (Holden 1994).

Statistics

The idea of using locally accurate models is also described in the statistical literature in (Cleveland *et al.* 1988) (see (Atkeson 1990) for a review of the literature of *local learning* methods in statistics), where local linear or quadratic models are weighted by smoothing functions. These methods store the entire data set and, for a given input point, form a locally weighted nonparametric representation of the system from the neighbouring data points. *Smoothing methods* such as Gaussian Kernel methods, as described in (Hastie and Tibshirani 1990) like other local averaging methods, suffer from the curse of dimensionality (Friedman 1991), and are computationally expensive. Such methods are applied to control problems in (Schaal and Atkeson 1994). The advantages of local representations are discussed in (Bottou and Vapnik 1992), where they suggest that a proper compromise between local and global methods will usually prove most effective as varying levels of complexity are required throughout the state space, although they claim that the 'local capacity' should match the data density, which is not necessarily true, as this would not relate model complexity to system complexity, but rather to data density. The more general goal of allocating local capacity is that the learning system should match the *local complexity* of the system.

Mixture Models used in statistics (Titterington *et al.* 1985, McLachlan and Basford 1988, Hastie and Tibshirani 1994) are created by mixing a number of probability distributions, and have many similarities to RBF nets (Bishop 1994, Nowlan 1991).

Operating regime based models could be viewed as a finite parameterisation of the state-dependent model. (Priestley 1988) describes *State Dependent Models* for non-linear time series which are basically linear models where the parameters depend in some way on the state. Tong's *Smooth Threshold Autoregressive* (STAR) models (Tong 1990) are also structurally equivalent to local model nets. In neither Priestley's or Tong's case, however, is much detail given about how to find the smooth weighting functions. (Billings and Voon 1987) also use a number of linear models to approximate a nonlinear system, but do not have smooth interpolation between basis functions.

Fuzzy models

The close relation of basis function nets to classes of *fuzzy logic systems* has also been discussed in (Jang and Sun 1993*b*), and (Brown and Harris 1994), where the similarity between membership functions and basis functions is pointed out. Some *fuzzy logic* systems can also be viewed as operating regime based models, e.g. the methods used in (Takagi and Sugeno 1985), where each operating regime is represented as a fuzzy set. This representation is appealing, since many systems change behaviour smoothly as a function of the operating point, and the soft transition between the regimes introduced by the fuzzy sets representations captures this feature in an elegant fashion. Algorithms for the identification of a decomposition into regimes on the basis of data are described by (Sugeno and Kang 1988) and several experimental and simulated examples are given (Takagi and Sugeno 1985, Sugeno and Kang 1986, Sugeno and Kang 1988, Nakamori *et al.* 1992). Similar applications are reported in (Sugeno and Kang 1988, Foss and Johansen 1993, Wang 1994) and (Harris *et al.* 1993). Related identification algorithms based on cluster analysis are described in (Bezdek *et al.* 1981*a*, Bezdek *et al.* 1981*b*, Hathaway and Bezdek 1993, Yoshinari *et al.* 1993, Nakamori and Ryoke 1994) and Chapters 2 and 13. Similar ideas are applied in chemometrics by (Næs and Isaksson 1991) and (Næs 1991). (Hunt *et al.* 1996*b*) discuss the similarity between fuzzy and basis function systems in more detail. Section 1.5.3 discusses the use of fuzzy methods in control.

Neural network models

Neural networks provide black-box nonlinear models and associated learning algorithms which have some analogies with biological information processing systems, and which have been increasingly applied to engineering problems in recent years. In terms of local model style architectures, the neural network community suggested the RBF network with parameterised local functions instead of constant coefficients in (Jones *et al.* 1989), followed up by (Stokbro *et al.* 1990*a*) and (Barnes *et al.* 1991), which describe approaches to modelling based on local linear models as generalisations of neural networks with localised receptive fields and radial basis-functions. The Adaptive Expert networks in (Jacobs *et al.* 1991) are also essentially local model systems, where the local models are called *expert networks* and the integration of the various *experts* is made by *gating networks*. These were developed into hierarchical models in (Jordan and Jacobs 1991, Jordan and Jacobs 1994). See also the similar approach in (Murray-Smith 1992). This probabilistic approach is found in Chapters 5 and 6. (Back and Tsoi 1991) describe the use of dynamic models as nodes in Multi-Layer Perceptrons – this could be seen as an MLP implementation of operating regime methods (with less locality).

Committee networks have received attention recently in the neural nets field, and have some relevance to multiple model methods. Here sub-models are combined in a committee (Perrone and Cooper 1993, Jacobs 1995) to improve performance. In the simplest case a uniform average of committee members is used, which is equivalent in structure to a local model net with uniform global basis functions – the difference in output of committee members is due to either differing types of committee members, or different local minima being found, if the optimisation methods used are nonconvex. More complex variations include differing weightings of committee members. Local model nets can be viewed as committee networks with 'more intelligent' weighting functions which demarcate the areas of validity of the 'committee members' – the local models/classifiers/controllers.[11]

Many other approaches to producing artificial systems with intelligent behaviour have developed sophisticated approaches to switching between 'behaviours' or 'skills', and much of this could prove relevant for multiple model approaches to modelling and control. We already mentioned the subsumption architectures (Brooks 1986), and arbitration networks (Yavnai 1993). Albus, in his wide-ranging paper (Albus 1991) states that *intelligence is the helmsman of behaviour*, a statement which fits in well with the idea of switching between or blending submodels or controllers.

1.4.5   Literature of structure identification methods

Parametric optimisation of weighting functions

A gradient descent optimisation technique for weighting function parameter optimisation is described in (Poggio and Girosi 1990). This method was applied in (Röscheisen *et al.* 1992) and also in (Hutchinson 1994), where some practical guidelines for clustering are given. Other RBF researchers used methods where a fixed number of basis functions was assumed, and the structure identification task was seen as the optimisation of the centres and widths. In the papers (Moody and Darken 1989, Sbarbaro 1992) clustering algorithms such as self-organising maps or k-means clustering were used to place the centres. A disadvantage of clustering algorithms is that they do not relate the location of the weighting functions to the complexity of the system being modelled, only to the location of data in the state space. Clustering methods such as k-means can be viewed as special cases of the EM (Expectation Maximisation) method (Dempster *et al.* 1977), which has been gaining increasing recognition in the neural net community (Nowlan 1991). In (Jordan and Jacobs 1994) the use of EM allowed the local model mismatch to be included in the optimisation of weighting functions.

The disadvantage of the techniques described above is that the user must still define the number of local models before identification starts, but as the complexity of the system is usually not fully understood, the optimal model structure is also unknown.

Constructive techniques

A hierarchical clustering technique based on a binary tree approach is used to recursively partition the operating range in (Stokbro *et al.* 1990a). The resulting model is 'flat', only the partitioning process is hierarchical. The partitioning is, however, not related to the local complexity of the system, but simply to the presence of data – a drawback of other similar algorithms.

Another option is to start off with a simple model, to estimate its parameters, determine where the representation is still unsatisfactory and to dynamically add new submodels to the model. This leads to a sequence of model structures followed by a parameter identification

---

[11] A Bayesian interpretation of such model averaging is discussed in section 10.7 of (Bishop 1995). Here, instead of just picking the most accurate model, all models are summed, weighted by the posterior probabilities of each model.

and confidence estimation stage. Constructive techniques which gradually enhance the model in this manner have a number of advantages: They automate the learning process by letting the model grow to fit the complexity of the system, but they do this robustly, by forcing growth to be guided by the availability of data and the complexity of the local models. This automatically determines the size of the model needed to approximate the system adequately, while preventing overfitting, e.g. (Murray-Smith and Gollee 1994, Murray-Smith and Hunt 1995) and (Murray-Smith 1992). (Waterhouse and Robinson 1996) use constructive methods together with EM to train hierarchical mixtures of experts models. See also page 87 for constructive approaches to fuzzy systems.

(Chen *et al.* 1991) used orthogonal least squares for the clustering task in developing radial basis function models. Their algorithm is a constructive one, which uses every data point as a candidate centre. Each time a basis function is added to the model, it chooses the best candidate centre by attempting to minimise the variance in the prediction error due to the model parameters. The *Resource Allocation Net* described in (Platt 1991) is a simple constructive algorithm, where when a data point is presented which causes an error larger than a given threshold a new basis function would be added at that point. Wynne-Jones suggests a constructive method where existing basis functions are split into two. The *Hierarchical Self-Organising Learning* (HSOL) algorithm, a hierarchical strategy for the construction of basis function models for classification, is described in (Lee and Kil 1991). Basis functions are added to the model in a coarse to fine strategy. (Carlin 1992) applied HSOL to modelling problems and similar coarse-to-fine ideas have been used for spline-based modelling applications (e.g. ASMOD in (Kavli 1993)). (Fritzke 1994) describes a constructive method based on a self-organising map framework. The heuristic local search algorithm (LSA) for operating regime based models splits the state space orthogonally to the axes of the state space (Johansen and Foss 1995). The spline-based MARS algorithm (Friedman 1991) mentioned earlier is also a constructive method.

The *Model Merging algorithm* described in (Omohundro 1991) attacks the problem in a different way, using a fine-to-coarse learning algorithm, where each data point is initially viewed as a model, and increasingly global models are created by merging the existing models. The bottom-up or top-down approaches both have advantages, and the choice will depend strongly on the application domain and the style of interpretation desired. See chapter 2 for a detailed example of such a procedure.

Hierarchical modelling and identification algorithms

The use of hierarchy in learning algorithms breaks down into two general camps. The decision tree methods started in the 1970s with *k-d trees* (Bently 1975), *Classification And Regression Trees* (CART) (Breiman *et al.* 1984), and *ID3* and *C4.5* (Quinlan 1993), and involve hierarchies of sharp partitions, dividing the input space into ever smaller areas. (Isaksson *et al.* 1991) and (Strömberg *et al.* 1991a) describe the use of such trees with dynamic models in the leaf nodes to model nonlinear dynamic systems. An alternative approach is the use of *soft partitions*, where the input space is no longer sharply partitioned, but there is a gradual transition between local models, allowing smooth interpolation between the local models. This also means that a point in the state space can activate local models in several leaves of the tree, with a differing level of membership to each.

Examples of this type of structure include *Basis-Function Trees* (Sanger 1991b, Sanger 1991a). The *Hierarchical Mixtures of Experts* (HME) structure, is a hierarchical structure (Jordan and Jacobs 1994) identified using Expectation Maximisation techniques (EM).

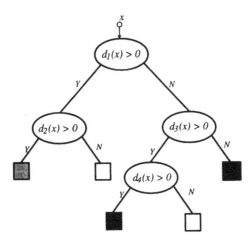

Figure 1.23 Decision tree structure. The conventional approach requires a number of discrete decisions which select leaves of the tree. These leaves can also be complex models or controllers, and the discrete nature of the node decisions can be altered to provide a smooth hierarchical blending structure.

(Miller and Rose 1996) presents more recent work which includes 'annealing' like characteristics in the learning process. Friedman used spline-based techniques to extend the CART ideas to soft splits for his *Multiple Adaptive Regression Splines* (MARS) algorithm (Friedman 1991). Quinlan also started to work with continuous systems modelling using *Model Trees* (Quinlan 1992). Links between wavelets and hierarchical networks (Bakshi and Stephanopoulos 1993) have also been investigated. Banan describes a constructive hierarchical method, with local linear models, which repeatedly partitions the operating range *randomly* in the areas producing errors. The 'average' network produced by the random splits is then the result of identification (Banan and Hjelmstad 1992). (Omohundro 1991) describes *Bump-* and *Balltrees* which have linear classifiers in the leaves of the tree. Constructive methods for hierarchical local model networks are described in (Murray-Smith 1992, Murray-Smith 1994).

The restriction of splits to hyperplanes orthogonal to the axes makes the method simple to implement and interpret, but it scales up poorly to higher dimensions and more complex models, especially when look-ahead search is used. Splits may be axis-oblique, i.e. the split depends on several variables, as shown in Figure 1.24, and they can also be crisp or soft (overlapping regimes). The use of smooth and oblique splits allows a variety of gradient-based optimisation algorithms to be applied to find the optimal partition, which is not possible in the classical decision tree methodology due to their crisp, axis-orthogonal partitions.

The longer term goal for hierarchical learning in multiple model systems is to make real use of hierarchy to make identification more efficient and tractable. A model can develop hierarchically to fit the data, and as the hierarchy develops and the representation of the system improves, decisions made earlier in the learning process (i.e. at higher levels in the decomposition process) can be re-evaluated, leading to gradual changes over a number of levels.

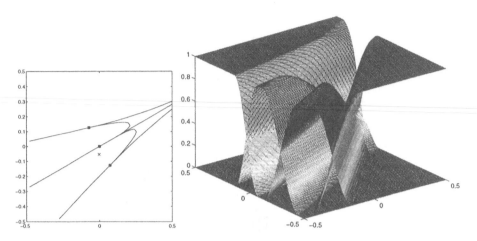

Figure 1.24 Contour and 3D representations of the basis functions of the leaves of the model tree for an axis-oblique partition. This shows graphically that the hierarchical structure can also be profitably viewed as a single layer partition.

1.4.6 Outlook for learning operating regime models

User-friendly methods for the easy integration of *a priori* knowledge into the model will become more and more important as the demands for accuracy, robustness and transparency of simulation models and control systems increase. There is a need for easy-to-use tools which allow the developer to creatively build engineering knowledge directly into the modelling process, leaving the learning algorithms to cope with the uncertainty in the process, and to warn the user when more information is needed, and support the user in obtaining this information.

Despite the improvements in tools, it is worth remembering that *'Thinking, intuition and insight cannot be made obsolete by automated model construction' (Ljung 1987).*

An immediately practical view of the methods described in this section is to see them as computationally- and data-intensive ways of supporting more traditional modelling methods, allowing the engineer to gain insight into the physical nature of the process by reproducing the behaviour with a learned model, understanding the behaviour and then creating a 'hand-built' simplified model which exhibits the essential behaviour of the structurally more complex empirically learned model. This is then more easily understood and validated, and is therefore more likely to be used in practical applications.

1.5 OPERATING REGIME BASED CONTROL – ALGORITHMS AND PROCEDURES

1.5.1 Gain scheduling like approaches

Traditionally, gain scheduling has been perhaps the most common systematic approach to control of strongly nonlinear systems in practice (Åström and Wittenmark 1989, Rugh 1991, Shamma and Athans 1992, Stein 1980, Whatley and Pott 1984, Gangsaas *et al.* 1986, Hyde and Glover 1993, Reichert 1992, Nichols *et al.* 1993). Even with the introduction of powerful nonlinear control strategies such as model predictive control (Garcia *et al.* 1989) and feedback linearisation (Isidori 1989), gain scheduling remains an attractive control strategy because it has some favourable engineering advantages such as simple design and tuning, and low computational complexity. It should be noted that by gain scheduling we mean a quite general class of control structures characterised by multiple local controllers, including Takagi–Sugeno fuzzy controllers (Takagi and Sugeno 1985).

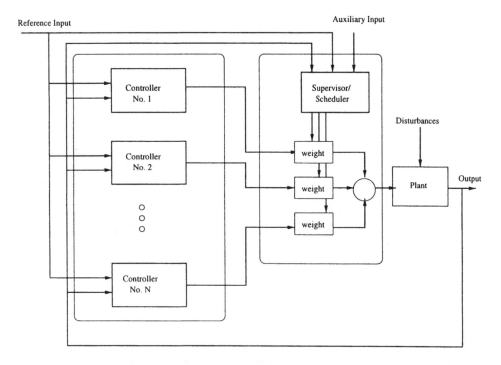

Figure 1.25 Typical structure of gain scheduled control system. Note that most of the literature of gain scheduled systems describes interpolation of parameters, but outputs can also be interpolated. See (Shorten 1996) for a discussion of the merits of parameter versus output interpolation.

A gain scheduling control system consists of two major components, see Figure 1.25: A family of controllers, and a scheduler that at each time instant assigns a controller (or a mixture of controllers) to be applied to the plant. One can think of the control system as having multiple local controllers that are applied in different operating regimes. In practice, gain scheduling can appear in different forms:

- A single controller from the family of controllers is selected. Alternatively, two or more controllers can be interpolated.
- One can interpolate local controllers' parameters (if they are parameterised in the same way), or the controllers' outputs.
- The family of controllers can be finite or infinite. In the infinite case the family of controllers is typically designed on the basis of a linearisation family of local models, e.g. (Wang and Rugh 1987).
- The scheduler can make its decision on the basis of a number of different variables and criteria. The variables can be states, inputs, measured disturbances, auxiliary variables, time etc.

■   Various control design methods can be applied and mixed to develop the family of controllers.

Local controller design

In virtually all design cases when explicit models are involved, the local controllers are designed on the basis of linear (or linearised) models about a family of equilibria. This is related to the fact that linear models and controllers are well understood and supported by software tools for identification, design and analysis. The gain scheduling framework in itself does not impose any restrictions on the local controller design algorithms, unless global stability and performance properties are taken directly into account. Hence, one can find examples of application of all the well known linear control design algorithms and heuristics.

Standard linear control design methods such as root locus (Tanaka and Sugeno 1992), pole placement (Shahruz and Behtash 1992) and frequency domain tools are widely applied. So are optimisation based methods such as LQR (Palm *et al.* 1995), generalised predictive control (GPC) (Chow *et al.* 1995), and optimal PID tuning (Jiang 1994). Recently, there has been a significant number of reports on the use of so-called robust and H_∞ control design methods in gain scheduling (Kellet 1991, Hyde and Glover 1993, Nichols *et al.* 1993). In particular, the use of Linear Matrix Inequalities for local control design with guaranteed global stability, robustness and performance are investigated in (Packard 1994, Apkarian *et al.* 1994, Apkarian and Ghainet 1995, Banerjee *et al.* 1995, Zhao *et al.* 1995*a*, Zhao *et al.* 1995*b*), see Chapters 12 and 13.

Scheduler design

Unlike local controller design, the scheduler design problem is not so well understood and there are fewer tools in the control engineer's toolbox to approach this design problem. The scheduler design problem consists of two major tasks: first to define the scheduling variables, and second to select a scheduling algorithm for selecting the local controller(s) to be applied.

Scheduling variables are usually selected on the basis of the following two heuristics (Shamma and Athans 1990, Shamma and Athans 1992):

■   The scheduling variables should capture the nonlinearities. After all, dealing with nonlinearities due to a wide operating range is usually the main reason for applying gain scheduled controllers.

■   The scheduling variables should be slowly time-varying compared to the desired bandwidth of the closed loop. This is motivated by the beneficial effect of separating timescales in complex and hierarchical control loops. It should, however, be noted that this heuristic is not always relevant. In particular, when the system is time-invariant and first order, this heuristic may not be necessary (Shamma and Athans 1990, Shamma and Athans 1992, Rugh 1991).

The simplest form of scheduling is hard switching between controllers or control parameters. Examples of this can be found in (Jiang 1994, Jacobs and Jordan 1993, Marcos and Artaza 1995, Morse 1995). Alternatively, the use of linear interpolation is very common (Hyde and Glover 1993, Nichols *et al.* 1993, Chow *et al.* 1995). As we have discussed in previous sections, operating regimes can conveniently be described as fuzzy sets and fuzzy inference can be applied to form a kind of interpolation (Takagi and Sugeno 1985, Tanaka and Sugeno 1992, Palm *et al.* 1995, Zhao *et al.* 1993, Zhao *et al.* 1995b, Cao *et al.* 1995). An advantage of interpolation in contrast to switching is that the control action will be smoother. If switching is applied, one should take explicit action to provide 'bumpless transfer' between the local controllers. Usually, this can be implemented by initialising the state of the controllers correctly after switching.

In general, one can interpolate the local controller outputs, or one can interpolate the local controller parameters (Kuipers and Åström 1994), see also Chapter 9. Of course, interpolation of controller parameters assumes that the controllers are homogeneous in structure, whereas interpolation of the local controller outputs does not impose any such constraints. Interpolation of outputs will increase controller order, whereas interpolation of parameters will not (Shorten 1996). This can be easily seen by considering the transfer function representation of each controller in the SISO case; in the presence of constant external scheduling vectors the transfer function of the control structure at intermediate equilibrium conditions is a convex sum of the individual controller transfer functions. The parameterisation of the controller may be of some importance if parameters are interpolated. Usually, the interpolation of zeros, poles and gains provides smoother and more robust control than the interpolation of polynomial coefficients in rational transfer functions (Chow *et al.* 1995, Shorten 1996).

Gain scheduled control system analysis

Analysis of the properties of the closed loop is of great importance. Somewhat surprisingly, gain scheduling control loop has been a neglected topic in the control theory literature until recently. In this section we will describe some results on stability and robustness of the closed loop, and discuss their relevance. Finally, we will briefly discuss performance.

Consider the problem of controlling a nonlinear system

$$x(t+1) \quad = \quad f(x(t), u(t)) \tag{1.13}$$

that can approximately be described as a weighted combination of a number of linear systems:

$$x(t+1) \quad = \quad \sum_{i=1}^{N} (x_i + A_i(x(t) - x_i) + B_i(u(t) - u_i))w_i(z(t)), \tag{1.14}$$

where the point (x_i, u_i) corresponds to an equilibrium, $z(t)$ is the scheduling variable, (A_i, B_i) is the controllable local linear model

$$A_i \quad = \quad \frac{\partial f}{\partial x}(x_i, u_i),$$

$$B_i \quad = \quad \frac{\partial f}{\partial u}(x_i, u_i).$$

Furthermore, consider a gain scheduled local linear state feedback

$$u(t) \quad = \quad u_i + \sum_{i=1}^{N} K_i(x(t) - x_i)w_i(z(t)). \tag{1.15}$$

The feedback matrix K_i can be designed using any design method, and the scheduler can be defined by arbitrary positive definite functions that satisfy

$$\sum_{i=1}^{N} w_i(z) \quad = \quad 1$$

for all z. It is clear that if there exists a matrix $P > 0$ such that $\tilde{A}_i^T P \tilde{A}_i - P = Q < 0$, then the linear control system

$$
\begin{aligned}
x(t+1) &= x_i + A_i(x(t) - x_i) + B_i(u(t) - u_i) \\
u(t) &= u_i + K_i(x(t) - x_i)
\end{aligned}
$$

with the closed-loop behaviour

$$
\begin{aligned}
x(t+1) - x_i &= \tilde{A}_i(x(t) - x_i) \\
\tilde{A}_i &= A_i + B_i K_i
\end{aligned}
$$

is asymptotically stable. This can easily be seen since the Lyapunov function $V(x) = x^T P x$ is decreasing along the system trajectory, e.g. (Khalil 1992). In (Tanaka and Sugeno 1992), this argument is extended to the nonlinear system (1.14) controlled by the gain scheduled controller (1.15). The stability condition is now that there must exist a common matrix $P > 0$ such that

$$
\tilde{A}_{ij}^T P \tilde{A}_{ij} - P < 0, \qquad \text{for all } i, j = 1, 2, ..., N \text{ where } w_i w_j \neq 0, \qquad (1.16)
$$

and $\tilde{A}_{ij} = (\tilde{A}_i + \tilde{A}_j)/2$, see also (Tanaka and Sano 1994). Again, this is seen quite easily from a Lyapunov argument, see also (Packard 1994, Apkarian et al. 1994, Apkarian and Ghainet 1995). In special cases, when for instance the system is first-order or in some canonical form, this is a quite strong result. However, one should be aware that these are only sufficient conditions for stability, that may be very conservative, in general. Hence, even if there does not exist a P matrix that satisfies the inequalities (1.16), then the closed loop may still be stable. Also notice that this result depends on how well the model (1.14) approximates the system (1.13). This is a topic that has been discussed in (Perev et al. 1995, Johansen and Foss 1993a), and in section 1.2.3.

In order to test the existence of the above-mentioned P matrix, it has been observed that (1.16) can be reformulated into a set of linear matrix inequalities (LMI) (Zhao et al. 1995a)

$$
\begin{pmatrix} P & P\tilde{A}_{ij} \\ \tilde{A}_{ij}^T P & P \end{pmatrix} > 0, \qquad \text{for all } i, j = 1, 2, ..., N \text{ where } w_i w_j \neq 0.
$$

LMIs can be studied and solved using the tools described in (Boyd et al. 1993, Boyd et al. 1994). If a P matrix is found, then Q can be computed and from the Lyapunov function it is possible to determine an upper bound on the convergence rate of the norm of the state, see Chapter 13. Related algorithms that guarantee H_∞ optimal performance are found in (Becker and Packard 1994, Apkarian et al. 1994, Apkarian and Ghainet 1995) and Chapter 12.

Robustness issues are studied in (Zhao et al. 1995b) and Chapter 13, who consider an uncertain model of the form

$$
x(t+1) = x_i + (A_i + \Delta A_i)(x(t) - x_i) + (B_i + \Delta B_i)(u(t) - u_i),
$$

where ΔA_i and ΔB_i represent the uncertainties. Related results can be found in (Tanaka and Sano 1994, Packard 1994, Becker and Packard 1994, Apkarian et al. 1994, Apkarian and Ghainet 1995).

With the formulation above, the closed loop can be written

$$
x(t+1) = \tilde{A}(z(t))x(t) \qquad (1.17)
$$

for some matrix $\tilde{A}$. If the local controllers are designed such that the poles of $A_i + B_i K_i$ are within the unit disc, then the system (1.17) will be asymptotically stable for any fixed value of the scheduling variable $z(t)$. However, it is well known that this is not a sufficient condition for uniform asymptotic stability of the system (1.17). If we consider $\tilde{A}(z(t)) = \tilde{A}(t)$ as an explicit function of time, then it is well known that a sufficient condition for uniform asymptotic stability of the system (1.17) is that the scheduling variables are sufficiently slowly time-varying (Desoer 1970), (Shamma and Athans 1990) and (Lawrence and Rugh 1990), in the sense that there exists a sufficiently small δ such that

$$\|\tilde{A}(t) - \tilde{A}(t-1)\| \leq \delta. \tag{1.18}$$

Hence, the heuristic to schedule on slow variables can be justified. However, it is important to bear in mind that when $\tilde{A}$ is viewed as a function of time, important structural information is lost when for instance it is partially a function of the state. Hence, the stability condition (1.18) may be conservative, and tight performance bounds and robustness margins may be hard to compute.

1.5.2 Multiple model adaptive control

In gain scheduling control systems, the scheduling algorithm is a deterministic function of the scheduling variables. The closely related Multiple Model Adaptive Control approach differs mainly from gain scheduling by the use of an estimator based scheduling algorithm used to weight the local controllers, see Chapter 11 or (Lainiotis 1976a, Lainiotis 1976b, Athans et al. 1977, Greene and Willsky 1980). A typical structure contains a model bank with Kalman filters with the twofold purpose of estimating the local model states as well as estimating local measures of accuracy. Sometimes a Markov model of the transition between the local models is also included (Sworder 1969, Blom and Bar-Shalom 1988). The local controllers are often LQG controllers (Chizeck et al. 1986, Griffiths and Loparo 1985), but other approaches exist. Closely related is also the approach in (Gendron et al. 1993, Narendra et al. 1995), where local model performance indices are used to select the local controller. One potential benefit of these approaches compared to gain scheduling is improved transient performance. The potential drawback is lack of robustness because there is a risk of selecting a poor controller when the system is not persistently excited. This is a general and well known problem with adaptive control approaches, e.g. (Narendra and Annaswamy 1989).

1.5.3 Fuzzy logic control

Within the area of fuzzy control there exists a wide variety of control structures and approaches. Somewhere in virtually all of these approaches one can find the idea of making a fuzzy partition of some region. Sometimes, this region can be viewed as the system's operating range, and the fuzzy partition can be interpreted as operating regimes.

The Takagi–Sugeno–Kang model has been discussed in section 1.4.4, and is used in Chapters 2 and 13, and the corresponding control structure is completely equivalent to gain scheduling based on local linear models and a fuzzy logic based scheduler. Therefore, this approach to fuzzy control was discussed in section 1.5.1 on gain scheduling, and we will not dwell on it here.

Fuzzy linguistic control is a control strategy based on linguistic (or rule-based) knowledge, see e.g. (Langari and Berenji 1992) or (Driankov et al. 1993) for a comprehensive introduction. The typical structure of a fuzzy linguistic controller is as shown in Figure 1.26.

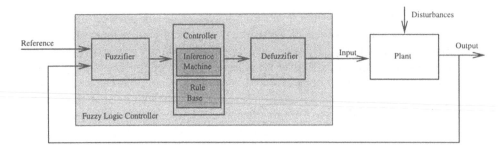

Figure 1.26   Typical structure of fuzzy logic control system.

Table 1.1   Example of a nonlinear PI-like fuzzy control rule table. The table shows the rate of change in control action as a function of the error and its derivative. The linguistic values are NL, negative large, NS, negative small, ZE, zero, PS, positive small, PL, positive large, which are being represented by fuzzy sets.

$\dot{e}/e$	NL	NS	ZE	PS	PL
NL	PL	PL	PS	ZE	ZE
NS	PL	PL	ZE	ZE	NS
ZE	PS	ZE	ZE	NS	NS
PS	ZE	ZE	ZE	NL	NL
PL	ZE	NS	NS	NL	NL

The effect of the components of the linguistic fuzzy logic controller is illustrated in Figure 1.27. In the first step the fuzzifier converts numerical information (like a measured output value, see Figure 1.27a) into a fuzzy set that the fuzzy inference mechanism can take as input, see Figure 1.27b. Uncertainty in the numerical information (such as measurement noise) can naturally be represented. The control rules in the rule base, Figure 1.27c, relate operating conditions (or the state) to control actions, such as

```
IF e is positive small AND ė is negative large THEN u̇ is
zero.
```

Clearly, the premise part of such control rules corresponds to a set of states that can be viewed as an operating regime, and the knowledge base can often be visualised and interpreted as being based on a (fuzzy) partitioning of the system's operating range, e.g. (Qin and Borders 1994). Often, the rule-base can be naturally represented as a table, see Table 1.1. The inference mechanism combines the fuzzy set representation of the measurements with the rule-base, and infers a fuzzy set representation of the control action, see Figure 1.27d. Finally, the defuzzification converts this fuzzy set into numerical information such as a command to the actuator, Figure 1.27e.

Hence, the design of such controllers consists of three main tasks:

1   Design of the rule-base on the basis of e.g. operators' or engineers' experience, knowledge and heuristics.
2   Design of membership functions that represents the fuzzy sets involved in the rule-base. This is often a considerable task due to the large number of numerical tuning parameters involved.

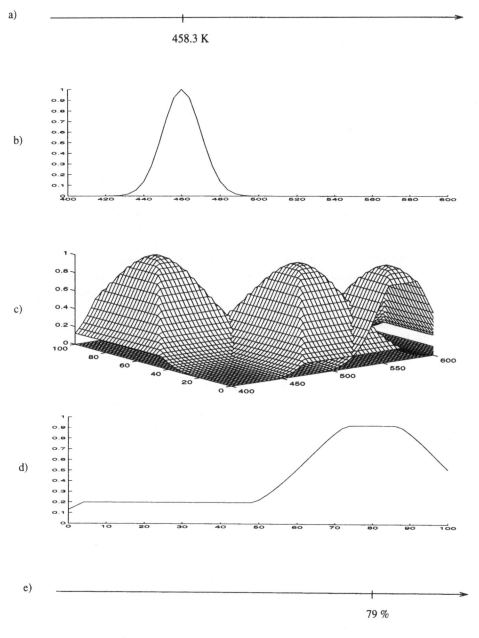

Figure 1.27 Example of fuzzy control inference.

3   Choice of inference mechanism, fuzzification and defuzzification algorithms. We will
    not discuss these issues here, but refer instead to (Driankov *et al.* 1993, Lee 1990*a*, Lee
    1990*b*).

The analysis of stability, robustness and performance of such control systems is still very
much an open problem, although some contributions exist. In (Langari and Tomizuka 1990)

it is assumed that the fuzzy logic controller is simply a static nonlinear function relating control error to control input. Applying a small gain type of argument, stability conditions related to the fuzzy logic controller parameters are then derived.

Another approach is taken by (Marin and Titli 1995) who assume the system is linear. It is suggested to compute the describing function of the nonlinear fuzzy logic controller numerically and use the Nyquist stability criterion to check stability of the closed loop. This technique is applied in (Olafsson 1995). Other approaches to stability analysis involve the circle criterion or conicity criterion (Driankov *et al.* 1993), and Interval Matrix method (Jamshidi and Titli 1995).

1.5.4   Global model-based control using local models

Instead of explicitly exploiting operating regimes directly in the control structure, one can apply operating regime approaches to develop an operating regime based model that can be applied in a model-based controller. This indirect approach has several advantages over the direct approaches described in previous sections. The main advantage is that a global controller is designed, which is expected to give better plant-wide control performance simply because global information can be applied to determine the control input at each time-step. This approach has been extensively studied using quite general model-based control design methods such as nonlinear model predictive control (Nakamori *et al.* 1992, Suzuki *et al.* 1992, Foss *et al.* 1995), feedback linearisation (including the adaptive control case) (Wang 1993, Johansen 1994*b*, Johansen 1994*a*), and sliding mode control (Ishigame *et al.* 1993).

With such model-based approaches it is evident that it is the quality of the particular model combined with the general properties of the model-based controllers that determines the stability, robustness and performance of the closed loop. Hence, the reader is referred to the general literature on such approaches (Rawlings *et al.* 1994, Isidori 1989, Sastry and Isidori 1989, Utkin 1992).

1.6   MODEL ANALYSIS, VALIDATION AND INTERPRETATION

Once a model structure and parameters have been identified from data it is necessary to validate the model's accuracy and properties. This is obviously a very important stage in a process which by its very nature has insufficient available intuition about the behaviour of the target system. Model validation is an issue which has often been given too little emphasis in the literature on nonlinear modelling and identification, but one which is very important in industrial situations. In 'local' methods, as with others, there will usually be a trade-off between flexibility and interpretability, the outcome of which will depend on their relative importance for a given application. It is therefore important to combine data-driven validation – is the model adequately accurate and robust for its purpose? – with more subjective validation, i.e. does the model behave in a way which seems physically plausible? Can the model be interpreted to give the engineer a better understanding of the system in question?

This can be rephrased as 'How do we extract *a posteriori* knowledge from the identified model?'. We have already discussed how the local representation, with its decomposition into operating regions, limited the interaction between model components and eased the introduction of *a priori* knowledge. It also, by the same logic, makes the extraction of knowledge *a posteriori* more simple.

Each operating regime can be handled more or less independently of the other operating regimes allowing one to solve complex nonlinear modelling or control problems by designing simple (e.g. linear) local models or controllers in each operating regime independently. This is important for validation, as the framework supports incremental modelling and simple model maintenance, because the modification of a single local model or operating regime has a more predictable effect on the global model.

Not surprisingly, this approach will in some cases *lead to spurious global behaviour of the model or controller even though the local behaviours and designs are well-behaved*, for various reasons. Hence it *is* important to analyse the global properties of the model or controller (see Chapter 9 for qualitative methods for simulating global behaviour). As we have discussed in section 1.5.1 and will discuss later, there exist some theoretical assumptions which relate the local properties to the global properties. However, in most practical situations one must rely on a less stringent analysis, based on extensive simulation of the global behaviour and transients, for example. This is partially due to conservativeness and lack of verifiability of the theoretical results.

One important aspect mentioned above, is that during modelling and identification the engineer is also gaining a better understanding of the system. This means that the *a priori* knowledge about the system in question will be added to, and can be used in a subsequent modelling iteration. The developer is still a vital part of the process, and the goal should be to enhance the power of the interactive software by automating the identification process wherever possible, but by always giving the engineer the freedom to intervene in any decisions. As knowledge about the system increases, more possibilities for better structuring and simplifying the identification task should become clearer. This then allows the automatic modelling and identification techniques to be used more successfully, and closes the loop in the modelling cycle seen in Figure 1.18, where the *a priori* knowledge is improved by the availability of the validated model.

Documentation properties of different representations

The final product, a working model, should therefore be seen as a contribution to the more general pool of engineering and scientific knowledge. The classical laws, rules or models we take as *a priori* knowledge today were also once poorly understood observed behaviour, which was then measured, analysed and turned into some simpler law or model. Kepler's laws of planetary motion were found only after painstaking acquisition of observed data, and the application of a variety of model structures to the data, estimation of the model parameters and validation of the models on new data!

Development of complex mathematical models is an expensive and time-consuming procedure. To lower costs and improve the maintainability of models, it is important that the model structures can be well documented for the use of future engineers. This means that the choice of local model structure, the variables used to schedule on, and the interactions between variables should all be easily documented. Local approaches are well suited to this, as the description of any complex system requires some sort of partitioning into understandable subsystems. They are certainly better suited to this than nonlinear black-box approaches such as neural networks. Also, if local models specific to a certain engineering discipline are used, it may often be possible to use many of the existing specialist tools for documenting the behaviour of such models.

By their very nature, graphical model structures such as Finite State Automata, Belief Networks or Markov Graphs have obvious advantages here. The interconnections between operating regimes are clear, and the scheduling/state transition variables can be clearly labelled, and if nonlinear transition functions are used, the same sort of visualisation techniques can be applied as for the basis functions in the next section. Decision trees are also graphical in their nature, and can be clarifying for some applications (e.g. diagnosis), but may not be ideal for intuitively describing dynamic systems.

Visualising the weighting functions

An advantage of operating regime approaches is that the model or controller structure can be visualised in terms of operating regimes, as we have seen in numerous figures above. This transfers the modelling or control design problem to a different level than thinking in terms of complex nonlinear equations. This framework is also engineering-friendly, since the operating regime concept is widely applied among engineers. This section discusses the interpretability of operating regimes and local models.

For one-dimensional problems it is relatively easy to get a clear picture of the model, as can be seen in Figure 1.28. Another interesting technique for one dimension is the effective kernel approach used in section 7.6.2, which gives insight into the data set and the learning algorithm, but this is unlikely to be of much use in other than toy examples.

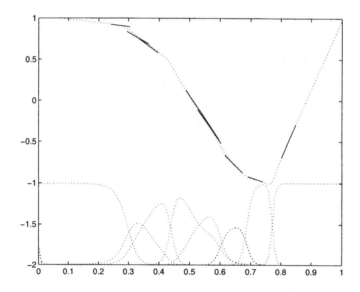

Figure 1.28   Local models, global model output and weighting functions.

In two and three dimensions it becomes harder, but it is still feasible to plot the weighting functions, as in Figure 1.29, from (Murray-Smith 1994). For higher dimensions it becomes difficult to get meaningful visualisation, other than 'slices' through the state space, unless a hierarchical representation is applied.

Note, however, that the weighting functions drawn above were unnormalised. When weighting functions are normalised their shape changes significantly (see, e.g. Figure 1.30). Chapter 8 deals more thoroughly with the effect of normalisation on the behaviour of the model.

A further aid to understanding the model is to go through the validation runs comparing the model output and actual output, while viewing the variables characterising the operating

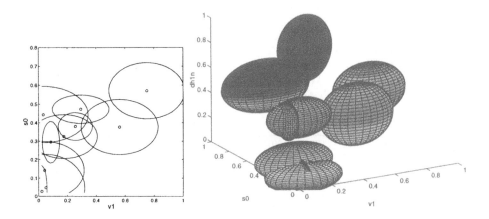

(a) 2-D basis function contour plot (contours drawn at 0.5).

(b) 3-D plot of weighting functions for roll mill model. The weighting functions are shown here in the unnormalised form, basically as ellipsoids representing the volume equivalent to that of the contour plot at 0.5. The hyper-ellipsoids in the figure correspond to the scales of the distance metrics of the weighting functions.

Figure 1.29 Visualisation of model operating regimes in two and three dimensions. A typical training run for the system (an aluminium mill) which is modelled by these operating regimes is shown in Figure 1.31.

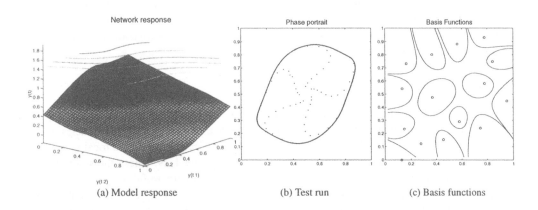

(a) Model response   (b) Test run   (c) Basis functions

Figure 1.30 Time series model with local linear models. This figure shows the nonlinear model response surface, time series data from a test run, and the normalised weighting functions.

regimes. The top plot in Figure 1.31 shows the model and actual output, while the plot below shows the characteristic variables for the run for the entire data set, with the portion being examined above marked as the darker area.

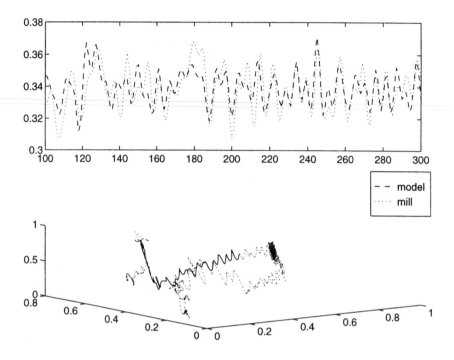

Figure 1.31 Model and system output, with the related area of the operating range. The visualisation tool allows you to examine the model residuals, while viewing the current position in the operating space below (the subset of the data corresponding to the model plot above is highlighted). More powerful tools would allow the use of such plots in conjunction with images such as Figure 1.29, where the local model closest to the data shown could be 'clicked' on and the structure and parameters viewed.

The use of such visualisation tools lets the developer examine the areas of the operating space responsible for large model deviations, to try and determine possible inadequacies in the model structure (e.g. too few local models in a particular region, ill-suited local model structures, etc.) with the aid of graphs such as shown in Figure 1.29. These tools are obviously limited to any three given dimensions, but can still provide useful insight in many cases. Further development of such tools would allow the user to select certain areas of the input space and find the nearest local models, so that their structures and parameters can be investigated, or gain more insight about the model workings. Axis-orthogonal partitions are produced by methods such as tensor product splines and fuzzy systems. These are easier to interpret, as in Figure 1.21. As described earlier, the hierarchical form of operating regime model can be viewed as a decision tree, a representation which, for binary trees with axis-orthogonal partitions (i.e. splitting on a single variable at a time), is often claimed to be especially suited to human interpretation.

For dynamic systems it is also helpful to plot the weighting functions against time, as in Figure 1.32.

The basis functions in a local model net are usually visualised as the individual ρ_k, but there are other ways of interpreting estimated models. Nonlinear data sets are often

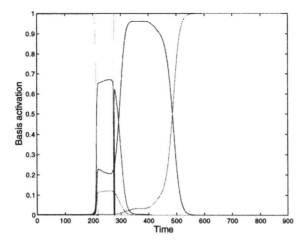

Figure 1.32   Plotting functions against time for a dynamic system.

modelled by simply *smoothing* the training data (Hastie and Tibshirani 1990) around the point of interest. This smoothing interpretation is described in section 7.6.2 on page 199, with graphical illustrations of such smooths.

Interpreting the local models

The local models in the operating regime approach are supposed to bring a greater transparency because they are *meaningful in their operating regimes*.[12] How far can we take micro-analysis of the component models, compared to macro-analysis of the combination of sub-models? The validity of the different methods of interpretation will depend on the training methods used. For example, linear local models estimated locally (see Chapter 7) around an equilibrium point (see Figure 1.33[13]) can be meaningfully treated as classical linearisations (although there will be differences from results produced by perturbation methods). If their parameters have been optimised globally however, it is no longer easy to do this if there is significant overlap with neighbouring local models.

A different method is to view the parameters of the current 'local model' of a homogeneous local model network as the weighted average of the local models' parameters . This can then be used to produce plots of important variables changing with the characterising variables. Examples of this include local models' poles and zeros (see Figure 3.7(b) on page 117 for an example of this), the eigenvalues of local state-space models, and parameter values themselves, if physically meaningful, with their local variance.[14]

The question of local interpretability becomes more complex when we deal with dynamic systems whose state trajectories pass through a number of local regions rapidly. Assume we have dynamic models that are supposed to be valid in some local regime i, i.e. $\dot{x} = f_i(x)$, where each local regime is parameterised by some weighting function ρ_i, leading to the

---

[12]We would like to thank Roger Bjørgan for contributing this section.

[13]We would like to thank K. Hunt and R. Haas for this diagram.

[14] Local estimates of accuracy are relatively easy to produce because of the partitioning of the state space inherent to the model structure, see (Murray-Smith 1994) for details.

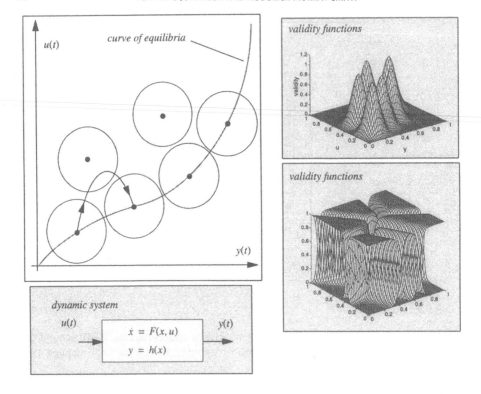

Figure 1.33 Interpreting local models. On-equilibrium local models can be interpreted classically, while off-equilibrium models can only be viewed as representing the dynamics on a part of a trajectory through the state space.

following global, dynamic model:

$$\dot{x} = \sum_{i=1}^{n_\mathcal{M}} \rho_i(x) f_i(x) = g(x). \tag{1.19}$$

If we introduce a small change to our notation:

$$\vec{\dot{x}} = \sum_{i=1}^{n_\mathcal{M}} \rho_i(x) \vec{f}_i(x) = \vec{g}(x), \tag{1.20}$$

for any trajectory of x, the vector $\vec{g}(x(t))$ is tangent to $\vec{x}(t)$. In Figure 1.34 we see a simple example of a vector field together with one trajectory produced by this vector field.

Having defined the local models and vector fields, we can return to the question of the interpretability of these models. The usual justification for local, linear models in dynamic systems theory is that if the operating trajectory is approaching an equilibrium point, then there is a regime around that equilibrium point where the linearised system describes the trajectory. This is the classical interpretation of local models, with a solid theoretical backing, producing valid approximations around the equilibrium point. If the system is adequately characterised by a linear (or affine) system in some region, even if it is not around

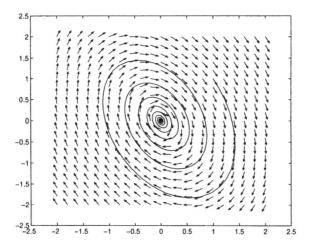

Figure 1.34   The vector field of a stable linear system and a sample trajectory.

an equilibrium point, we can still interpret the behaviour in this regime by using the local, linear model, *provided that the trajectory stays in this regime.*

If the trajectory doesn't stay in a single regime, a new vector field will be patched together by the local models. The linearity of each of the individual local models does not *necessarily* imply that the overall behaviour of the full system will be easy to interpret. To illustrate this we use two of the linear systems shown in Figure 1.34, and give them different offset terms (see equation (1.21)), and smooth validity functions to switch between the regimes

$$\dot{x} = \rho_1(x)\,[Ax + b] + \rho_2(x)\,[Ax - b]\,, \tag{1.21}$$

$$A = \begin{bmatrix} 2 & 5 \\ -10 & -3 \end{bmatrix} \qquad b = \begin{bmatrix} -10 \\ 10 \end{bmatrix}. \tag{1.22}$$

Two trajectories created by this system from the two following, slightly different initial conditions,

$$x_0 = \begin{bmatrix} 0 \\ 3.047427004 \end{bmatrix} \qquad x_0 = \begin{bmatrix} 0 \\ 3.047427005 \end{bmatrix} \tag{1.23}$$

are visualised in Figure 1.35(b): One trajectory is attracted to the upper equilibrium point, the other is pulled away into a large limit cycle. Naturally, this can be described as the usual sensitivity to initial conditions that often occur in nonlinear, dynamic systems. The main point is that even though the different regimes may be simple linear systems, if the trajectory doesn't stay in one of the regimes, the overall response of the interpolated system will be hard to interpret in terms of individual regimes. The new dynamic system can only be understood as a whole, hence making the complete system less interpretable than we may initially have hoped for.

Also note that in our simple example we can visualise the vector field for the full model, but even for this simple case it is difficult to predict which trajectory an initial condition would produce. A driving input to the system or a higher-order system would further complicate matters.

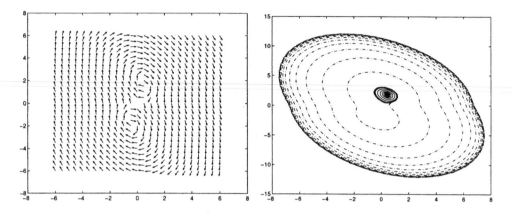

(a) The vector field of two stable linear systems.

(b) The response of two slightly different initial conditions. Dashed line goes to limit cycle, solid to an equilibrium point.

Figure 1.35   Vector field, and actual trajectories. Although the vector field appears smooth, the actual trajectories followed can quickly diverge.

Local behaviour around an equilibrium point

Treating the model as a linear parameter varying system,

$$\dot{x} = \sum_{i=1}^{n_{\mathcal{M}}} \rho_i(x)(A_i x + d_i) = A(x)x + d(x) = g(x), \tag{1.24}$$

we can linearise in the usual way to find the local behaviour around an equilibrium point x_0 of this system, $A(x_0)x_0 = -d(x_0)$:

$$\Delta\dot{x} = \frac{dg(x_0)}{dx}\Delta x = \left(A(x_0) + \frac{dA(x_0)}{dx}x_0 + \frac{dd(x_0)}{dx}\right)\Delta x. \tag{1.25}$$

Finding the partial derivatives for our particular structure and substituting we get

$$\Delta\dot{x} = \sum_{i=1}^{n_{\mathcal{M}}} \left((A_i x_0 + d_i)\frac{d\rho_i(x_0)}{dx}^T + \rho_i(x_0)A_i\right)\Delta x. \tag{1.26}$$

It follows immediately that in order to interpret the local model parameters,

$$A(x_0) \quad = \quad \sum_{i=1}^{n_{\mathcal{M}}} A_i \rho_i(x_0), \tag{1.27}$$

as the linearisation of the system (in the classical sense), it is required that

$$\sum_{i=1}^{n_{\mathcal{M}}} (A_i x_0 + d_i)\frac{d\rho_i(x_0)}{dx}^T \quad = \quad 0. \tag{1.28}$$

Notice that

$$\sum_{i=1}^{n_\mathcal{M}} \frac{d\rho_i(x_0)}{dx} = 0 \qquad (1.29)$$

for any x_0, and by the nature of the weighting functions, $\frac{d\rho_i(x_0)}{dx} \neq 0$ only for regions i that belong to the neighbourhood of x_0:

$$\mathcal{N}(x_0) = \{i \mid \rho_i(x_0) > 0\}.$$

Hence, (1.28) is equivalent to

$$\sum_{i \in \mathcal{N}(x_0)} \left((A_i - A_j)x_0 + (d_i - d_j) \right) \frac{d\rho_i(x_0)}{dx}^T = 0, \qquad (1.30)$$

where j is any element in $\mathcal{N}(x_0)$. Hence, it is clear that interpreting $A(x_0)$ as a linearisation in areas with non-negligible basis function gradients requires that neighbouring local model parameters are similar, i.e. the system changes behaviour smoothly, and the granularity of the decomposition is sufficiently fine. Moreover, it is desirable to have smooth weighting functions.

The problems of partitioning dynamic systems outlined in this section highlight the limits of sensible application of partitioning approaches to modelling and control. Local approaches can represent complex nonlinear systems, but if this results in local models which cannot be seen as classical approximations around an equilibrium point, then some of the hoped for interpretation advantages are lost.

1.7 CONCLUDING REMARKS

The purpose of this chapter is twofold. First, as the contributed chapters in the book represent a number of different approaches, each with its own language, insight and interpretations, this chapter should serve as an introduction to the basic concepts and ideas in operating regime and multiple model approaches to nonlinear modelling and control. This is basically about the motivation and the methods for partitioning a nonlinear dynamic system such that it can be represented by simpler subsystems, and can be more easily understood by humans.

The second half of the chapter gave a survey and overview of the current state of the art in terms of available procedures, algorithms, tools and applications. Here we have seen that there is a wide range of competing approaches. The main differences lie in the ease with which the various methods can incorporate external knowledge, and their suitability for use with learning algorithms.

There was also some discussion of methods for the interpretation of multiple model systems. This extends from simple visualisation of the operating regions, to results about model stability, and as in any approach to modelling or control of complex systems, the development of user-friendly tools will be a decisive factor in determining its popularity and success. There is already a great deal of applicable theory in this area, and practising engineers have always instinctively applied similar techniques. We believe that the theory and experience now needs to be translated into intuitively understandable design techniques and associated software tools which can integrate heterogeneous model structures, provide graphical interpretations and automatically diagnose potential problems with the given design – all the features needed for it to be regularly applied by nn-researchers to real world problems.

REFERENCES

AIZERMANN, M., E. BRAVERMAN AND L. RONZONOER (1964) 'Theoretical foundations of the potential function method in pattern recognition learning'. *Automatika i Telemekhanika* **25**, 147–169.

ALBUS, J. S. (1975) 'A new approach to manipulator control: The cerebellar model articulation controller (CMAC)'. *Trans. ASME. Jnl. Dyn. Sys. Meas. and Control* **63**(3), 220–227.

ALBUS, J. S. (1991) 'Outline of a theory of intelligence?'. *IEEE Transactions on Systems, Man, and Cybernetics* **21**(3), 473–509.

ANTSAKLIS, P. J., K. M. PASSINO AND S. J. WANG (1991) 'An introduction to autonomous control systems'. *IEEE Control Systems Magazine* **11**, 5–13.

AOYAMA, A. AND V. VENKATASUBRAMANIAN (1993) Integrating neural networks with first-principles knowledge for bioreactor modeling and control. Paper 147i, 1993 Annual AIChE Meeting, November, St. Louis.

APKARIAN, P. AND P. GHAINET (1995) 'A convex characterization of gain scheduled H_∞ controllers'. *IEEE Trans. Automatic Control* **40**, 853–864.

APKARIAN, P., P. GHAINET AND G. BECKER (1994) Self-scheduled H_∞ control of linear parameter-varying systems. In 'Proc. American Control Conference'. pp. 856–860.

ÅSTRÖM, K. J. AND B. WITTENMARK (1989) *Adaptive Control*. Addison-Wesley.

ÅSTRÖM, K. J. AND T. HÄGGLUND (1988) *Automatic tuning of PID controllers*. Instrument Society of America.

ÅSTRÖM, K. J. AND T. J. MCAVOY (1992) 'Intelligent control'. *J. Process Control* **2**, 115–125.

ÅSTRÖM, K. J., J. J. ANTON AND K. E. ÅRZÈN (1986) 'Expert control'. *Automatica* **22**, 277–286.

ATHANS, M., D. CASTANOU, K.-P. DUNN, C. S. GREENE, W. H. LEE, N. R. SANDELL AND A. S. WILLSKY (1977) 'The stochastic control of the F-8C aircraft using a multiple model adaptive control (MMAC) method – Part 1: Equilibrium flight'. *IEEE Trans. Automatic Control* **22**, 768–780.

ATKESON, C. G. (1990) Memory-based approaches to approximating continuous functions. In 'Workshop on Nonlinear Modelling and Forecasting, Santa Fe Institute'. Addison-Wesley.

BACK, A. D. AND A. C. TSOI (1991) 'FIR and IIR synapses, a new neural network architecture for time series modeling'. *Neural Computation* **3**(3), 375–385.

BADR, A., W. GHARIEB AND Z. BINDER (1991) Multimodel variable structure control systems. In 'Proc. IEEE Conf. Decision and Control, Brighton, UK'. pp. 2068–2073.

BAILEY, J. E. AND D. F. OLLIS (1986) *Biochemical Engineering Fundamentals*. McGraw-Hill, Singapore.

BAKSHI, B. R. AND G. STEPHANOPOULOS (1993) 'Wave-Net: a multiresolution, hierarchical neural neural network with localized learning'. *AIChE Journal* **39**(1), 57.

BANAN, M. R. AND K. D. HJELMSTAD (1992) Self-organization of architecture by simulated hierarchical adaptive random partitioning. In 'IJCNN, Baltimore'. Vol. III. pp. 823–828.

BANERJEE, A., Y. ARKUN, R. PEARSON AND B. OGUNNAIKE (1995) H_∞ control of nonlinear processes using multiple linear models. In 'Proc. European Control Conference, Rome'. pp. 2671–2676.

BARNES, C., S. BROWN, G. FLAKE, R. JONES, M. O'ROURKE AND Y. C. LEE (1991) Applications of neural networks to process control and modelling. In 'Artificial Neural Networks, Proceedings of 1991 Internat. Conf. Artif. Neur. Nets'. Vol. 1. pp. 321–326.

BARTON, P. I. AND C. C. PANTELIDES (1994) 'Modeling of combined discrete/continuous processes'. *AIChE Journal* **40**, 966–979.

BECKER, G. AND A. PACKARD (1994) 'Robust performance of linear parametrically varying systems using parametrically-dependent linear feedback'. *Systems and Control Letters* **23**, 205–215.

BELLMAN, R. E. (1961) *Adaptive Control Processes*. Princeton University Press. Princeton, NJ.

BENAIM, M. (1994) 'On functional approximation with normalised gaussian units'. *Neural Computation* **6**(1), 319–333.

BENCZE, W. J. AND G. F. FRANKLIN (1995) 'A separation principle for hybrid control system design'. *IEEE Control Systems Magazine* **15**(2), 80–85.

BENTLY, J. L. (1975) 'Multidimensional binary search trees used for associative searching'. *Communications of the ACM* **18**(9), 509–517.

BERNARDO, J. M. AND A. F. M. SMITH (1994) *Bayesian Theory*. John Wiley. New York.

BEZDEK, J. C., C. CORAY, R. GUNDERSON AND J. WATSON (1981a) 'Detection and characterization of cluster substructure. II. Fuzzy c-varieties and complex combinations thereof'. *SIAM J. Applied Mathematics* **40**, 352–372.

BEZDEK, J. C., C. CORAY, R. GUNDERSON AND J. WATSON (1981b) 'Detection and characterization of cluster substructure. I. Linear structure: Fuzzy c-lines'. *SIAM J. Applied Mathematics* **40**, 339–357.

BILLINGS, S. A. AND S. CHEN (1989) 'Extended model set, global data and threshold model identification of severely non-linear systems'. *Int. J. Control* **50**, 1897–1923.

BILLINGS, S. A. AND W. S. G. VOON (1987) 'Piecewise linear identification of non-linear systems'. *Int. J. Control* **46**, 215–235.

BISHOP, C. (1995) *Neural Networks for Pattern Recognition*. Clarendon Press, Oxford.

BISHOP, C. M. (1994) Training with noise is equivalent to Tikhonov Regularization. Submitted for publication.

BLOM, H. A. P. AND Y. BAR-SHALOM (1988) 'The interacting multiple model algorithm for systems with Markovian switching coefficients'. *IEEE Trans. Automatic Control* **33**, 780–783.

BOTTOU, L. AND V. VAPNIK (1992) 'Local learning algorithms'. *Neural Computation* **4**(6), 888–900.

BOX, G. E. P. AND G. M. JENKINS (1970) *Time Series Analysis: Forecasting and Control*. Holden-Day, San Francisco, CA.

BOYD, S., L. EL GHAOUI, E. FERON AND V. BALAKRISHNAN (1994) *Linear Matrix Inequalities in System and Control Theory*. SIAM, Philadelphia.

BOYD, S., V. BALAKRISHNAN, E. FERON AND L. EL GHAOUI (1993) Control system analysis and synthesis via linear matrix inequalities. In 'Proc. American Control Conference'. pp. 2147–2154.

BRANICKY, M. (1994) Stability of switched and hybrid systems. In 'Proceedings of Conference on Decision and Control, pp. 3498-3503'.

BREIMAN, L. AND W. S. MEISEL (1976) 'General estimates of the intrinsic variability of data in nonlinear regression models'. *J. American Statistical Association* **71**, 301–307.

BREIMAN, L., J. H. FRIEDMAN, R. A. OLSHEN AND C. J. STONE (1984) *Classification and Regression Trees*. Wadsworths & Brooks, Monterey, CA.

BROCKETT, R. (1993) 'Hybrid models for motion control systems'. In Hybrid Systems, Lecture Notes in Computer Science.

BROOKS, R. A. (1986) 'A robust layered control system for a mobile robot'. *IEEE Journal of Robotics and Automation*.

BROOMHEAD, D. S. AND D. LOWE (1988) 'Multivariable functional interpolation and adaptive networks'. *Complex Systems* **2**, 321–355.

BROWN, M. AND C. HARRIS (1994) *Neurofuzzy Adaptive Modelling and Control*. Prentice Hall. Hemel Hempstead, UK.

BUNTINE, W. (1994) 'Operations for learning using graphical models'. *Journal of Artificial Intelligence Research*.

CANNON, M. AND J.-J. SLOTINE (1995) 'Space-frequency localized basis function networks for nonlinear system estimation and control'. *Neurocomputing* **9**(3), 293–342.

CAO, S. G., N. W. REES AND G. FENG (1995) Stability analysis and design for a class of continuous-time fuzzy control systems. Preprint.

CARLIN, M. (1992) Radial Basis Function networks and nonlinear data modeling. In 'Neuro–Nimes 92'. pp. 623–633.

CHEN, S. AND S. A. BILLINGS (1994) Neural networks for nonlinear dynamic system modelling and identification. In H. C. J. (Ed.). 'Advances in Intelligent Control'. Taylor & Francis. London. pp. 85–112.

CHEN, S., C. F. N. COWAN AND P. M. GRANT (1991) 'Orthogonal least squares learning algorithm for radial basis function networks'. *IEEE Transactions on Neural Networks*.

CHIZECK, H. J., A. S. WILLSKY AND D. CASTANON (1986) 'Discrete-time Markovian jump linear quadratic optimal control'. *Int. J. Control*, 213–231.

CHOW, C.-M., A. G. KUZNETSOV AND D. W. CLARKE (1995) Using multiple models in predictive control. In 'Proc. European Control Conference, Rome'. pp. 1732–1737.

CLEVELAND, W. S., S. J. DEVLIN AND E. GROSSE (1988) 'Regression by local fitting'. *Journal of Econometrics* **37**, 87–114.

CYROT-NORMAND, D. AND H. D. V. MIEN (1980) Non-linear state-affine identification methods: Application to electrical power plants. In 'Proc. IFAC Symposium on Automatic Control in Power Generation, Distribution and Protection'. pp. 449–462.

DEMPSTER, A. P., N. M. LAIRD AND D. B. RUBIN (1977) 'Maximum likelihood from incomplete data via the EM algorithm'. *J. Royal Statistical Society Series B* **39**, 1–38.

DESOER, C. A. (1970) 'Slowly varying discrete system $x_{t+1} = a_t x_t$'. *Electronics Letters* **6**, 339–340.

DOROFEYUK, A. A., A. D. KASAVIN AND I. S. TORGOVITSKY (1970) Application of automatic classification methods to process identification in industry. In 'Preprints 2nd IFAC Symposium on Identification and Process Parameter Identification, Prague, Czechoslovakia'. p. 4.1.

DRIANKOV, D., H. HELLENDOORN AND M. REINFRANK (1993) *An Introduction to Fuzzy Control*. Springer-Verlag, Berlin.

FARMER, J. D. AND J. J. SIDOROWICH (1987) 'Predicting chaotic time series'. *Physical Review Letters* **59**(8), 845–848.

FEDOROV, V. V. (1972) *Theory of Optimal Experiments*. Academic Press. New York.

FOSS, B. A. AND T. A. JOHANSEN (1993) On local and fuzzy modelling. In '3rd Int. Conf. on Industrial Fuzzy Control and Intelligent Systems'. Houston, Texas.

FOSS, B. A., T. A. JOHANSEN AND A. V. SØRENSEN (1995) 'Nonlinear predictive control using local models – applied to a batch fermentation process'. *Control Engineering Practice* **3**, 389–396.

FRIEDMAN, J. H. (1991) 'Multivariate Adaptive Regression Splines'. *Annals of Statistics* **19**, 1–141.

FRITZKE, B. (1994) Supervised learning with growing cell structures. In J. D. Cowan, G. Tesauro and J. Alspector (Eds.). 'Advances in Neural Information Processing Systems 6'. Morgan Kaufmann Publishers. San Franciso, CA. pp. 255–262.

GANGSAAS, D., K. V. BRUCE, J. D. BLIGHT AND U. L. LY (1986) 'Applications of modern synthesis of aircraft control. Three case studies'. *IEEE Trans. Automatic Control* **31**, 995–1014.

GARCIA, C. E., D. M. PRETT AND M. MORARI (1989) 'Model predictive control: Theory and practice – A survey'. *Automatica* **25**, 335–348.

GAWTHROP, P. J. (1995) Continuous-time local state local model networks. In 'Proc. IEEE Conf. Systems, Man and Cybernetics, Vancouver, Canada'.

GAWTHROP, P. J. (1996) Continuous-time local model networks. In R. W. Żbikowski and K. Kunt (Eds.). 'Neural Adaptive Control Technology'. Vol. 1. World Scientific.

GEMAN, S., E. BIENENSTOCK AND R. DOURSAT (1992) 'Neural networks and the bias/variance dilemma'. *Neural Computation* **4**(1), 1–58.

GENDRON, S., M. PERRIER, J. BARETTE, M. AMJAD, A. HOLKO AND N. LEGAULT (1993) 'Deterministic adaptive control of SISO processes using model weighting adaptation'. *Int. J. Control* **58**, 1105–1123.

GIROSI, F., M. JONES AND T. POGGIO (1993) Priors, stabilizers and basis functions: from regularization to radial, tensor and additive splines. MIT AI Memo 1430. MIT.

GOODWIN, G. C. AND R. L. PAYNE (1977) *Dynamic System Identification: Experiment Design and Data Analysis.* Academic Press, New York.

GREENE, C. S. AND A. S. WILLSKY (1980) An analysis of the multiple model adaptive control algorithm. In 'Proc. of the IEEE Conference on Decision and Control'. pp. 1142–1145.

GRIFFITHS, B. E. AND K. A. LOPARO (1985) 'Optimal control of jump linear quadratic Gaussian systems'. *Int. J. Control* **42**, 791–819.

HABER, R., I. VAJK AND L. KEVICZKY (1982) Nonlinear system identification by "linear" systems having signal-dependent parameters. In 'Preprints 6th IFAC Symp. on Identification and System Parameter Identification'. Washington D.C.. pp. 421–426.

HALME, A. (1989) Expert system approach to recognize the state of fermentation and to diagnose faults in bioreactors. In N. M. Fish, R. I. Fox and N. F. Thornhill (Eds.). 'Computer Applications in Fermentation Technology'. Elsevier.

HARRIS, C., C. G. MOORE AND M. BROWN (1993) *Intelligent Control: Aspects of Fuzzy Logic and Neural Nets.* World Scientific.

HARTMAN, E. AND J. D. KEELER (1991) 'Predicting the future: Advantages of semilocal units'. *Neural Computation* **3**(4), 566–578.

HASTIE, T. AND R. TIBSHIRANI (1994) Discriminant analysis by Gaussian mixtures. Technical report. AT&T Bell Laboratories. Murray Hill, NJ.

HASTIE, T. J. AND R. J. TIBSHIRANI (1990) *Generalized Additive Models.* Monographs on Statistics and Applied Probability 43. Chapman and Hall. London.

HATHAWAY, R. J. AND J. C. BEZDEK (1993) 'Switching regression models and fuzzy clustering'. *IEEE Trans. Fuzzy Systems* **1**, 195–204.

HILHORST, R. A. (1992) Supervisory Control of Mode-Switch Processes. PhD thesis. University of Twente – Electrical Engineering Department.

HILHORST, R. A., J. VAN AMERONGEN AND P. LÖHNBERG (1991) Intelligent adaptive control of mode-switch processes. In 'Proc. IFAC International Symposium on Intelligent Tuning and Adaptive Control, Singapore'.

HLAVÁCKOVÁ, K. AND R. NERUDA (1993) 'Radial Basis Function networks'. *Neural Network World* **1**, 93–101.

HOLDEN, S. B. (1994) On the Theory of Generalization and Self-Structuring in Linearly Weighted Connectionist Networks. PhD thesis. Cambridge University.

HUNT, K. J., D. SBARBARO, R. ŻBIKOWSKI AND P. J. GAWTHROP (1992) 'Neural networks for control systems – A survey'. *Automatica* **28**, 1083–1112.

HUNT, K. J., J. C. KALKKUHL, H. FRITZ AND T. A. JOHANSEN (1996a) 'Constructive empirical modelling of longitudinal vehicle dynamics using local model networks'. *Control Engineering Practice* **4**(2), 167–178.

HUNT, K. J., R. HAAS AND M. BROWN (1995) 'On the functional equivalence of fuzzy inference systems and spline-based networks'. *Int. J. Neural Systems* **6**, 171–184.

HUNT, K. J., R. HAAS AND R. MURRAY-SMITH (1996b) 'Extending the functional equivalence of radial basis function networks and fuzzy inference systems'. *Trans. IEEE on Neural Networks.*

HUTCHINSON, J. M. (1994) Radial Basis Function Approach to Financial Time Series Analysis. PhD thesis. Massachusetts Institute of Technology, Dept. EECS.

HYDE, R. A. AND K. GLOVER (1993) 'The application of scheduled H_∞ controllers to a VSTOL aircraft'. *IEEE Trans. Automatic Control* **38**, 1021–1039.

ISAKSSON, A. J., L. LJUNG AND J.-E. STRÖMBERG (1991) On recursive construction of trees as models of dynamical systems. In '30th Conf. on Decision & Control, Brighton'. pp. 1686–1687.

ISHIGAME, A., T. FURUKAWA, S. KAWAMOTO AND T. TANIGUCHI (1993) 'Sliding mode controller design based on fuzzy inference for nonlinear systems'. *IEEE Trans. Ind. Electronics* **40**, 64–70.

ISIDORI, A. (1989) *Nonlinear Control Systems*. Springer-Verlag.

IVAKHNENKO, A. G. (1970) 'Heuristic self-organization in problems of engineering cybernetics'. *Automatica* **6**, 207–219.

IVAKHNENKO, A. G. (1971) 'Polynomial theory of complex systems'. *IEEE Transactions on Systems, Man, and Cybernetics* **1**(4), 364–378.

JACOBS, R. A. (1995) 'Adaptive mixtures of local experts'. *Neural Computation* **7**, 867–888.

JACOBS, R. A. AND M. I. JORDAN (1993) 'Learning piecewise control strategies in a modular neural network architecture'. *IEEE Trans. Systems, Man, and Cybernetics* **23**, 337–345.

JACOBS, R. A., M. I. JORDAN, S. J. NOWLAN AND G. E. HINTON (1991) 'Adaptive mixtures of local experts'. *Neural Computation* **3**(1), 79–87.

JAMSHIDI, M. AND A. TITLI (1995) Stability of fuzzy control system via interval matrix method. In 'Proc. 3rd IEEE Mediterranean Symp. New Directions in Control and Automation, Cyprus'.

JANG, J.-S. R. AND C.-T. SUN (1993a) 'Functional equivalence between radial basis function networks and fuzzy inference systems'. *IEEE Trans. Neural Networks* **4**, 156–159.

JANG, J. S. R. AND C. T. SUN (1993b) 'Functional equivalence between radial basis function networks and fuzzy inference systems'. *IEEE Transactions on Neural Networks* **4**(1), 156–158.

JENSEN, F. V. (1996) *An introduction to Bayesian Networks*. UCL Press.

JIANG, J. (1994) 'Optimal gain scheduling controller for a diesel engine'. *IEEE Control Systems Magazine* **14**(4), 42–48.

JOHANSEN, T. A. (1994a) Adaptive control of MIMO non-linear systems using local ARX models and interpolation. In 'IFAC ADCHEM 94'. Kyoto, Japan.

JOHANSEN, T. A. (1994b) 'Fuzzy model based control: Stability, robustness, and performance issues'. *IEEE Trans. Fuzzy Systems* **2**, 221–234.

JOHANSEN, T. A. (1994c) Operating regime based process modeling and identification. Technical Report 94-109-W. Dr. Ing. thesis. Department of Engineering Cybernetics, Norwegian Institute of Technology, Trondheim, Norway. http://www.itk.unit.no/ansatte/Johansen_Tor.Arne/dring_taj.ps.gz.

JOHANSEN, T. A. (1995) On the optimality of the Takagi-Sugeno-Kang fuzzy inference mechanism. Accepted for the 4th IEEE Conf. on Fuzzy Systems, Yokohama, Japan.

JOHANSEN, T. A. AND B. A. FOSS (1993a) 'Constructing NARMAX models using ARMAX models'. *Int. J. Control* **58**, 1125–1153.

JOHANSEN, T. A. AND B. A. FOSS (1993b) State-space modeling using operating regime decomposition and local models. In 'Preprints 12th IFAC World Congress, Sydney, Australia'. Vol. 1. pp. 431–434.

JOHANSEN, T. A. AND B. A. FOSS (1995) 'Identification of non-linear system structure and parameters using regime decomposition'. *Automatica* **31**, 321–326.

JONES, R. D. AND CO-WORKERS (1991) Nonlinear adaptive networks: A little theory, a few applications. Technical Report 91-273. Los Alamos National Lab., NM.

JONES, R. D., Y. C. LEE, C. W. BARNES, G. W. FLAKE, K. LEE, P. S. LEWIS AND S. QIAN (1989) Function approximation and time series prediction with neural networks. Technical Report 90-21. Los Alamos National Lab., New Mexico.

JORDAN, M. AND R.A. JACOBS (1994) 'Hierarchical mixtures of experts and the EM algorithm'. *Neural Computation* **6**, 181–214.

JORDAN, M. I. AND R. A. JACOBS (1991) Hierarchies of adaptive experts. In J. E. Moody, S. J. Hanson and R. P. Lippmann (Eds.). 'Advances in Neural Information Processing Systems 4'. Morgan Kaufmann Publishers. San Mateo, CA.

KASAVIN, A. D. (1972) 'Adaptive piecewise approximation algorithms in the identification problem'. *Automation and Remote Control* **33**, 2001–2006.

KAVLI, T. (1993) 'ASMOD – an algorithm for adaptive spline modelling of observation data'. *International Journal of Control* **58**(4), 947–967.

KELLET, M. G. (1991) Continuous scheduling of H_∞ controllers for a MS760 Paris aircraft. In P. H. Hammond (Ed.). 'Robust Control System Design Using H_∞ and Related Methods'. Institute of Measurement and Control, London.

KHALIL, H. (1992) *Nonlinear Systems*. Macmillan.

KOLMOGOROFF, A. (1936) 'Sulla theoria di Volterra della lotta per l'esistenza'. *G. Istit. Ital. Degli Attuari* **7**, 74–80.

KONSTANTINOV, K. AND T. YOSHIDA (1989) 'Physiological state control of fermentation processes'. *Biotechnology and Bioengineering* **33**, 1145–1156.

KONSTANTINOV, K. B. AND T. YOSHIDA (1991) 'A knowledge-based pattern recognition approach for real-time diagnosis and control of fermentation processes as variable structure plants'. *IEEE Trans. Systems, Man, and Cybernetics* **21**(4), 908–914.

KOSKO, B. (1994) 'Fuzzy systems as universal approximators'. *IEEE Trans. Computers* **43**, 1329–1333.

KRAMER, M. A., M. L. THOMPSON AND P. M. PHAGAT (1992) Embedding theoretical models in neural networks. In 'Proc. American Control Conference, Chicago, Il.'. pp. 475–479.

KUIPERS, B. AND K. ÅSTRÖM (1994) 'The composition and validation of heterogeneous control laws'. *Automatica* **30**, 233–249.

KURCOVA, V. (1992) Universal approximation using feedforward neural networks with Gaussian Bar units. In 'Proc. ECAI'92, Vienna'. pp. 193–197.

LAINIOTIS, D. (1976a) 'Partitioning: A unifying framework for adaptive systems. I: Estimation'. *Proc. IEEE* **64**, 1126–1143.

LAINIOTIS, D. (1976b) 'Partitioning: A unifying framework for adaptive systems. II: Control'. *Proc. IEEE* **64**, 1144–1161.

LANE, S. H., D. A. HANDELMAN AND J. J. GELFAND (1991) Higher order CMAC neural networks - theory and practice. In 'Proc. American Control Conference, Boston, USA'. pp. 1579–1585.

LANE, S. H., D. A. HANDELMAN AND J. J. GELFAND (1992) 'Theory and development of higher-order CMAC neural networks'. *IEEE Control System Magazine* **12**(2), 23–30.

LANGARI, G. AND M. TOMIZUKA (1990) Stability of fuzzy linguistic control systems. In 'Proc. 29th IEEE Conf. Decision and Control, Honolulu, Hawaii'. pp. 2185–2190.

LANGARI, R. AND H. R. BERENJI (1992) Fuzzy logic in control engineering. In D. A. White and D. A. Sofge (Eds.). 'Handbook of Intelligent Control'. Van Nostrand Reinhold, New York.

LAWRENCE, D. A. AND W. J. RUGH (1990) 'On a stability theorem for nonlinear systems with slowly-varying input'. *IEEE Trans. Automatic Control* **35**, 860–864.

LEE, C. C. (1990a) 'Fuzzy logic in control systems: Fuzzy logic controller – Part I'. *IEEE Trans. Systems, Man, and Cybernetics* **20**, 404–418.

LEE, C. C. (1990b) 'Fuzzy logic in control systems: Fuzzy logic controller – Part II'. *IEEE Trans. Systems, Man, and Cybernetics* **20**, 419–435.

LEE, S. AND M. R. KIL (1991) 'A Gaussian potential function network with hierarchically self-organizing learning'. *Neural Networks* **4**, 207–224.

LIND, M. (1988) Representing goals and functions of complex systems: An introduction to multi-level flow modelling. Technical Report, Institute of Automatic Control Systems, Technical University of Denmark.

LIND, M. (1991) Representations and abstractions for interface design using multilevel flow modelling. In G. R. S. Weir and J. R. Alty (Eds.). 'Human-Computer Interaction and Complex Systems'. Academic Press, London. pp. 223–244.

LJUNG, L. (1987) *System Identification — Theory for the User*. Prentice Hall. Englewood Cliffs, New Jersey, USA.

LUND, E. J., J. G. BALCHEN AND B. A. FOSS (1991) Multiple model estimation with inter-residual distance feedback. In 'Preprints IFAC/IFORS Symposium on Identification and System Parameter Estimation, Budapest'. pp. 1512–1517.

LUND, E. J., J. G. BALCHEN AND B. A. FOSS (1992) Monitoring processes with structurally different operating regimes. In 'Preprints IFAC Symposium on On-line Fault Detection and Supervision in the Chemical Process Industries, Newark, Delaware'. pp. 126–131.

MACKAY, D. J. C. (1991) Bayesian Methods for Adaptive Models. PhD thesis. California Institute of Technology, Pasadena, CA.

MARCOS, M. AND F. ARTAZA (1995) A methodology for the design of rule-based supervisors. In 'Proc. European Control Conference'. pp. 1619–1624.

MARIN, J. P. AND A. TITLI (1995) On the use of describing function to analyse and design fuzzy controllers. In 'Proc. IFSA-95, Sao Paulo, Brazil'.

MASON, J. C. AND P. C. PARKS (1992) Selection of neural network architectures: Some approximation theory guidelines. In 'In: K. Warwick, G. W. Irwin, K. J. Hunt (Eds), Neural networks for control and systems'. Peter Peregrinus. pp. 151–180.

MAVROVOUNIOTIS, M. L. AND S. CHANG (1992) 'Hierarchical neural networks'. *Comp. Chem. Engr.* **16**, 347–369.

MCLACHLAN, G. AND K.E. BASFORD (1988) *Mixture Models: Inference and Applications to Clustering.* Marcel Dekker.

MCMICHAEL, D. W. (1995) Bayesian growing and pruning strategies for MAP-optimal estimation of Gaussian mixture models. In '4th IEE Intern. Conf. on Artificial Neural Networks'. pp. 364–368.

MESAROVIC, M. D., D. MACKO AND Y. TAKAHARA (1970) *Theory of Hierarchical, Multilevel Systems.* Academic Press.

MILLER, D. AND K. ROSE (1996) 'Hierarchical, unsupervised learning with growing via phase transitions'. *Neural Computation* **8**(2), 425–450.

MOODY, J. AND C. DARKEN (1989) 'Fast-learning in networks of locally-tuned processing units'. *Neural Computation* **1**, 281–294.

MORSE, A. S. (1995) Control using logic-based switching. In A. Isidori (Ed.). 'Trends in Control'. Springer-Verlag, London. pp. 69–114.

MURRAY-SMITH, R. (1992) A Fractal Radial Basis Function network for modelling. In 'Inter. Conf. on Automation, Robotics and Computer Vision, Singapore'. Vol. 1. pp. NW–2.6.1–NW–2.6.5.

MURRAY-SMITH, R. (1994) A Local Model Network Approach to Nonlinear Modelling. PhD Thesis. Department of Computer Science, University of Strathclyde, Glasgow, Scotland.

MURRAY-SMITH, R. AND H. GOLLEE (1994) A constructive learning algorithm for local model networks. In 'Proc. IEEE Workshop on Computer-intensive methods in control and signal processing, Prague, Czech Republic'. pp. 21–29.

MURRAY-SMITH, R. AND K. J. HUNT (1995) Local model architectures for nonlinear modelling and control. In K. J. Hunt, G. R. Irwin and K. Warwick (Eds.). 'Neural Network Engineering in Dynamic Control Systems'. Advances in Industrial Control. Springer-Verlag. pp. 61–82.

NÆS, T. (1991) 'Multivariate calibration when data are split into subsets'. *J. Chemometrics* **5**, 487–501.

NÆS, T. AND T. ISAKSSON (1991) 'Splitting of calibration data by clustering analysis'. *J. Chemometrics* **5**, 49–65.

NAKAMORI, Y. AND M. RYOKE (1994) 'Identification of fuzzy prediction models through hyperellipsoidal clustering'. *IEEE Trans. Systems, Man, and Cybernetics* **24**, 1153–1173.

NAKAMORI, Y., K. SUZUKI AND T. YAMANAKA (1992) A new design of a fuzzy model predictive control system for nonlinear processes. In T. Terano, M. Sugeno, M. Mukaidono and K. Shigemasu (Eds.). 'Fuzzy Engineering Toward Human Friendly Systems'. IOS Press, Amsterdam. pp. 788–799.

NARENDRA, K. S. AND A. M. ANNASWAMY (1989) *Stable Adaptive Systems.* Prentice-Hall, Englewood Cliffs, NJ.

NARENDRA, K. S. AND K. PARTHASARATHY (1990) 'Identification and control of dynamical systems using neural networks'. *IEEE Transaction on Neural Networks* **1**(1), 4–27.

NARENDRA, K. S., J. BALAKRISHNAN AND M. K. CILIZ (1995) 'Adaptation and learning using multiple models, switching and tuning'. *IEEE Control Systems Magazine* **15**(3), 37–51.

NICHOLS, R. A., R. T. REICHERT AND W. J. RUGH (1993) 'Gain scheduling for H_∞ controllers for a MS760 Paris aircraft'. *IEEE Trans. Control Systems Technology* **1**, 69–75.

NOWLAN, S. J. (1991) Soft Competitive Adaptation: Neural Network Learning Algorithms based on Fitting Statistical Mixtures. CMU-CS-91-126. School of Computer Science, Carnegie Mellon University. Pittsburgh, PA.

OLAFSSON, J. F. (1995) Fuzzy control of water desalination plants. Diploma thesis, Department of Engineering Cybernetics, Norwegian Institute of Technology, Norway.

OMOHUNDRO, S. M. (1987) 'Efficient algorithms with neural network behavior'. *J. Complex Systems* **1**, 273–347.

OMOHUNDRO, S. M. (1991) Bumptrees for efficient function, constraint and classification learning. In R. P. Lippmann, J. E. Moody and D. S. Touretzky (Eds.). 'Advances in Neural Information Processing Systems 3'. Morgan Kaufmann Publishers. San Francisco, CA. pp. 693–699.

OPOITSEV, V. I. (1970) 'Identification of static plants by means of piecewise linear functions'. *Automation and Remote Control* **31**, 809–815.

PACKARD, A. (1994) 'Gain scheduling via linear fractional transformations'. *Systems and Control Letters* **22**(1), 79–92.

PALM, R., D. DRIANKOV AND U. REHFUESS (1995) Lyapunov linearization based design of Takagi-Sugeno fuzzy controllers. Preprint.

PAO, Y.-H. (1992) 'Functional link net computing'. *Computer* pp. 76–79.

PARK, J. AND I. W. SANDBERG (1991) 'Universal approximation using radial-basis-function networks'. *Neural Computation* **3**(2), 246–257.

PARK, J. AND I. W. SANDBERG (1993) 'Approximation and Radial-Basis-Function networks'. *Neural Computation* **5**(3), 305–316.

PEARL, J. (1988) *Probabilistic Reasoning in Intelligent Systems*. Morgan Kaufmann.

PEREV, K., C. JACOBSON AND B. SHAFAI (1995) 'An algorithm for identification of smooth nonlinear systems based on parameterized linearization families'. *Int. J. Control* **61**, 1013–1043.

PERRONE, M. P. AND L. N. COOPER (1993) When networks disagree: ensemble methods for hybrid neural networks. In R. J. Mammone (Ed.). 'Artificial Neural Networks for Speech and Vision'. Chapman and Hall. London.

PETRIDIS, Y. AND A. KEHAGIAS (1996) 'Modular neural networks for map classification of time series and the partition algorithm'. *IEEE Trans. Neural Networks* **7**, 73–86.

PETTIT, N. AND P. E. WELLSTEAD (1993) Piecewise-linear systems with logic control: A state-space representation. In 'Proc. European Control Conference, Groningen'. pp. 1581–1586.

PLATT, J. (1991) 'A Resource-Allocating Network for function interpolation'. *Neural Computation* **3**(2), 213–225.

POGGIO, T. AND F. GIROSI (1990) Networks for approximation and learning. In 'Proceedings of the IEEE'. Vol. 78. pp. 1481–1497.

POTTMANN, M., H. UNBEHAUEN AND D. E. SEBORG (1993) 'Application of a general multi-model approach for identification of highly nonlinear processes – a case study'. *Int. J. Control* **57**(1), 97–120.

POWELL, M. J. D. (1987) Radial Basis Functions for multivariable interpolation: A review. In 'Algorithms for Approximation'. Clarendon Press. Oxford. pp. 143–167.

PRIESTLEY, M. B. (1988) *Non-linear and Non-stationary Time Series Analysis*. Academic Press.

PSICHOGIOS, D. C. AND L. H. UNGAR (1992) 'A hybrid neural network – first principles approach to process modeling'. *AIChE J.* **38**, 1499–1511.

QIN, S. J. AND G. BORDERS (1994) 'A multiregion fuzzy logic controller for nonlinear process control'. *IEEE Trans. Fuzzy Systems* **2**, 74–89.

QUINLAN, J. (1992) Learning with continuous classes. In 'Australian AI Conf'. pp. 343–348.

QUINLAN, J. (1993) *C4.5 Programs for Machine Learning*. Morgan Kaufmann.

RAJBMAN, N. S., A. A. DOROFEYUK AND A. D. KASAVIN (1981) Identification of nonlinear processes by piecewise approximation. In P. Eykhoff (Ed.). 'Trends and Progress in System Identification'. Pergamon Press, Oxford. pp. 185–238.

RASMUSSEN, J. (1986) *Information Processing and Human-Machine Interaction: An Approach to Cognitive Engineering*. North-Holland.

RAWLINGS, J. B., E. S. MEADOWS AND K. R. MUSKE (1994) Nonlinear model predictive control: A tutorial and survey. In 'Preprints IFAC Symposium ADCHEM, Kyoto, Japan'. pp. 203–214.

REICHERT, R. T. (1992) 'Dynamic scheduling of modern-robust-control autopilot designs for missiles'. *IEEE Control Systems Magazine* **12**(5), 35–42.

RÖSCHEISEN, M., R. HOFMANN AND V. TRESP (1992) Neural control for rolling mills: Incorporating domain theories to overcome data deficiency. In J. E. Moody, S. J. Hanson and R. P. Lippmann (Eds.). 'Advances in Neural Information Processing Systems 4'. Morgan Kaufmann. San Mateo, CA. pp. 659–666.

RUGH, W. (1991) 'Analytical framework for gain scheduling'. *IEEE Control Systems Magazine* **11**(1), 79–84.

RUGH, W. J. (1981) *Nonlinear System Theory: The Volterra-Wiener Approach*. Johns Hopkins Press.

SANGER, T. D. (1991a) 'A tree-structured adaptive network for function approximation in high-dimensional spaces'. *IEEE Trans. on Neural Networks* **2**(2), 285–293.

SANGER, T. D. (1991b) 'A tree-structured algorithm for reducing computation in networks with separable basis functions'. *Neural Computation* **3**(1), 67–78.

SASTRY, S. S. AND A. ISIDORI (1989) 'Adaptive control of linearizable systems'. *IEEE Trans. Automatic Control* **34**, 1123–1131.

SBARBARO, D. G. (1992) A comparative study of different learning algorithms for Gaussian networks. In 'IFAC symposium on Intelligent Components and Intruments for Control Applications. Malaga, Spain'. pp. 301–305.

SCHAAL, S. AND C. ATKESON (1994) 'Robot juggling: Implementation of memory based learning'. *IEEE Control Systems Magazine* **14**(1), 57–71.

SCOTT, G. M. AND W. H. RAY (1993) 'Creating efficient nonlinear neural network process models that allow model interpretation'. *J. Process Control* **3**, 163–178.

SHAFER, G. AND PEARL, J. (EDS.) (1990) *Readings in Uncertain Reasoning*. Morgan Kaufmann. San Mateo, CA.

SHAHRUZ, S. M. AND S. BEHTASH (1992) 'Design of controllers for linear parameter-varying systems by gain scheduling technique'. *J. Mathematical Analysis and Applications* **168**, 195–217.

SHAMMA, J. S. AND M. ATHANS (1990) 'Analysis of gain scheduled control for nonlinear plants'. *IEEE Trans. Automatic Control* **35**, 898–907.

SHAMMA, J. S. AND M. ATHANS (1992) 'Gain scheduling: Potential hazards and possible remedies'. *IEEE Control Systems Magazine* **12**(3), 101–107.

SHORTEN, R. (1996) A Study of Hybrid Dynamical Systems with Application to Automotive Control. PhD thesis. Department of Electrical Engineering, University College Dublin, Republic of Ireland, June 1996.

SIMON, H. A. (1976) *Administrative Behavior 3rd ed.*. Free Press.

SIMONYI, K. E., N. K. LOH AND R. E. HASKELL (1989) An application of expert hierarchical control to piecewise linear systems. In 'Proc. 28th IEEE Conf. Decison and Control, Tampa, FL.'. pp. 822–827.

SJÖBERG, J., H. HJALMARSSON AND L. LJUNG (1994) Neural networks in system identification. In 'Preprints 10th IFAC Symp. System Identification, Copenhagen'. Vol. 2. pp. 49–72.

SKEPPSTEDT, A., L. LJUNG AND M. MILLNERT (1992) 'Construction of composite models from observed data'. *Int. J. Control* **55**(1), 141–152.

SØRHEIM, E. (1990) 'A combined network architecture using ART2 and back propagation for adaptive estimation of dynamical processes'. *Modeling, Identification and Control* **11**, 191–199.

SÖDERMAN, U., J. TOP AND J.-E. STRÖMBERG (1993) The conceptual side of mode switching. In 'Proc. IEEE Conf. Systems, Man, and Cybernetics, Le Touquet, France'. pp. 245–250.

SPECHT, D. F. (1991) 'A general regression neural network'. *IEEE Transactions on Neural Networks.*

ŠTECHA, J. AND V. HAVLENA (1994) Parameter tracking with alternative noise models. In 'Proc. IEEE Workshop CMP'94'. pp. 171–175.

STEIN, G. (1980) Adaptive flight control — a pragmatic view. In K. S. Narandra and R. V. Monopoli (Eds.). 'Applications of Adaptive Control'. Academic Press, New York.

STOKBRO, K., D. K. UMBERGER AND J. A. HERTZ (1990a) 'Exploiting neurons with localized receptive fields to learn chaos'. *Complex Systems* **4**(3), 603–622.

STOKBRO, K., J. A. HERTZ AND D. K. UMBERGER (1990b) 'Exploiting neurons with localized receptive fields to learn chaos'. *J. Complex Systems* **4**, 603.

STRÖMBERG, J.-E., F. GUSTAFSSON AND L. LJUNG (1991a) Trees as black-box model structures for dynamical systems. In 'Proc. European Control Conference, Grenoble'. pp. 1175–1180.

STRÖMBERG, J.-E., F. GUSTAFSSON AND L. LJUNG (1991b) Trees as black-box model structures for dynamical systems. In 'European Control Conference, Grenoble'. pp. 1175–1180.

SU, H.-T., N. BHAT AND T. J. MCAVOY (1992) Integrated neural networks with first principles models for dynamic modeling. In Preprints IFAC DYCORD+ '92, College Park, Maryland.

SUGENO, M. AND G. T. KANG (1986) 'Fuzzy modelling and control of multilayer incinerator'. *Fuzzy Sets and Systems* **18**, 329–346.

SUGENO, M. AND G. T. KANG (1988) 'Structure identification of fuzzy model'. *Fuzzy Sets and Systems* **26**, 15–33.

SUGENO, M. AND T. YASUKAWA (1993) 'A fuzzy-logic-based approach to qualitative modeling'. *IEEE Trans. Fuzzy Systems* **1**, 7–31.

SUZUKI, K., Y. NAKA AND K. BITO (1992) Fuzzy multi-model control of a high-purity distillation system. In T. Terano, M. Sugeno, M. Mukaidono and K. Shigemasu (Eds.). 'Fuzzy Engineering Toward Human Friendly Systems'. IOS Press, Amsterdam. pp. 684–693.

SWORDER, D. D. (1969) 'Feedback control of a class of linear systems with jump parameters'. *IEEE Trans. Automatic Control* **14**, 9–14.

TAKAGI, T. AND M. SUGENO (1985) 'Fuzzy identification of systems and its applications for modeling and control'. *IEEE Trans. on Systems, Man and Cybernetics* **15**(1), 116–132.

TANAKA, K. AND M. SANO (1994) 'A robust stabilizing problem of fuzzy controller systems and its application to backing up control of a truck-trailer'. *IEEE Trans. Fuzzy Systems* **2**, 119–134.

TANAKA, K. AND M. SUGENO (1992) 'Stability analysis and design of fuzzy control systems'. *Fuzzy Sets and Systems* **45**, 135–156.

THOMPSON, M. L. AND M. A. KRAMER (1994) 'Modeling chemical processes using prior knowledge and neural networks'. *AIChE J.* **40**, 1328–1340.

TITTERINGTON, D., A.F.M. SMITH AND U.E. MAKOV (1985) *Statistical Analysis of Finite Mixture Distributions*. John Wiley & Sons. Chichester.

TONG, H. (1990) *Non-linear Time Series: A Dynamical System Approach*. Oxford University Press. Oxford Statistical Science Series 6.

UTKIN, V. I. (1977) 'Variable structure systems with sliding modes'. *IEEE Trans. Automatic Control* **22**, 212–222.

UTKIN, V. I. (1992) *Sliding Modes in Control and Optimization*. Springer-Verlag, Berlin.

WAHBA, G. (1990) Spline models for observation data. In 'Regional Conference Series in Applied Mathematics'. SIAM. Philadelphia, PA.

WAHBA, G. (1992) Multivariate function and operator estimation, based on smoothing splines and reproducing kernels. In M. Casdagli and S. Eubank (Eds.). 'Nonlinear Modeling and Forecasting, SFI Studies in the Sciences of Complexity'. Vol. XII. Addison-Wesley.

WAHLBERG, B. (1991) 'System identification using Laguerre models'. *IEEE Trans. Automatic Control* **36**, 551–562.

WANG, J. AND W. J. RUGH (1987) 'Feedback linearization families for nonlinear systems'. *IEEE Trans. Automatic Control* **32**, 935–940.

WANG, L.-X. (1993) 'Stable adaptive fuzzy control of nonlinear systems'. *IEEE Trans. Fuzzy Systems* **1**, 146–155.

WANG, L.-X. (1994) *Adaptive Fuzzy Systems and Control: Design and Stability Analysis*. Prentice Hall.

WATERHOUSE, S., D. MACKAY AND T. ROBINSON (1996) Bayesian methods for mixtures of experts. Neural Information Processing Systems, **8**, 351–357, MIT press. Cambridge, MA.

WATERHOUSE, S. R. AND A. J. ROBINSON (1996) Constructive algorithms for hierarchical mixtures of experts. Submitted to NIPS'95. Neural Information Processing Systems, **8**, 351–357, MIT press. Cambridge, MA.

WHATLEY, M. J. AND D. C. POTT (1984) 'Adaptive gain improves reactor control'. *Hydrocarbon Processing* pp. 75–78.

WHITTAKER, J. (1990) *Graphical Models in Applied Multivariate Statistics*. John Wiley & Sons.

WIENER, N. (1958) *Nonlinear problems in random theory.*. The Technology Press, MIT and John Wiley and Sons Inc. New York.

YAVNAI, A. (1993) Arbitration network: A new approach for combining reflexive behaviors and reasoning-driven behaviors in intelligent autonomous systems. In 'Intelligent Autonomous Systems 3'. IOS Press, Washington. pp. 428–438.

YOSHINARI, Y., W. PEDRYCZ AND K. HIROTA (1993) 'Construction of fuzzy models through clustering techniques'. *Fuzzy Sets and Systems* **54**, 157–165.

ZADEH, L. A. (1973) 'Outline of a new approach to the analysis of complex systems and decision processes'. *IEEE Trans. Systems, Man, and Cybernetics* **3**, 28–44.

ZENG, X.-J. AND M. G. SINGH (1994) 'Approximation theory of fuzzy systems – SISO case'. *IEEE Trans. Fuzzy Systems* **2**, 162–176.

ZHANG, X.-C., A. VISALA, A. HALME AND P. LINKO (1994) 'Functional state modelling approach for bioprocesses: Local models for aerobic yeast growth processes'. *J. Process Control* **4**, 127–134.

ZHAO, J., V. WERTZ AND R. GOREZ (1995a) Design a stabilizing fuzzy and/or non-fuzzy state feedback controller using LMI method. In 'Proc. European Control Conference, Rome'. pp. 1201–1206.

ZHAO, J., V. WERTZ AND R. GOREZ (1995b) Linear TS fuzzy model based robust stabilizing controller design. In 'IEEE Conf. Decision and Control, New Orleans'.

ZHAO, Z.-Y., M. TOMIZUKA AND S. ISAKA (1993) 'Fuzzy gain scheduling of PID controllers'. *IEEE Trans. Systems, Man, and Cybernetics* **23**, 1392–1398.

PART TWO

Modelling

CHAPTER TWO

Fuzzy Set Methods for Local Modelling and Identification

R. BABUŠKA and H. B. VERBRUGGEN

There are several methods for modelling nonlinear systems. A main distinction can be made between global and local models. In this chapter we concentrate on approximation of a nonlinear system by a set of local linear models. Each local model is valid for a certain range of operating conditions and an interpolative scheduling mechanism combines the outputs of the local models into a continuous global output. Such a model structure can be conveniently represented by means of fuzzy If-Then rules. Construction of a rule-based fuzzy model requires identification of the antecedent and consequent structure, of the membership functions for different operating regions and estimation of the consequent regression parameters. While the latter task can be solved using linear estimation techniques, the construction of the membership functions is a nonlinear optimisation problem. Several existing constructive techniques are reviewed in this chapter and a method based on fuzzy clustering is described in more detail. The presented approach does not require any prior knowledge about the operating regimes and also an appropriate number of rules can be determined automatically. If a sufficiently rich identification data set covering the operating ranges of interest is not available, the rules obtained from data can be combined with prior knowledge transformed into the membership functions for the relevant operating regions and the local models. The models provided by the user can also be nonlinear (semi)mechanistic models based on first principles. Examples are provided to illustrate the basic concepts of the described fuzzy modelling and identification techniques.

2.1 INTRODUCTION

There exist many different models of a system and the choice of a particular modelling and identification method depends on the aim of modelling. A model can be developed for analysing the system's behaviour, for design purposes, for changing a system's parameters in order to improve its performance, for bottleneck analysis, controller design, etc. In this chapter we emphasise the use of nonlinear models for control which requires modelling and identification techniques that lead to simple, transparent and mathematically tractable models. In many applications a framework is needed for combining quantitative information and numerical data with qualitative and heuristic knowledge. To achieve this, user-friendly methods are needed for efficient translation of the knowledge into a computer-manageable form, for interfacing qualitative information with numerical data and for appropriate validation of the models. Modelling based on the use of fuzzy sets in combination with local regression techniques has a great potential to achieve these goals. The theory of fuzzy modelling and fuzzy model-based control has been developed in recent years (Takagi and Sugeno 1985, Pedrycz 1993, Yager and Filev 1994, Brown and Harris 1994) and a number of successful applications to complex industrial systems have been reported (Sugeno and Kang 1986, Sugeno and Tanaka 1991, Babuška and Verbruggen 1994, Kaymak 1994). Important synergy also exists between the methods based on fuzzy sets and artificial neural networks (Brown and Harris 1994, Buckley and Hayashi 1994, Pedrycz 1995). These techniques accompanied by a suitable control design method will certainly enrich the set of tools available for the design of complex industrial control systems.

2.2 FUZZY MODELLING

Fuzzy If-Then rules are used not only to incorporate human knowledge in fuzzy expert systems and controllers (Gupta *et al.* 1985, Driankov *et al.* 1993), but also can be applied to modelling of nonlinear dynamic systems. In that case fuzzy sets are used to partition the continuous domains of the system variables into overlapping regions. A fuzzy model describes the system either by establishing direct relations between the input and output fuzzy regions in the form of linguistic rules or by using a simple local model for each region. The overlap of the fuzzy regions ensures a smooth transition among the rules' outputs. In this chapter we concentrate mainly on rules with fuzzy antecedents and functional consequents, usually defined as local linear regression models (Takagi and Sugeno 1985). The structure of this rule-based model and the related inference mechanism are explained in this section.

2.2.1 Fuzzy models for nonlinear regression

Fuzzy models can be used as a tool for nonlinear regression, i.e. modelling the dependence of a response variable $y \in \mathcal{Y} \subset \mathbb{R}$ on the regression vector $\varphi = [\varphi_1, \ldots, \varphi_d]^T$ over some domain $\varphi \in \mathcal{D} \subset \mathbb{R}^d$ containing the available data. The system that generated the data is presumed to be described by:

$$y = f(\varphi) + \epsilon . \tag{2.1}$$

The deterministic function f captures the dependence of y on φ and the additive stochastic component ϵ reflects the dependence of y on quantities other than φ. A common approach to identification of nonlinear dynamic systems is to transform the time functions in a static nonlinear regression problem (Leonaritis and Billings 1985). For input–output models, a relation between the past input–output data and future outputs is sought. A finite number of past inputs $u(i)$ and outputs $y(i)$ is collected into the regression vector $\varphi(k)$

$$\varphi(k) = [y(k), y(k-1), \ldots, y(k-n+1), u(k), u(k-1), \ldots, u(k-m+1)]^T .$$

The problem is then to infer the unknown function $y(k+1) = f(\varphi(k))$ from a set of input–output measurements. Similarly, a nonlinear state-space model can be represented as a concatenation of two static functions:

$$
\begin{aligned}
x(k+1) &= g(x(k), u(k)), \\
y(k) &= h(x(k)).
\end{aligned}
$$

The state transition vector function g maps the current state and input into the new state. The output (scalar) function h maps the current state into the output. If the state is measured directly on the system or is reconstructed, both g and h can be approximated using nonlinear regression techniques. Above we assumed that y is a scalar, i.e. the system under study is a multi-input, single-output (MISO) system. This is no restriction, since a multi-input, multi-output (MIMO) system can be always represented as a set of coupled MISO systems.

In nonlinear regression (2.1), the aim is to use the available data to construct a function $\hat{f}(\varphi)$ that can serve as a reasonable approximation to $f(\varphi)$ not only for the given data but over the entire domain $\mathcal{D}$. The definition of "reasonable approximation" may depend on the purpose for which the model is constructed. If the aim of modelling is only to obtain a rule for predicting y given values of φ, accuracy is the only relevant criterion. Often, however, beside accurate predictions one wants to obtain such $\hat{f}$ that can be used to analyse and understand the properties of the true underlying function f, i.e. the real system that generated the data. This property can be called the *transparency* of the model. In the following sections we show how fuzzy models can be used for transparent modelling of nonlinear systems.

2.2.2 Takagi–Sugeno fuzzy model

The fuzzy rule-based model used in our approach for nonlinear regression was introduced by Takagi and Sugeno (1985) and in fuzzy modelling literature it is usually referred to as Takagi–Sugeno (TS) model. The most general form of a TS fuzzy rule is:

$$\textbf{If } \underbrace{\varphi \text{ is } A_i}_{\text{antecedent}} \textbf{ then } \underbrace{y_i = f_i(\varphi)}_{\text{consequent}}. \tag{2.2}$$

The first part of the rule, called the antecedent, is defined as a fuzzy proposition "φ is A_i" where φ is a crisp (nonfuzzy) vector and A_i is a fuzzy set defined by its (multivariate) membership function

$$\mu_{A_i}(\varphi): \mathbb{R}^d \to [0, 1].$$

The index $i = 1, \ldots, K$ denotes the ith rule, where K is the number of rules in the rule base. The degree of fulfilment (degree of truth) β_i of the antecedent proposition for a given value of the regression vector φ is evaluated as the degree of membership of this vector into the set A_i: $\beta_i = \mu_{A_i}(\varphi)$. By means of the fuzzy sets, the regressor space is partitioned into smaller regions, in which the regression problem becomes more tractable. The overlap of the membership functions results in gradual changes of the membership degrees for the adjacent rules and in a continuous transition from one consequent to another. A certain meaning is usually associated with each of the antecedent fuzzy sets. In a univariate case one may think of examples like "temperature is HIGH," "pressure is LOW," etc. The meaning of *linguistic terms* HIGH and LOW is reflected in the position and shape of their membership functions in the respective domains. Parameterised membership functions are usually used, such as triangular, trapezoidal, Gaussian or sigmoidal.

Since it is difficult to assign a linguistic meaning to multidimensional fuzzy sets, the antecedent proposition in (2.2) is usually expressed as a combination of simple propositions with univariate fuzzy sets defined for the individual components of φ. The propositions are combined using conjunction and disjunction, i.e. logical connectives **and** and **or** respectively. The complement **not** can be used as well. In this way, propositions like

"**If** temperature is **not** HIGH **or** pressure is LOW **and** ..."

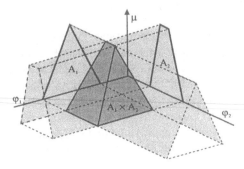

Figure 2.1 Cartesian product of two fuzzy sets using the minimum intersection operator.

can be created. The degree of fulfilment of the antecedent is computed as a combination of the membership degrees of the individual propositions using the fuzzy logic operators for conjunction (t-norms such as the minimum or product), disjunction (t-conorms such as the maximum or bounded sum) and the complement. For instance, in rules with the most common *conjunctive* form of the antecedent:

$$\text{If } \varphi_1 \text{ is } A_{i,1} \text{ and } \varphi_2 \text{ is } A_{i,2} \text{ and } ,\ldots, \text{ and } \varphi_d \text{ is } A_{i,d} \text{ then } y_i = f_i(\varphi) \qquad (2.3)$$

the degree of fulfilment $\beta_i(\varphi)$ is calculated as:

$$\beta_i(\varphi) = \mu_{A_{i,1}}(\varphi_1) \wedge \mu_{A_{i,2}}(\varphi_2) \wedge \ldots \wedge \mu_{A_{i,d}}(\varphi_d) \qquad (2.4)$$

when minimum ($\wedge$) is used for conjunction. For the product operator we obtain:

$$\beta_i(\varphi) = \mu_{A_{i,1}}(\varphi_1) \cdot \mu_{A_{i,2}}(\varphi_2) \cdot \cdots \cdot \mu_{A_{i,d}}(\varphi_d) . \qquad (2.5)$$

Notice that $\beta_i(\varphi)$ in (2.4) and (2.5) is in fact a membership function of a multivariate fuzzy set created by intersecting the univariate sets on the Cartesian product space of the components of φ.[1] The shape and support of the resulting multidimensional fuzzy set depends on the particular intersection operator used. Figure 2.1 shows an example with the minimum operator.

The second part of the rule (2.2), called the consequent, is some fixed function of the regression vector. The rules may in general differ both in the antecedent fuzzy sets and the consequent functions. However, in practice, f_is are instances of a suitable parameterised function, whose structure remains unchanged for all the rules and only its parameters vary. A simple and practically useful parameterisation is an affine form, resulting in the following rules:

$$\text{If } \varphi \text{ is } A_i \text{ then } y_i = a_i^T \varphi + b_i, \qquad (2.6)$$

where a_i is a parameter vector and b_i is a scalar offset. This model will be called an *affine TS model*. When $b_i = 0$, $i = 1,\ldots,K$, the model is called a *linear TS model*. The consequents of the affine TS model are hyperplanes in the product space $(\mathcal{D} \times \mathcal{Y}) \subset \mathcal{R}^{d+1}$. For a linear TS model all the hyperplanes contain the origin. The antecedents define in which region of the regression space the hyperplanar consequent model is valid. In this way, the affine TS models can approximate the regression surface piecewise by hyperplanes. A special case of (2.6) exists for constant functions $y_i = b_i$, $i = 1,\ldots,K$. This model will be called a *singleton model* since a constant is a singleton fuzzy set.

---

[1] This approach can be compared to generating multivariate spline basis functions as a tensor product of univariate basis functions (Brown and Harris 1994).

2.2.3 Inference in the TS model

Inference in fuzzy rule-based systems is a process of deriving an output fuzzy set given the rules and the known inputs, using the so called compositional rule of inference (Driankov *et al.* 1993). If a numerical output is needed, the resulting fuzzy set must be defuzzified, i.e. replaced by a single number $y \in \mathcal{Y}$ that is representative for the entire set. The mode (maximum) and the centre of gravity methods are commonly used. Since the consequents in the TS model are not fuzzy sets but crisp functions, the inference and defuzzification reduce to the weighted mean:

$$y = \frac{\sum_{i=1}^{K} \beta_i(\varphi) y_i}{\sum_{i=1}^{K} \beta_i(\varphi)} . \tag{2.7}$$

Here $\beta_i(\varphi)$ is the degree of fulfilment of the antecedent of the ith rule and y_i is the output of the local consequent model of that rule. Denoting the normalised degree of fulfilment $\bar{\beta}_i(\varphi) = \beta_i(\varphi) / \sum_{j=1}^{K} \beta_j(\varphi)$, for the affine model (2.6) the global output can be written as:

$$y = (\sum_{i=1}^{K} \bar{\beta}_i(\varphi) a_i^T) \varphi + \sum_{i=1}^{K} \bar{\beta}_i(\varphi) b_i. \tag{2.8}$$

An affine TS model with a common consequent structure can be written as a global model with input-dependent parameters

$$y = a^T(\varphi)\varphi + b(\varphi), \tag{2.9}$$

where the "parameters" $a(\varphi)$, $b(\varphi)$ are computed as convex linear combinations of the consequent parameters a_i and b_i, i.e.:

$$a(\varphi) = \sum_{i=1}^{K} \bar{\beta}_i(\varphi) a_i, \quad b(\varphi) = \sum_{i=1}^{K} \bar{\beta}_i(\varphi) b_i . \tag{2.10}$$

Notice, however, that the output of the global model is not a convex function, which results in some undesirable properties of the TS interpolation mechanism as shown in the following example. Consider a TS model of the following form:

$$\textbf{If } x \text{ is } A_i \textbf{ then } y_i = a_i x + b_i, \quad i = 1, 2, \tag{2.11}$$

where the affine consequents were designed as local linearisation of a nonlinear function $y = f(x)$. The antecedent sets A_1, A_2 are defined by trapezoidal membership functions shown in the bottom part of Figure 2.2.

The top part of this figure presents the function $y = f(x)$ along with the consequents $y_1 = a_1 x + b_1$, $y_2 = a_2 x + b_2$ and the global output of the TS model y_{TS} calculated as the weighted mean (2.8). Notice that the model output is not a smooth function, due to the piecewise linear antecedent membership functions. Further it significantly differs from the convex function $y = f(x)$ which the designer might want to approximate by the TS structure. The consequent parameters are often estimated from the system input–output data by minimising the prediction error using least squares. Due to the undesirable interpolation properties, the least square optimisation may favour rule consequents that are *not* local linear approximations of the global nonlinear system, depending on the antecedent membership functions. Therefore, these membership functions should be carefully chosen

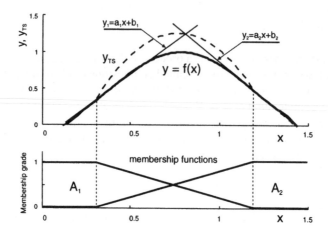

Figure 2.2   Interpolation in the TS model.

or a suitable identification technique must be used to extract them from data. One such technique is presented in Section 2.3.3. More details can be found in (Babuška *et al.* 1994).

The use of membership functions (validity functions, basis functions) for partitioning the operating space of the system into overlapping regions is not restricted to fuzzy modelling. A similar approach is applied for instance by Johansen (1994) or Banerjee *et al.* (1995). Analogies also can be found in nonlinear statistical regression using splines (de Boor 1978, Friedman 1991) or radial basis functions (Jang and Sun 1993). The main difference of the latter two techniques from the TS modelling is that they use constant functions (singletons) as local models instead of linear or nonlinear functions.

2.3   DEVELOPING FUZZY MODELS

A TS model can be designed in several ways, using prior knowledge, data or a combination of both. Some of the techniques are reviewed in this section.

2.3.1   Linearisation of nonlinear white-box models

Prior knowledge may be involved in the design of the antecedent membership functions, and in the choice of the consequent structure. The consequent parameters can be computed by local linearisation of a known mechanistic model or can be estimated from data. This approach is illustrated on the well-known problem of balancing a pole on a cart. The differential equations for the pole motion are:

$$\dot{x}_1 = x_2$$
$$\dot{x}_2 = \frac{g\sin(x_1) - amlx_2^2\sin(2x_1)/2 - a\cos(x_1)u}{4l/3 - aml\cos^2(x_1)}. \tag{2.12}$$

Here x_1 and x_2 denote the angle and angular velocity of the pole respectively, g is the gravity constant, l is the distance from pivot to pole's centre of mass, u is the force applied to the cart and $l = 1/(m + M)$ where m is the mass of the pole and M the mass of the cart.

A simple TS model for $x_1 \in (-\pi/2, \pi/2)$ can be designed by linearising (2.12) around $|x_1| = 0$ and $|x_1| = \pi/2$. The corresponding TS rules are:

$$\textbf{If } |x_1| \text{ is AROUND ZERO } \textbf{then } \dot{x} = A_1 x + B_1 u$$
$$\textbf{If } |x_1| \text{ is not AROUND ZERO } \textbf{then } \dot{x} = A_2 x + B_2 u. \tag{2.13}$$

The fuzzy set AROUND ZERO can be defined using for instance a triangular membership function: $\mu(x_1) = 1 - 2|x_1|/\pi$. The membership degree of the complement **not** AROUND ZERO is $1 - \mu(x_1)$. Since $\pm\pi/2$ is not an equilibrium point, linearisation of (2.12) around this point would give additional constant terms in (2.13) and hence an affine TS model would be obtained. Since this model is less suitable for controller design and analysis than the linear state-space TS model (2.13), Wang *et al.* (1995) use the following approximation of the matrices A_1, B_1, A_2 and B_2:

$$A_1 = \begin{bmatrix} 0 & 1 \\ g/(4l/3 - aml) & 0 \end{bmatrix}, \qquad B_1 = \begin{bmatrix} 0 \\ a/(4l/3 - aml) \end{bmatrix},$$

$$A_2 = \begin{bmatrix} 0 & 1 \\ 2g/(\pi(4l/3 - amlc^2)) & 0 \end{bmatrix}, \qquad B_2 = \begin{bmatrix} 0 \\ ac/(4l/3 - amlc^2) \end{bmatrix},$$

where c should be $c = \cos(\pi/2)$. Since this model would not be controllable, $c = \cos(\pi/1.95)$ is used. Derivation of matrices B_1 and B_2 follows directly from the linearisation of (2.12) around $|x_1| = 0$ and $|x_1| = \pi/2$ respectively. Element $a_1(2, 1)$ is obtained by neglecting the second term in the numerator of (2.12). This is clearly an approximation since this term can be neglected only for small angular velocities. Linearisation of (2.12) around $\pi/2$ results in a constant term only, say c_2, where $c_2 = g/(4l/3 - amlc^2)$. For $|x_1| = \pi/2$ we can substitute c_2 and x_1 into the state equation $\dot{x}_2 = a_2(2, 1)x_1$ to obtain $c_2 = a_2(2, 1)\pi/2$ and thus $a_2(2, 1) = 2c_2/\pi$. This is again an approximation valid in a close neighbourhood of $|x_1| = \pi/2$.

As shown by Wang *et al.* (1995), this model can be easily extended to the entire working range $x_1 \in [-\pi, \pi]$ using four rules. A feedback Takagi-Sugeno controller with guaranteed stability can be designed that uses the same antecedents as the model and the consequents are control laws of the form $u = -P_i x$, $i = 1, \ldots, 4$, where P_is are designed using standard methods such as pole placement.

2.3.2 Templates for membership functions

For many practical systems a complete and exact mathematical model is not known and also no particular knowledge may be available about the partition of the operating space. In such a case, a small number of equally spaced and shaped membership functions can be used for each antecedent variable and the consequent parameters can be estimated from data using least squares (note that (2.8) is linear in the parameters a_i and b_i). In literature, this approach is sometimes denoted as template-based modelling. A simple example illustrates this approach.

Example. Consider a nonlinear dynamic system described by a first-order difference equation:

$$y(k + 1) = y(k) + u(k)e^{-3|y(k)|} . \tag{2.14}$$

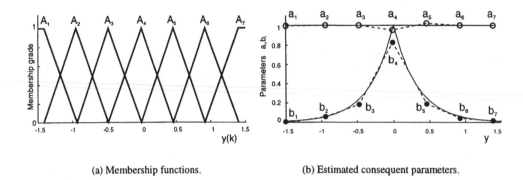

(a) Membership functions.   (b) Estimated consequent parameters.

Figure 2.3 (a) Equidistant triangular membership functions designed for the output $y(k)$; (b) comparison of the true system nonlinearity (solid line) and its approximation in terms of the estimated consequent parameters (dashed line).

A set of 300 simulated input–output data shown in Fig. 2.4(a) are used for identification. Knowing the structure of the "true" system we choose the following TS model structure:

$$\textbf{If } y(k) \text{ is } A_i \textbf{ then } y_i(k + 1) = a_i y(k) + b_i u(k), \qquad (2.15)$$

where only the output $y(k)$ is used in the rule antecedents since the system nonlinearity is caused by the dependence of the system gain on this variable. Assuming that no further prior knowledge is available, we design 7 equally spaced triangular membership functions A_1 to A_7 on the domain of $y(k)$, see Figure 2.3(a). The number 7 is chosen quite arbitrarily, models with 5 or 3 membership functions work as well, 9 membership functions gives even more accurate results.

The consequent parameters estimated by least squares are plotted in Figure 2.3(b) against the cores of the antecedent fuzzy sets as empty circles for a_i and filled circles for b_i, $i = 1, \ldots, 7$. Plotted is also the interpolation between the parameters (dashed line) and the true system nonlinearity (solid line). The interpolation between a_i and b_i is linear since the membership functions are piecewise linear (triangular). One can observe that the dependence of the consequent parameters on the antecedent variable approximates quite accurately the system's nonlinearity, which gives the model a certain transparency. Also the individual consequent parameters correspond to local linearisations of the original nonlinear system around the cores of the fuzzy sets, i.e. the points where the rules are valid to the highest degree. The values of the parameters $a = [1.00, 1.00, 1.00, 0.97, 1.01, 1.00, 1.00]^T$ and $b = [0.01, 0.05, 0.20, 0.81, 0.20, 0.05, 0.01]^T$ indicate the strong nonlinearity of the u-part (a nonlinear gain) and the linearity of the y-part (linear dynamics) of (2.14). Validation of the model in simulation using fresh data is shown in Figure 2.4(b).

2.3.3 Product space clustering

This section explains how fuzzy clustering can be used for constructing affine TS fuzzy models from data. The approach is based on an assumption that the identification data is a representative sample of the regression surface (surface defined by equation $y = f(\varphi)$). A fuzzy clustering algorithm based on an adaptive distance measure is applied to partition this data into several (hyperellipsoidal) clusters. These clusters are projected on to

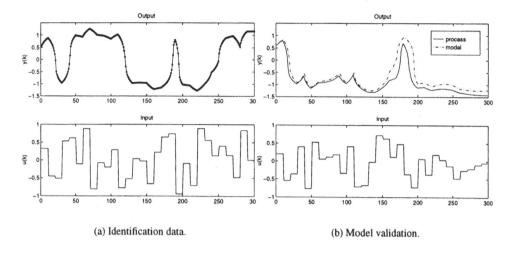

(a) Identification data.   (b) Model validation.

Figure 2.4   (a) Identification data set; (b) validation of the model on a fresh data set.

the antecedent variables to form the membership functions and from the fuzzy covariance matrices of the clusters the consequent parameters are derived. Cluster validity measures and cluster merging techniques can be applied to find an appropriate number of rules. The following paragraphs describe this method in a greater detail.

Geometric interpretation of nonlinear regression

This approach is based on a geometric interpretation of the nonlinear regression problem (2.1). In the product space of the regression vector φ and the response variable y: $(\mathcal{D} \times \mathcal{Y}) \subset \mathbb{R}^{d+1}$ the unknown nonlinear function $y = f(\varphi)$ represents a nonlinear hypersurface, i.e. a subspace of dimension d, called the regression surface. The goal is to find the parameters of the TS model such that it approximates this hypersurface in a piecewise manner by hyperplanes. This problem can be regarded as a pattern recognition task similar to straight line or contour extraction in image processing for which fuzzy clustering techniques have been successfully applied (Dave 1992, Krishnapuram and Freg 1992). The algorithms used in pattern recognition are based on hyperplanar prototypes (fuzzy linear varieties and fuzzy c-elliptotypes (Bezdek 1981)), i.e. the cluster prototypes are defined as hyperplanes. The distance function measures the distance of the data points from the hyperplanes. Yoshinari *et al.* (1993) also applied the fuzzy c-elliptotypes algorithm to fitting univariate functions. The approach presented here is different since it uses fuzzy clustering with an adaptive distance measure. The cluster prototypes are defined as points but the distance measure adapts automatically to the shape of the cluster using an estimate of the cluster covariance matrix (Gustafson and Kessel 1979). The identification technique based on the Gustafson–Kessel (GK) algorithm was presented in our previous articles (Babuška and Verbruggen 1994, Babuška and Verbruggen 1995*a*) and similar approaches have been used also by Nakamori and Ryoke (1994) and Zhao *et al.* (1994). On the other hand, many clustering-based identification methods proposed in literature (Sugeno and Yasukawa 1993, Pedrycz 1984, Pedrycz 1993, Berenji and Khedar 1993) are conceptually different from our approach since they use different model structures, different clustering algorithms and except Berenji and Khedar (1993) do not apply clustering in the product space of the model variables.

The Gustafson–Kessel algorithm

A set of N data pairs (φ_j, y_j), $j = 1, \ldots, N$ is available. Denoting $z_j = [\varphi_j, y_j]^T$ we can write this data set in a matrix form

$$Z = [z_1, \ldots, z_N], \quad z_j \in R^{d+1}. \tag{2.16}$$

The vectors z_j will be partitioned into K clusters with prototypes

$$v_i = [v_{i,1}, \ldots, v_{i,d+1}]^T \in R^{d+1}, \quad i = 1, \ldots, K. \tag{2.17}$$

The K-tuple of the cluster prototypes is denoted: $V = [v_1, \ldots, v_K]$. The partitioning of the data is defined by means of a fuzzy partition matrix

$$U = [\mu_{ij}]_{K \times N} \tag{2.18}$$

where $\mu_{i,j} \in [0, 1]$ represents the membership degree of the data vector z_j in the ith cluster with the prototype v_i. The GK algorithm finds the partition matrix and the cluster prototypes by minimising the following objective function:

$$J(Z, V, U) = \sum_{i=1}^{K} \sum_{j=1}^{N} \mu_{i,j}^m d^2(z_j, v_i) \tag{2.19}$$

subject to:

$$\sum_{i=1}^{K} \mu_{i,j} = 1, \quad j = 1, \ldots, N \text{ and} \tag{2.20}$$

$$0 < \sum_{j=1}^{N} \mu_{i,j} < N, \quad i = 1, \ldots, K. \tag{2.21}$$

Here $m > 1$ is a parameter that controls fuzziness of the clusters, with higher values of m the clusters overlap more. Typically $m = 2$ is used. The function $d(z_j, v_i)$ is the distance of the data vector z_j from the cluster prototype v_i. The constraint (2.20) avoids the trivial solution $U = 0$ and the constraint (2.21) guarantees that clusters are neither empty nor contain all the points to degree 1. This is a standard definition of clustering with a fuzzy objective function for which iterative algorithms that minimise (2.19) subject to (2.20) and (2.21) have been derived (Bezdek 1981). The shape of the clusters is determined by the particular distance measure $d(z_j, v_i)$ involved. Gustafson and Kessel (1979) generalised the algorithm for an adaptive distance measure:

$$d^2(z_j, v_i) = (z_j - v_i)^T M_i(z_j - v_i) \tag{2.22}$$

where M_i is a positive definite matrix adapted according to the actual shapes of the individual clusters approximately described by the cluster covariance matrices F_i:

$$F_i = \frac{\sum_{j=1}^{N} \mu_{i,j}^m (z_j - v_i)(z_j - v_i)^T}{\sum_{j=1}^{N} \mu_{i,j}^m}. \tag{2.23}$$

The distance inducing matrix M_i is calculated as a normalised inverse of the cluster covariance matrix:

$$M_i = \det(F_i)^{\frac{1}{d+1}} F_i^{-1} \, . \tag{2.24}$$

The normalisation by the determinant of F_i is involved in order to avoid a trivial solution of (2.19) for $M_i = 0$.

Algorithm: Given a data set Z, the number of clusters K and an initial (random) partition U, the cluster prototypes V and the final partition matrix U are found by repeating the following steps:

1 Compute the cluster prototypes (cluster means): $v_i = \dfrac{\sum_{j=1}^{N} \mu_{i,j}^m z_j}{\sum_{j=1}^{N} \mu_{i,j}^m} \, .$

2 Calculate the cluster covariance matrices F_i using (2.23).

3 Compute the matrices M_i using (2.24).

4 Calculate the squared distances $d^2(z_j, v_i)$ as given in (2.22).

5 Update the fuzzy partition matrix U: $\mu_{i,j} = \dfrac{d^2(z_j, v_i)^{-1/(m-1)}}{\sum_{l=1}^{K} d^2(z_j, v_l)^{-1/(m-1)}}$

if $d^2(z_j, v_i) = 0$ for some $i = k$, set $\mu_{k,j} = 1$ and $\mu_{i,j} = 0, \forall i \neq k$.

until a specified convergence criterion is satisfied, e.g. $\|U_l - U_{l-1}\| < \epsilon$ where $\|\cdot\|$ is a suitable matrix norm, l is the iteration step and ϵ is a termination tolerance, typically $\epsilon = 0.01$.

After convergence, the partition matrix U, the cluster prototypes v_i and covariance matrices $F_i, i = 1, \ldots, K$ are obtained. The following section explains how the TS rules are derived from this information.

Extracting rules from the clusters

The distance measure (2.22) defines the clusters as hyperellipsoids whose shape is roughly described by the eigenstructure of the cluster covariance matrix F_i. The eigenvectors of F_i define the orientation of the hyperellipsoid axes and the length of the axes is given by the corresponding eigenvalues of F_i. Clustering the data spread around the regression surface results in flat hyperellipsoids that can be seen locally as hyperplanes. The eigenvector $\Phi_i = [\phi_{i,1}, \ldots, \phi_{i,d+1}]$ corresponding to the smallest eigenvalue of the ith cluster's covariance matrix F_i is the normal vector of the hyperplane, see Figure 2.5. Recalling that v_i is the cluster prototypical point we can directly write an implicit form of the affine TS consequents:

$$\Phi_i^T \cdot ([\varphi, y] - v_i) = 0 \, . \tag{2.25}$$

After an algebraic manipulation we obtain the parameters a_i and b_i of the explicit form (2.6):

$$a_i = -\frac{1}{\phi_{i,d+1}} [\phi_{i,1}, \ldots, \phi_{i,d}]^T, \quad b_i = \frac{1}{\phi_{i,d+1}} \Phi_i^T \cdot v_i \, , \tag{2.26}$$

and thus the clusters found by the GK algorithm can be represented by a set of TS rules (2.6).

The antecedent fuzzy sets A_i are obtained by projecting the fuzzy set defined pointwise in the product space (by the ith row of the partition matrix U, $\mu_i: (\mathcal{D} \times \mathcal{Y}) \to [0, 1]$) on to the regression vector space, see the bottom part of Figure 2.5:

$$\mu_{A_i}(\varphi_l) = \max_{j:\varphi_j=\varphi_l,\ j>l} \mu_{i,j}, \quad \forall l \in \{1, \dots, N\}. \tag{2.27}$$

Projection can be seen as a reverse operation to the composition of a multivariate fuzzy set from univariate fuzzy sets by means of intersection on the Cartesian product space, depicted in Figure 2.1.

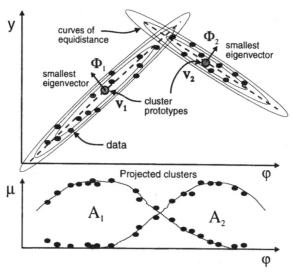

Figure 2.5 The eigenvector Φ_i corresponding to the smallest eigenvalue of F_i determines the normal vector of the hyperplane. The antecedent membership functions are obtained by projecting the partition on to the regressor.

Since the fuzzy sets A_i are defined pointwise for the identification data, the TS model representation (2.6) can be directly used only for predicting values contained in the identification data set. For new data, a way must be found to compute the membership degrees in the sets A_i. In the univariate case, this can be accomplished simply by interpolating the known membership degrees or by fitting them with a parameterised membership function. For multivariate regression, a direct interpolation in the regression space becomes a nontrivial problem, since the data is usually not on a hyperrectangular lattice but scattered around the space. At the same time, multidimensional fuzzy sets are difficult to interpret. Therefore, it is more convenient to decompose the multidimensional antecedent fuzzy sets into the conjunctive form of univariate fuzzy sets (2.3). This is achieved by means of projection, analogous to the projection of the clusters to the regression space (2.27). A drawback of this approach is that a decomposition error results for fuzzy sets that are oblique to the axes. On the other hand, the conjunctive form is more transparent and allows for a linguistic interpretation of the model.

The consequent parameters a_i, b_i also can be estimated by solving the set of equations (2.8) in the least squares sense. The membership degrees obtained by clustering can be directly substituted for the normalised degrees of fulfilment $\bar{\beta}_i$, since the constraint (2.20) guarantees that the sum of the degrees over all the rules equals one. The parameters obtained in this way are optimal w.r.t. the model output y and also can partially compensate for the decomposition error that may occur due to antecedent decomposition. If the model will serve as a numerical predictor, the least squares approach is usually preferred.

Determining the number of rules

The number of clusters, i.e. the number of rules in the resulting TS model, must be specified before clustering. So far we have assumed that the appropriate number of rules is known in advance. However, in practice this is usually not the case even though some general guidelines exist. The number of rules is related to the type of nonlinearity the system is expected to exhibit. The more rules the model contains, the finer approximation can be obtained, but also more parameters must be estimated and their variance is therefore higher due to overfitting. The number of clusters also depends on the dimension of φ and on the number of measurements available. Few data points cannot be partitioned into many clusters because in such a case the hyperplanes are not sufficiently determined. Empirical relations among the number of clusters K, the dimension of the regression vector d and the number of data points required have been suggested (Jain and Dubes 1988). If no particular knowledge about the type of the process nonlinearity is available, automated procedures for determining the number of rules can be applied. In connection with fuzzy clustering two main approaches are used: validity measures and compatible cluster merging.

Validity measures are numerical indicators that assess the qualities of the clusters, such as fuzzy hypervolume, total intracluster distance, fuzzy partition density, etc., and mostly were introduced in the pattern recognition setting (Krishnapuram and Freg 1992). The data must be clustered several times, each time with a different number of clusters. The number of clusters that minimises (maximises) the validity measure is selected. A dedicated validity measure was proposed for the clustering-based identification of nonlinear systems (Babuška and Verbruggen 1995b) that combines a measure of the cluster flatness with the mean prediction error. This measure prefers a few flat clusters to a larger number of small ones if both settings lead to approximately the same prediction error. This approach conceptually resembles the use of information criteria in linear system identification (Akaike 1974) and can be used also for selecting the structure of the model.

Cluster merging approaches start with a high number of clusters and proceed by gradually merging similar clusters (Krishnapuram and Freg 1992, Kaymak and Babuška 1995). The initial number of clusters must be set sufficiently high such that the nonlinearity of the regression hypersurface can be captured accurately enough. The number of clusters is iteratively reduced by merging clusters that are sufficiently close and approximately parallel, see Figure 2.6. Two clusters i and j are approximately parallel if the dot product of their

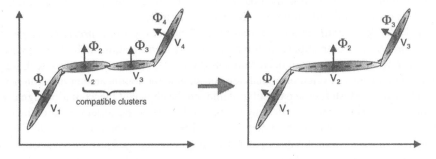

Figure 2.6 Merging compatible clusters.

normal vectors is close to one:

$$|\Phi_i \cdot \Phi_j| \geq k_1, k_1 \text{ close to } 1 .\tag{2.28}$$

The distance of the clusters is measured as the Euclidean distance of the cluster prototypical points:

$$\|v_i - v_j\| \leq k_2, k_2 \text{ close to } 0 .\tag{2.29}$$

By evaluating these cluster compatibility criteria for all pairs of clusters, one obtains two matrices whose elements indicate the degree of similarity between the ith and jth clusters measured according to the corresponding criterion. An automated fuzzy decision-making procedure combines the two criteria (allowing for compensation between them). A user-defined threshold parameter is used to determine which clusters should be merged. This threshold is related to the complexity/accuracy trade-off of the resulting TS model. Merging means that a new partition matrix is created with less rows (clusters). The membership degrees of the combined clusters are computed by adding up the membership degrees of the compatible clusters, i.e. the new partition matrix also fulfils the conditions (2.20) and (2.21). The GK algorithm is initialised with this new partition matrix and run again. Because of this good initialisation, the convergence of the algorithm is now much faster compared to a random initialisation.

The benefits from using the cluster merging technique are two-fold. 1) The number of rules can be optimised without testing all the cluster numbers from two to the specified upper limit, as with the cluster validity measures. This may significantly reduce the computational effort. 2) By decreasing the number of clusters gradually, a better solution can be found since small regions with a few data points can be captured using initially a higher number of clusters, see (Kaymak and Babuška 1995) for details. In practice, some experiments are needed for finding the right threshold, especially for noisy data.

Another approach to rule base simplification has been developed recently by Setnes (1995), using a measure of similarity among the membership functions obtained by cluster projections. Similar membership functions are replaced by one common function, reducing the number of linguistic terms needed and thus also the number of rules. In this way, a more transparent and understandable rule base is obtained. Experimental results show that both the cluster merging and the similarity-based approach lead to good results.

Summary of the identification procedure

In this section the identification method described above is summarised and a simulation example is presented for illustration of the individual steps.

Step 1: *Design identification experiments and collect a set of representative process measurements.* This is an important initial step for any identification method, since it determines the information content of the identification data set. As opposed to linear techniques, pseudo-random binary excitation signals are not suitable for nonlinear identification in general and for fuzzy clustering in particular. Although the choice of the excitation signal may be problem-dependent, the input data should preferably excite the system in the entire range of the considered variables both in amplitude and frequency. The pseudo-random binary signal is not suitable since it only contains two amplitude levels. Data recorded during the routine process operation can be used as well, provided they fulfil the above requirements. If the identification data set does not cover the entire operating range, additional rules can be supplied by the user, see also Step 6.

Step 2: *Choose the model structure.* The relevant system variables must be selected for the model inputs and when identifying dynamic systems the order of the model must be chosen. After the appropriate structure is defined, the matrix Z (2.16) is constructed from the identification data.

Step 3: *Cluster the data.* The initial number of clusters K, and the clustering parameters m and ϵ must be defined before clustering. The choice of K has been discussed in the previous section, $m = 2$ and $\epsilon = 0.01$ are good settings for most problems. Note that for large data sets, the clustering procedure may be computationally demanding.

Step 4: *Determine the number of clusters.* Using a cluster validity measure or the compatible cluster merging technique an appropriate number of clusters can be found. This step obviously involves several repetitions of Step 3 with different K and a different initial partition matrix U.

Step 5: *Generate the rules.* By projecting the clusters onto the antecedent variables, the membership functions are obtained. The consequent parameters are extracted from the eigenstructure of the cluster covariance matrix or estimated by least squares.

Step 6: *Validate the model.* The validation of fuzzy models has several facets, namely a standard validation through numerical simulations and comparisons with the process data, analysis of the linear consequent models (stability, step responses, gains, time-constants, non-minimum phase behaviour, etc.) and analysis of the coverage of the input space by the combinations of the antecedent membership functions. For an incomplete rule base, additional rules can be provided based on prior knowledge or local linearisation of first-principle models.

A numerical example

The identification method will be illustrated using a simulation example from the literature (Ikoma and Hirota 1993). An artificial time series is generated by a nonlinear autoregressive system:

$$x(k+1) = f(x(k)) + \epsilon(k), \quad f(x) = \begin{cases} 2x - 2, & 0.5 \leq x, \\ -2x, & -0.5 < x < 0.5, \\ 2x + 2, & x \leq -0.5, \end{cases} \quad (2.30)$$

where $\epsilon(k)$ is an independent random variable of $N(0, \sigma^2)$ with $\sigma = 0.3$. From the generated data $x(k)$ $k = 0, \ldots, 200$, with an initial condition $x(0) = 0.1$, the first 100 points are used for identification and the rest for model validation. Ikoma and Hirota (1993) used this data for estimating a fuzzy relational model with nine equally-spaced triangular reference fuzzy sets. We show that by means of fuzzy clustering, a TS affine model with three reference fuzzy sets can be obtained and that both the membership functions and the consequent parameters reflect accurately the underlying piecewise linear system. It is assumed that the only prior knowledge is that the data was generated by a nonlinear autoregressive system:

$$x(k+1) = f(x(k), x(k-1), \ldots, x(k-p+1)), \quad (2.31)$$

where p is the system order. Here $\varphi(k) = [x(k), x(k-1), \ldots, x(k-p+1)]^T$ is the regression vector and $x(k+1)$ is the response variable. The matrix Z is constructed from the identification data:

$$Z = \begin{bmatrix} x(p) & x(p+1) & \cdots & x(N-1) \\ \cdots & \cdots & \cdots & \cdots \\ x(1) & x(2) & \cdots & x(N-p) \\ x(p+1) & x(p+2) & \cdots & x(N) \end{bmatrix}. \qquad (2.32)$$

To identify the system we need to find the order p and to approximate the function f by a TS affine model. In this example we demonstrate how the order of the system, and the number of clusters can be determined by means of a cluster validity measure. For flat hyperellipsoids, one of the axes is much shorter than the others, thus one of the eigenvalues should be significantly smaller than the remaining ones. Denoting λ_i^s and λ_i^L the smallest and the largest eigenvalue of F_i respectively, a cluster flatness index t_i can be defined as their ratio $t_i = \lambda_i^s / \lambda_i^L$. Aiming simultaneously at a low approximation error $e = 1/N \sum_{j=1}^{N} (x(j) - \hat{x}(j))^2$, where $\hat{x}$ is the model output, we define the validity measure as:

$$v = t \cdot e,$$

where t is computed as the average of t_is, $i = 1, 2, \ldots, K$. This validity measure was calculated for a range of model orders $p = 1, 2 \ldots, 5$ and number of clusters $K = 2, 3 \ldots, 7$. The results are shown in a matrix form in Figure 2.7(b). The minimal v (printed in boldface) was obtained for $p = 1$ and $K = 3$ which corresponds to (2.30). In Figure 2.7(a) v is also plotted as a function of K for orders $p = 1, 2$. Note that this function may have several local minima, of which the first is usually chosen in order to obtain a simpler model.

Figure 2.8(a) shows the projection of the obtained clusters onto the variable $x(k)$ for the correct system order $p = 1$ and the number of clusters $K = 3$. Figure 2.8(b) shows also the cluster prototypes:

$$V = \begin{bmatrix} -0.772 & -0.019 & 0.751 \\ 0.405 & 0.098 & -0.410 \end{bmatrix}.$$

From the cluster covariance matrices given below one can already see that the variance in one direction is higher than in the other one, thus the hyperellipsoids are flat and the model can be expected to represent a functional relationship between the variables involved in clustering:

$$F_1 = \begin{bmatrix} 0.057 & 0.099 \\ 0.099 & 0.249 \end{bmatrix}, \quad F_2 = \begin{bmatrix} 0.063 & -0.099 \\ -0.099 & 0.224 \end{bmatrix}, \quad F_3 = \begin{bmatrix} 0.065 & 0.107 \\ 0.107 & 0.261 \end{bmatrix}.$$

This is confirmed by examining the eigenvalues of the covariance matrices:

$$\lambda_{1,1} = 0.015, \qquad \lambda_{1,2} = 0.291,$$
$$\lambda_{2,1} = 0.017, \qquad \lambda_{2,2} = 0.271,$$
$$\lambda_{3,1} = 0.018, \qquad \lambda_{3,2} = 0.308.$$

One can see that for each cluster one of the eigenvalues is an order of magnitude smaller that the other one. The eigenvectors corresponding to the smallest eigenvalue of each covariance matrix are:

$$\Phi_1 = [-0.921, 0.389]^T, \quad \Phi_2 = [0.904, 0.428]^T, \quad \Phi_3 = [-0.915, 0.404]^T. \qquad (2.33)$$

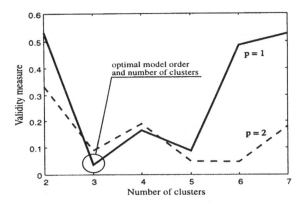

(a) Validity measure v as function of p and K.

model order

v	1	2	3	4	5
2	0.5312	0.3307	0.5076	4.6473	1.2753
3	**0.0367**	0.0893	0.0797	0.2162	2.4587
4	0.1660	0.1903	5.6228	0.3617	1.6028
5	0.0885	0.0493	0.0630	0.1819	0.2764
6	0.4852	0.0477	0.4380	0.2248	0.5174
7	0.5287	0.1806	2.0788	0.1264	0.1310

number of clusters

(b) v calculated for $p = 1, 2 \ldots, 5$ and $K = 2, 3 \ldots, 7$.

Figure 2.7 The cluster validity measure for different model orders and number of clusters.

By applying (2.26) we derive the parameters a_i and b_i of the affine TS model shown below. Parameterised piecewise exponential membership functions were used to define the antecedent fuzzy sets:

$$\mu(x(k); \sigma_1, \alpha_1, \sigma_2, \alpha_2) = \begin{cases} e^{-\left(\frac{x(k)-\alpha_1}{2\sigma_1}\right)^2} & \text{if } x(k) < \alpha_1 \\ e^{-\left(\frac{x(k)-\alpha_2}{2\sigma_2}\right)^2} & \text{if } x(k) > \alpha_2 \\ 1 & \text{otherwise.} \end{cases} \quad (2.34)$$

They were fitted to the projected clusters A_1 to A_3 by numerically optimising the parameters $\sigma_1, \alpha_1, \sigma_2$ and α_2. The result is shown by dashed lines in Figure 2.8(a). After labelling these fuzzy sets NEGATIVE, ABOUT ZERO and POSITIVE, the obtained TS models can be written as:

If $x(k)$ is NEGATIVE   then $x(k + 1) = 2.371x(k) + 1.237$
If $x(k)$ is ABOUT ZERO   then $x(k + 1) = -2.109x(k) + 0.057$
If $x(k)$ is POSITIVE   then $x(k + 1) = 2.267x(k) - 2.112$

The estimated consequent parameters correspond approximately to the definition of the line segments in the deterministic part of (2.30). Also the partition of the antecedent domain is in agreement with the definition of the system.

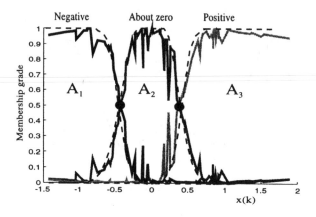

(a) Fuzzy partition projected on $x(k)$.

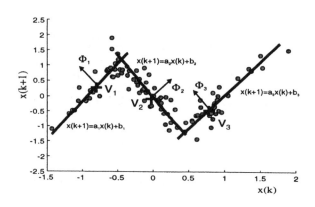

(b) Local linear models extracted from the clusters.

Figure 2.8 Result of fuzzy clustering for $p = 1$ and $K = 3$. Part (a) shows the membership functions obtained by projecting the partition matrix onto $x(k)$. Part (b) gives the cluster prototypes v_i, the orientation of the eigenvectors Φ_i and the direction of the affine consequent models (lines).

Converting an affine TS model into a singleton model

As shown previously, an affine TS model derived by fuzzy clustering approximates a non-linear hypersurface piecewise by hyperplanes. A piecewise linear model also can be obtained using a singleton model, i.e. a special case of the affine model (2.6) where $a_i = 0$ and thus $y_i = b_i, i = 1, \ldots K$. It is easy to show that in order to obtain linear interpolation between the constant consequents b_i, the antecedent fuzzy sets must be defined by triangular membership functions such that $\sum_{i=1}^{K} \mu_i = 1$, i.e. they are in fact piecewise linear basis splines of order 2 (Brown and Harris 1994) and the product intersection operator must be used (2.5). The cores of the membership functions (i.e. knots of the basis functions)

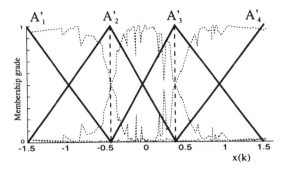

Figure 2.9   Triangular membership functions for the singleton model.

coincide with the intersection points of the adjacent membership functions of the affine TS model. Extra sets must be added at the extreme points of the domain. The construction of the triangular membership functions for the above example is shown in Figure 2.9. Note that the intersection points are in the neighbourhood of -0.5 and 0.5, which agrees with (2.30). In our example, the singleton model consists of the following rules:

> If $x(k)$ is A_1' then $x(k+1) = -1.5000$
> If $x(k)$ is A_2' then $x(k+1) = 0.4291$
> If $x(k)$ is A_3' then $x(k+1) = -0.4264$
> If $x(k)$ is A_4' then $x(k+1) = 1.5000$

where the consequent parameters are estimated from the data using least squares. In the multivariate case, the triangular membership functions are constructed for each antecedent variable separately. A smooth regression surface can be obtained by using higher-order (e.g. piecewise quadratic) splines. The advantage of this approach is that via fuzzy clustering one can find unevenly spaced knots reflecting the local (non)linearity of the regression surface, thus the number of rules needed for approximating the surface with a desired accuracy can be significantly reduced. Another advantage of the singleton model is that it can be easily inverted, since the consequents do not depend on the inputs of the model.

2.3.4   Other identification methods

Beside the methods presented in the previous sections, other identification techniques for fuzzy models have been introduced in literature. The approach of Sugeno and his co-workers (Takagi and Sugeno 1985, Sugeno and Kang 1986, Sugeno and Kang 1988, Sugeno and Tanaka 1991) is based on independent identification of the antecedent and consequent parts of the rules. The consequent parameters are estimated by least squares and the antecedent space is *a priori* partitioned using a small number of triangular membership functions (usually two per an antecedent variable) whose parameters are adjusted by using numerical optimisation. Other authors, such as (Jang 1993) apply error backpropagation methods to optimise the antecedent parameters of a fuzzy model in a similar way as the weights in a neural network.

   Although with these techniques one can arrive to numerically accurate nonlinear "black-box" models, in our view they are less suitable for fuzzy modelling since they destroy the interpretability of the model unless some other constraints are imposed. In many presented application examples, for instance in (Takagi and Sugeno 1985, Sugeno and Tanaka 1991), the output of the resulting fuzzy model over the considered domain is a consequence of the interpolation among several rules, and not of the local models themselves.

2.4 APPLICATION EXAMPLES

Fuzzy modelling techniques have been successfully applied to various complex nonlinear systems. This section presents two problems from different application domains, for which fuzzy models were built by using the identification method based on fuzzy clustering. Some other applications from literature are briefly reviewed as well.

2.4.1 Pressure modelling and control

An affine TS model describing the pressure dynamics in a laboratory fermentation tank is estimated from a set of input–output measurements. The system has two inputs (inlet air flow-rate u_2 and outlet valve position u_1) and one output (pressure in the fermenter headspace y). Experiments were designed to collect the identification and validation data (van Can *et al.* 1995). The inputs of the system were excited with multi-sinusoidal signals, such that the entire operating range was covered, see Figure 2.10. The process is modelled as a nonlinear ARX model $y(k + 1) = f(y(k), u_1(k), u_2(k))$ where f is approximated by an affine TS model (2.6).

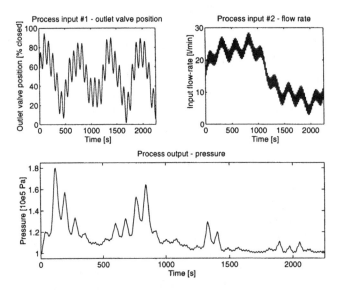

Figure 2.10   Process measurements used for identification.

The GK clustering algorithm with compatible cluster merging (see Section 2.3.3) was applied to the data (400 samples), initially with 9 clusters. After merging four and then two clusters, the procedure arrived to the final partition that divides the space of the regression vector $[y(k), u_1(k), u_2(k)]^T$ into three fuzzy regions. Membership functions for each of the regressors are obtained by fitting the cluster projections by parametric exponential functions (2.34). Finally, after estimating the consequent parameters by least squares, the following rules were obtained:

> **If pressure** $y(k)$ is LOW **and valve** $u_1(k)$ is OPEN
>
> **and flow-rate** $u_2(k)$ is LOW
>
> **then** $y(k + 1) = 0.52y(k) + 5 \cdot 10^{-5}u_1(k) + 3.2 \cdot 10^{-3}u_2(k) + 0.44$
>
> **If pressure** $y(k)$ is MEDIUM **and valve** $u_1(k)$ is HALF CLOSED
>
> **and flow-rate** $u_2(k)$ is MEDIUM
>
> **then** $y(k + 1) = 0.76y(k) + 2.4 \cdot 10^{-4}u_1(k) + 4.9 \cdot 10^{-3}u_2(k) + 0.04$
>
> **If pressure** $y(k)$ is HIGH **and valve** $u_1(k)$ is CLOSED
>
> **and flow-rate** $u_2(k)$ is HIGH
>
> **then** $y(k + 1) = 0.85y(k) + 7.4 \cdot 10^{-4}u_1(k) + 5.3 \cdot 10^{-3}u_2(k) - 0.45$

The membership functions for the pressure $y(k)$ and valve position $u(k)$ are shown in Figure 2.11.

The three consequent models are in a good agreement with linear models obtained by identifying discrete-time first order models from local step-responses measured on the process. Also the global output of the model obtained by simulation on a fresh data set follows the process output accurately, see Figure 2.12.

The TS fuzzy model is incorporated into a standard nonlinear predictive control scheme. The sample time is 5 s and the prediction horizon is one sample. The rate of change of the valve position is constrained to 10% per sample time. Real-time control results given in Figure 2.13 show that the model-based controller can accurately follow different reference signals. In these control experiments, the inlet flow-rate is kept constant.

2.4.2 Identification of respiratory dynamics

In this example, fuzzy clustering is used to estimate parameters of the respiratory dynamics from pressure–flow records from mechanically ventilated subjects. Figure 2.14 shows a data set consisting of 11 respiratory cycles, 5737 data samples in total. A physical model of the respiratory dynamics describes the relation among the pressure P, the lungs volume V (integrated flow-rate) and the flow-rate Q:

$$P = P_0 + E \cdot V + R \cdot Q. \tag{2.35}$$

The respiratory elastance E, the resistance R and the elastic recoil pressure P_0 (pressure when the volume V equals zero) are parameters that can be easily estimated from data using for instance least squares. From the values of these parameters the condition of the subject can be detected.

Since the parameters during the different phases of the mechanical ventilation may differ (Peslin *et al.* 1992), it is interesting to apply fuzzy clustering in order to automatically find the segments of data that yield reliable linear parameter estimates. The data samples $[V, Q, P]^T$ were partitioned into four clusters (the number of clusters was chosen based on prior knowledge) The parameters E, R and P_0 were extracted from the eigenstructure of the clusters (see section 2.3.3). Figure 2.15 shows the partitioning of the respiratory cycle into four segments for which different local models were found. The relation between the parameters for inspiration (phase 1) and expiration (phase 3) are in a good agreement with the results from the literature (Peslin *et al.* 1992). Numerical details of this study will be published elsewhere.

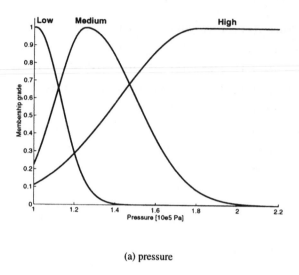

(a) pressure

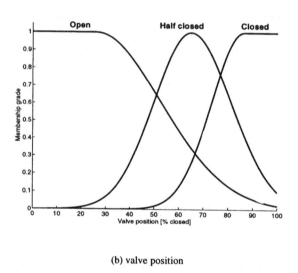

(b) valve position

Figure 2.11 Membership functions for pressure and valve position.

Other applications of fuzzy modelling to complex systems have been presented in lit-erature. Sugeno and his co-workers demonstrated their identification methods on mod-elling of a multilayer incinerator (Sugeno and Kang 1986), prediction of river water flow (Sugeno and Tanaka 1991), modelling and control of a steel-making process (Takagi and Sugeno 1985), and other systems. Takagi–Sugeno fuzzy rules are also used for hierar-chical control of an unmanned helicopter (Sugeno *et al.* 1993) and a model car (Sugeno and Nishida 1985). Nakamori and Ryoke (1994) describe an application to river water

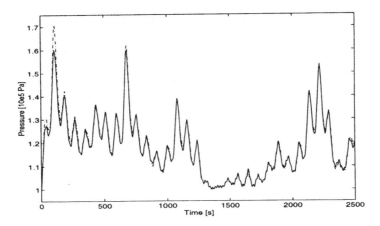

Figure 2.12   Model validation (solid line: process, dash-dotted line: model).

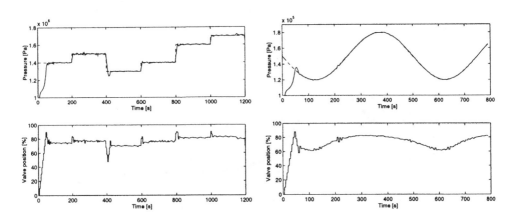

Figure 2.13   Real-time control response of a nonlinear predictive controller based on the TS fuzzy model for a stepwise and sinusoidal reference signal. Dashed line is the reference and solid line the process output.

quality prediction and (Zhao *et al.* 1994) modelled a glass melting furnace. In (Babuška *et al.* 1996) fuzzy model is developed for describing in qualitative terms the enzyme kinetics in a Penicillin–G conversion as a function on the concentrations of the components involved in the conversion. Kaymak (1994) applied fuzzy set techniques to modelling of enzymatic soil removal in a washing process. Also methods for controller design based on a fuzzy model have been developed and applied to cold-rolling process, automatic train operation (Terano *et al.* 1994) and combustion control for a refuse incineration plant (Sugeno and Kang 1986).

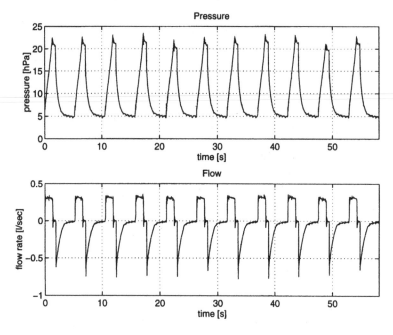

Figure 2.14   Pressure–flow measurements.

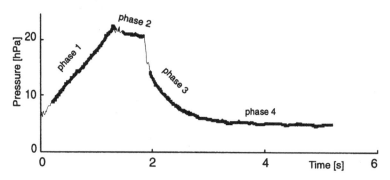

Figure 2.15   Partitioning of the respiratory cycle.

2.5   CONCLUDING REMARKS

In this chapter we described some methods for nonlinear system identification and modelling with the help of fuzzy sets. Main attention is paid to the Takagi–Sugeno class of fuzzy models and an identification technique based on fuzzy clustering is described in more detail. Simulation and real-world examples illustrate the potential of the presented approach.

Despite a number of successful applications, fuzzy modelling still cannot be considered a mature field. The available techniques for constructing and tuning fuzzy models are rather specialised and complex and their use requires skills and knowledge of fuzzy modelling experts. Rather than as a fully automated identification technique, fuzzy modelling should be seen as an interactive method facilitating an active participation of the user in a computer-assisted modelling session. For this purpose also special software tools need to be developed.

REFERENCES

AKAIKE, H. (1974) 'A new look at the statistical model identification'. *IEEE Transactions on Automatic Control* **19**(6), 716–723.

BABUŠKA, R. AND H.B. VERBRUGGEN (1994) Applied fuzzy modeling. In 'Proceedings IFAC Symposium on Artificial Intelligence in Real Time Control'. Valencia, Spain. pp. 61–66.

BABUŠKA, R. AND H.B. VERBRUGGEN (1995a) Identification of composite linear models via fuzzy clustering. In 'Proceedings European Control Conference'. Rome, Italy. pp. 1207–1212.

BABUŠKA, R. AND H.B. VERBRUGGEN (1995b) New approach to constructing fuzzy relational models from data. In 'Proceedings 3rd European Congress on Fuzzy and Intelligent Technologies'. Aachen, Germany. pp. 583–587.

BABUŠKA, R., H.B. VERBRUGGEN AND H.J.L. VAN CAN (1996) Fuzzy modeling of enzymatic Penicillin–G conversion. Submitted to IFAC World Congress.

BABUŠKA, R., R. JAGER AND H.B. VERBRUGGEN (1994) Interpolation issues in Sugeno–Takagi reasoning. In 'Proceedings IEEE World Congress on Computational Intelligence'. Orlando, FL. pp. 859–863.

BANERJEE, A., Y. ARKUN, R. PEARSON AND B. OGUNNAIKE (1995) H_∞ control of nonlinear processes using multiple linear models. In 'Proceedings European Control Conference'. Rome, Italy. pp. 2671–2676.

BERENJI, H. AND P.S. KHEDAR (1993) Clustering in product space for fuzzy inference. In 'Proceedings of Second International Conference on Fuzzy Systems'. San Francisco, CA. pp. 1402–1407.

BEZDEK, J. (1981) *Pattern Recognition with Fuzzy Objective Function*. Plenum Press, New York.

BROWN, M. AND C. HARRIS (1994) *Neurofuzzy Adaptive Modelling and Control*. Prentice Hall. New York.

BUCKLEY, J. J. AND Y. HAYASHI (1994) Fuzzy neural networks. In R. R. Yager and L. A. Zadeh (Eds.). 'Fuzzy Sets, Neural Networks and Soft Computing'. van Nostrand Reinhold. New York. pp. 233–249.

DAVE, R. (1992) Boundary detection through fuzzy clustering. In 'IEEE International Conference on Fuzzy Systems'. San Diego, CA. pp. 127–134.

DE BOOR, C. (1978) *A Practical Guide to Splines*. Springer-Verlag, New York.

DRIANKOV, D., H. HELLENDOORN AND M. REINFRANK (1993) *An Introduction to Fuzzy Control*. Springer, Berlin.

FRIEDMAN, J. (1991) 'Multivariate adaptive regression splines'. *The Annals of Statistics* **19**(1), 1–141.

GUPTA, M., KANDEL, A., BANDLER, W. AND KISZKA, J. (EDS.) (1985) *Approximate Reasoning in Expert Systems*. Elsevier Science Publishers. Amsterdam, The Netherlands.

GUSTAFSON, D. AND W.C. KESSEL (1979) Fuzzy clustering with a fuzzy covariance matrix. In 'Proc. IEEE CDC'. San Diego, CA. pp. 761–766.

IKOMA, N. AND K. HIROTA (1993) 'Nonlinear autoregressive model based on fuzzy relation'. *Information Sciences* **71**, 131–144.

JAIN, A. K. AND R. C. DUBES (1988) *Algorithms for Clustering Data*. Prentice Hall. Englewood Cliffs, NJ.

JANG, J.-S. (1993) 'ANFIS: Adaptive-network-based fuzzy inference systems'. *IEEE Transactions on Systems, Man & Cybernetics* **23**(3), 665–685.

JANG, J.-S. AND C.-T. SUN (1993) 'Functional equivalence between radial basis function networks and fuzzy inference systems'. *IEEE Transactions on Neural Networks* **4**(1), 156–159.

JOHANSEN, T. (1994) Operating Regime Based Process Modelling and Identification. PhD dissertation. The Norwegian Institute of Technology – University of Trondheim. Trondheim, Norway.

KAYMAK, U. (1994) Application of fuzzy methodologies to a washing process. Chartered designer thesis. Delft University of Technology. Control Lab., Faculty of El. Eng., Delft.

KAYMAK, U. AND R. BABUŠKA (1995) Compatible cluster merging for fuzzy modeling. In 'Proceedings FUZZ-IEEE/IFES'95'. Yokohama, Japan. pp. 897–904.

KRISHNAPURAM, R. AND CHIN-PIN FREG (1992) 'Fitting an unknown number of lines and planes to image data through compatible cluster merging'. *Pattern Recognition* 25(4), 385–400.

LEONARITIS, I. AND S.A. BILLINGS (1985) 'Input-output parametric models for non-linear systems'. *International Journal of Control* 41, 303–344.

NAKAMORI, Y. AND M. RYOKE (1994) 'Identification of fuzzy prediction models through hyperellipsoidal clustering'. *IEEE Transactions on Systems, Man and Cybernetics* 24(8), 1153–73.

PEDRYCZ, W. (1984) 'An identification algorithm in fuzzy relational systems'. *Fuzzy Sets and Systems* 13, 153–167.

PEDRYCZ, W. (1993) *Fuzzy Control and Fuzzy Systems (second, extended edition)*. John Wiley and Sons. New York, NY.

PEDRYCZ, W. (1995) *Fuzzy Sets Engineering*. CRC Press. Boca Raton, FL.

PESLIN, R., F. DA SILVA, C. DUVIVIER AND F. CHABOT (1992) 'Respiratory mechanics studied by multiple linear regression in unsedated ventilated patients'. *Eur. Respir. J.* 5, 871–878.

SETNES, M. (1995) Fuzzy Rule Base Simplification Using Similarity Measures. Master's thesis. Delft University of Technology. Delft, The Netherlands.

SUGENO, M. AND G.T. KANG (1986) 'Fuzzy modelling and control of multilayer incinerator'. *Fuzzy Sets and Systems* 18, 329.

SUGENO, M. AND G.T. KANG (1988) 'Structure identification of fuzzy model'. *Fuzzy Sets and Systems* 28, 15–33.

SUGENO, M. AND K. TANAKA (1991) 'Successive identification of a fuzzy model and its application to prediction of a complex system'. *Fuzzy Sets and Systems* 42, 315–334.

SUGENO, M. AND M. NISHIDA (1985) 'Fuzzy control of model car'. *Fuzzy Sets and Systems* 16, 103–113.

SUGENO, M. AND T. YASUKAWA (1993) 'A fuzzy-logic-based approach to qualitative modeling'. *IEEE Transactions on Fuzzy Systems* 1, 7–31.

SUGENO, M., M.F. GRIFFIN AND A. BASTIAN (1993) Fuzzy hierarchical control of an unmanned helicopter. In 'Proceedings of the Fifth IFSA World Congress'. Seoul, Korea. pp. 179–181.

TAKAGI, T. AND M. SUGENO (1985) 'Fuzzy identification of systems and its application to modeling and control'. *IEEE Transactions on Systems, Man and Cybernetics* 15(1), 116–132.

TERANO, T., K. ASAI AND M. SUGENO (1994) *Applied Fuzzy Systems*. Academic Press, Inc.. Boston.

VAN CAN, H., H.A.B. TE BRAAKE, C. HELLINGA, A.J. KRIJGSMAN, H.B. VERBRUGGEN AND K.CH.A.M. LUYBEN (1995) 'Design and real-time testing of a neural model predictive controller for a nonlinear system'. *Chemical Engineering Science*. To appear.

WANG, H., K. TANAKA AND M. GRIFFIN (1995) Parallel distributed compensation of nonlinear systems by Takagi–Sugeno fuzzy model. In 'Proceedings FUZZ-IEEE/IFES'95'. Yokohama, Japan. pp. 531–538.

YAGER, R. AND D.P. FILEV (1994) *Essentials of Fuzzy Modeling and Control*. John Wiley. New York.

YOSHINARI, Y., W. PEDRYCZ AND K. HIROTA (1993) 'Construction of fuzzy models through clustering techniques'. *Fuzzy Sets and Systems* 54, 157–165.

ZHAO, J., V. WERTZ AND R. GOREZ (1994) A fuzzy clustering method for the identification of fuzzy models for dynamical systems. In '9th IEEE International Symposium on Intelligent Control'. Columbus, OH.

CHAPTER THREE

Modelling of Electrically Stimulated Muscle

H. GOLLEE, K. J. HUNT, N. DE N. DONALDSON and J. C. JARVIS

Human muscle can be made to contract by electrical stimulation. There are many well-established uses for this technique and new applications are being investigated. We are interested in two uses: restoration of function to paralysed limbs and reconfiguration of one of the back muscles to assist a failing heart.

Muscle modelling has been widely studied in the past. Model types range from biophysical models, which are based on the structure and processes of actual muscles, through analogue models, to purely mathematical descriptions. The latter are derived only from the input (neural activation) and the output (mechanical response) signals. We are searching for mathematical models, which can represent the complex nonlinear behaviour of muscle, so that the contraction can be better controlled.

We have developed a number of dynamic muscle models based on force measurements taken from isometric contracting muscles stimulated by an irregular pulse train. The results of experimental investigations have shown that a simple linear transfer-function model of muscle response can be rather inaccurate over the full operational range. The muscle response is significantly nonlinear with respect to the level of stimulation. However, the experiments with linear system identification provide results which can be used to choose structure parameters of a dynamic model.

Nonlinear models based on local model networks (LMN) are able to capture the nonlinear effects and provide accuracy over a wide operational range. We investigate the use of Radial Basis Function (RBF) networks and LMN with local linear models. We describe the types of network models used, the training algorithms, and show some typical experimental results based on real measured data.

3.1 INTRODUCTION

We are interested in obtaining models of electrically activated muscle which could be used in stimulator controllers for cardiac assistance from skeletal muscles (Salmons and Jarvis 1992), or for functional electrical stimulation of limb muscle in paralysed patients (Hambrecht 1992). For both applications, a model of the muscle is essential to allow development of algorithms for its controlled stimulation.

A variety of different approaches to employ skeletal muscle for cardiac assistance is discussed in (Salmons and Jarvis 1992). One possible setup uses a muscle from the patient's back transformed into a Skeletal Muscle Ventrical (SMV) which is wrapped around a part of the artery and functions as a pumping device. Two control loops have to be considered:

1 The pumping power should support the weak heart in such a way that only the power actually needed is generated. Thus, the required pumping activity depends on parameters of the circulation system such as blood pressure and flow.

2 The skeletal muscle ventricle should generate the pumping power required to perform the functions described above, in phase with the heart pulse. On top of that, the long-term changes in the skeletal muscle due to the artificial stimulation should affect its characteristics in a desired way.

Naturally, control task (1) requires a model of the human circulatory system, whereas task (2) needs a model of artificially stimulated skeletal muscle.

In this section, we first give a brief introduction to the field of artificial electrical stimulation of muscle. Following that, the application considered in this work is presented.

3.1.1 Muscle modelling

Muscle physiology

In order to apply artificial electrical stimulation to muscle, it is essential to take some basic physiological properties of the neuro-muscular system into account; an introduction to this can be found in standard physiology textbooks, e.g. in (Silbernagl and Despopoulos 1991).

In skeletal muscle, extrafusal muscle fibres are the primary unit of contraction. They are stimulated by axons of α-motoneurons, which are situated in the spinal cord. The transmission of information takes place in the form of impulse trains; a stimulation pulse can create an action potential in the neuron if it exceeds a certain threshold. A new activation of this neuron is only possible after a certain recreation period, thus avoiding over-stimulation. The neural information may be regarded as being encoded in the inter-pulse intervals (IPIs), or in the instantaneous pulse frequency, of the pulse train.

One motoneuron activates 5 to 1000 muscle fibres simultaneously. All the fibres activated by the same motoneuron can be distributed over the entire muscle and form a motor unit, which represents the 'unit' of muscle force in a normally innervated muscle. Two types of motor units can be distinguished: fast and slow units. Slow motor units are more fatigue resistant and thus are able to generate a certain force for a longer time, whereas fast motor units can produce a higher force in a shorter time. The ratio of fast and slow motor units in a muscle has a great influence on its characteristics.

All the motoneurons going from the spinal cord to the same muscle form a nerve. Each motoneuron can be stimulated selectively by the Central Nervous System (CNS), enabling a graduated muscle activation. Here, those motoneurons which belong to slow, fatigue resistant motor units are recruited first, and fast motor units are only recruited if high force is necessary. The contractile force of each motor unit can be increased by applying a stimulation with a higher pulse frequency.

Artificial muscle activation can take place by applying electrical impulses to the motoneurons, thus generating action potentials which are transmitted to the corresponding muscle fibres. With current chronic implanted electrodes, single motoneurons are not stimulated directly. The electrodes are relatively large and stimulate many neurons. Thus, with artificial electrical stimulation the muscle activation can be varied by

(a) the energy of the electric pulse, which defines the number of recruited motor units, and
(b) the pulse frequency, or the inter-pulse interval (IPI), which determines the contraction of the recruited muscle fibres.

Due to the fact that fast motor units have larger motoneurons, these units are recruited first, which is contrary to the way the recruitment takes place when the muscle is stimulated by the CNS. Moreover, all recruited motor units are stimulated synchronously, as opposed to the natural stimulation which can take place asynchronously. An asynchronous recruitment allows a smoother muscle contraction and some muscle fibres might be at rest, which increases fatigue resistance.

Model structures

Different types of muscle models are used for different purposes. The range extends from analytical models which are based on physical properties of the muscle, either on a microscopic or on a macroscopic level, to parametric models which are purely mathematical descriptions.

Analytical models are popular in biology and biomechanics as they are based on physical properties of the neuro-muscular system, either on a microscopic level, like the cross bridge model (Huxley 1957), or by describing entire muscle, as in the Hill-type model based on the approach described in (Hill 1938). Their advantages are mainly that their parameters are in some way related to physiological properties of the modelled system.

For use in implantable muscle stimulator devices, different model characteristics from the ones inherent to the group of analytical models become important, namely that the model

■ should be easy to adapt to the muscle using data from standard experiments which do not damage the muscle tissue;
■ should be controller orientated, i.e. controller design based on such a model should be possible. The designed controller must not be computationally expensive as it should be implemented in the form of an implantable device.

For these reasons, most real-world implementations are based on simple linear muscle models. Often, the contraction is assumed to be isometric,[1] which gives a single-input (stimulation) single-output (force) system. Experiments have shown that a second order dynamic linear model with delay can be applied (Zahalak 1992). To take the nonlinear recruitment of motor units into account, the linear structure is extended by a nonlinear static recruitment curve, which leads to a *Hammerstein* model as shown in Figure 3.1 (Haber and Unbehauen 1990).

However, such a model does not cope with nonlinear effects other than those related to recruitment. In particular, the doublet effect (a more than linear increase in muscle force when two input pulses with very short IPI are applied) is a nonlinear effect not covered by common parametric models. We therefore address this problem by introducing more general nonlinear parametric model structures.

---

[1] During isometric contraction the muscle length is constant.

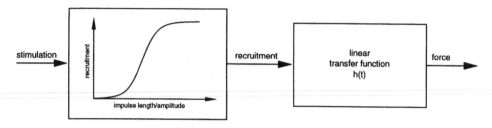

Figure 3.1 Hammerstein cascade structure.

NARMAX structure The most general form of input-output model is the well-known NARMAX[2] structure (Leontaritis and Billings 1985),

$$y(t) = f(u(t - k), \ldots u(t - k - n_u), y(t - 1), \ldots y(t - n_y), e(t - 1), \ldots e(t - n_e))$$
$$+ e(t). \quad (3.1)$$

Here, $f()$ is a generally nonlinear, smooth function, $y(t)$ is the system output, $u(t)$ is the system input and $e(t)$ describes a white noise disturbance. The discrete time is expressed by index t, and k is an input delay. If we restrict ourselves to single-input–single-output systems, then $y(t) \in \mathbb{Y} \subset \mathbb{R}$, $u(t) \in \mathbb{U} \subset \mathbb{R}$ and $e(t) \in \mathbb{E} \subset \mathbb{R}$.

The arguments of $f()$ can be written as a single vector $\underline{\psi}$,

$$\underline{\psi}(t) = [u(t - k), \ldots u(t - k - n_u), y(t - 1), \ldots y(t - n_y),$$
$$e(t - 1), \ldots e(t - n_e)]^T \quad (3.2)$$
$$\in \mathbb{U}^{n_u+1} \times \mathbb{Y}^{n_y} \times \mathbb{E}^{n_e} \subset \mathbb{R}^{n_u+n_y+n_e+1} .$$

Then, equation (3.1) becomes

$$y(t) = f(\underline{\psi}(t)) + e(t) . \quad (3.3)$$

NARX structure It is only necessary to consider the delayed noise terms in the data vector $\underline{\psi}$, as in equation (3.2), if the disturbances are correlated. However, often disturbances with white noise characteristics can be assumed, in which case the data vector can be simplified,

$$\underline{\psi}(t) = [u(t - k), \ldots u(t - k - n_u), y(t - 1), \ldots y(t - n_y)]^T \quad (3.4)$$
$$\in \mathbb{U}^{n_u+1} \times \mathbb{Y}^{n_y} \subset \mathbb{R}^{n_u+n_y+1} .$$

When this data vector is used in equation (3.3), the structure becomes a NARX[3] model.

When identifying a system one aims at finding a parameterised structure which approximates the unknown function $f()$ in equation (3.3). In section 3.2 we look at linear functions for $f()$. Nonlinear approaches are described in section 3.3. In both cases we restrict ourselves to (N)ARX structures, i.e. the data vector (3.4) is used.

---

[2]Nonlinear AutoRegressive Moving Average with eXogenous inputs.
[3]Nonlinear AutoRegressive with eXogenous inputs.

3.1.2 Experiments

In this paper we use data from experiments with a *tibialis anterior* rabbit muscle which is stimulated by irregular supramaximal pulse trains. The term *supramaximal* refers to the fact that the amplitude and length of the stimulation pulses are chosen such that all motoneurons of the muscle are recruited. This corresponds to the upper right region of the recruitment curve shown in Figure 3.1. The muscle stimulation is varied by the irregularity of the IPI, which introduces additional dynamical nonlinearities as mentioned above.

The experimental protocol is described in detail in (Jarvis 1993). Briefly, the following conditions applied to the experiments:

- The muscle was stimulated using electrical impulses with 200 microseconds duration and an amplitude three times the threshold for muscle stimulation, which ensured supramaximal stimulation.
- The contractive force of the muscle was measured and recorded, while the muscle length was kept constant, i.e. the muscle was contracting in an isometric way.
- The inter-pulse intervals (IPIs) were varied randomly between 1 and 70 milliseconds.
- The duration of one pulse train did not exceed 300 milliseconds. Together with periods of rest of 30 seconds between the pulse trains, this ensured that fatigue did not affect the recorded data.
- A constant-frequency burst of impulses (25 milliseconds IPI) was delivered every 5 minutes to check that the preparation did not show progressive deterioration during the experiment.

A total of 60 data sets, containing the input pulses and the contractive force, each of 590 milliseconds duration, was recorded at 1 kHz sampling frequency. A number of typical records are shown in Figure 3.2.

3.2 LINEAR SYSTEM IDENTIFICATION

3.2.1 Linear model structures

The classical approach to find a linear model is based on the expansion of the general NARX description of the system, equation (3.3) with (3.4), into a Taylor series around some operating point $[u^o, y^o]$. For small deviations around this point higher-order terms can be ignored and we obtain a linear approximation,

$$
\begin{aligned}
\tilde{y}(t) &= y(t) - f(\underline{\psi}^o) \\
&= -a_1\tilde{y}(t-1) - \cdots - a_{n_y}\tilde{y}(t-n_y) + \\
&\quad + b_0\tilde{u}(t-k) + \cdots + b_{n_u}\tilde{u}(t-k-n_u) + e(t) ,
\end{aligned} \tag{3.5}
$$

where the coefficients a_i, b_j are defined by

$$
a_i = \left. \frac{-\partial f}{\partial y(t-i)} \right|_{u^o, y^o} , \qquad\qquad i = 1, \ldots n_y \tag{3.6}
$$

$$
b_j = \left. \frac{\partial f}{\partial u(t-k-j)} \right|_{u^o, y^o} , \qquad\qquad j = 0, \ldots n_u \tag{3.7}
$$

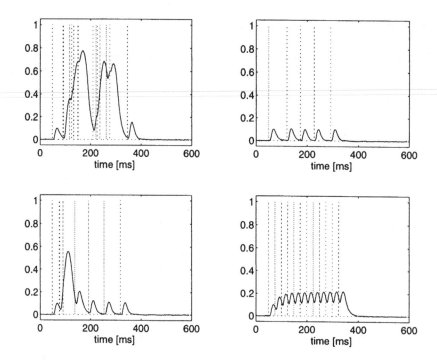

Figure 3.2 Some typical data records. The input pulses are shown dotted, the muscle force solid. The graph on the bottom right shows a record of a tetanic contraction as a result of a constant-frequency burst. On the three other graphs, records with random stimulation patterns are depicted.

The deviation variables are defined by

$$\tilde{y}(t) = y(t) - f(\underline{\psi}^o) = y(t) - y^o ,$$
$$\tilde{u}(t) = u(t) - u^o ,$$
$$\text{and} \quad \underline{\tilde{\psi}}(t) = \underline{\psi}(t) - \underline{\psi}^o , \tag{3.8}$$
$$\text{where} \quad \underline{\psi}^o = [u^o, \dots u^o, \ y^o, \dots y^o]^T .$$

Introducing the parameter vector $\underline{\Theta} \in \mathbb{R}^{n_u + n_y + 1}$,

$$\underline{\Theta} = [\Theta_0, \dots \Theta_{n_u}, \ \Theta_{n_u+1}, \dots \Theta_{n_u+n_y}]^T$$
$$= [b_0, \dots b_{n_u}, \ -a_1, \cdots -a_{n_y}]^T , \tag{3.9}$$

equation (3.5) becomes,

$$\tilde{y}(t) = \underline{\tilde{\psi}}^T(t)\underline{\Theta} + e(t) . \tag{3.10}$$

Defining polynomials A and B as

$$A(q^{-1}) = 1 + a_1 q^{-1} + \cdots + a_{n_y} q^{-n_y}$$
$$B(q^{-1}) = b_0 + b_1 q^{-1} + \cdots + b_{n_u} q^{-n_u} , \tag{3.11}$$

equation (3.10) can be rewritten as a transfer function,

$$A(q^{-1})\,\bar{y}(t) = q^{-k}\,B(q^{-1})\,\bar{u}(t) + e(t)\,. \qquad (3.12)$$

Here q^{-1} is the time shift operator, i.e. $q^{-n}\,y(t) = y(t-n)$.

Often it can be assumed that the number of delayed inputs corresponds to the number of delayed outputs,

$$n = n_u + 1 = n_y\,, \qquad (3.13)$$

which simplifies the data vector (3.4) to

$$\underline{\psi}_{k,n}(t) = [u(t-k),\dots u(t-k-n+1),\ y(t-1),\dots y(t-n)]^T\,. \qquad (3.14)$$

Using data vector (3.14) the linear ARX structure is defined by only two design parameters, n and k.

3.2.2 Sampling period

The experimental data introduced in section 3.1.2 are sampled with a period $T_{exp} = 1$ ms, which is sufficiently short to represent all system dynamics in the discrete samples. We are interested in increasing this period to an 'optimal' value which is

1   not too long, so that no information about the dynamics of the real system is lost due to a violation of the sampling theorem, and
2   not too short, to prevent the system from modelling fast disturbances (noise).

Furthermore, a larger sampling period is desirable because this goes with a reduction of the size of the data set and hence leads to a decreased computational effort.

The effects related to a changed sampling period are different for the pulse-like input signal and for the continuous-in-value output signal. In the following sections they are discussed separately.

Sampling of the input signal

For the input signal, a change of the sampling period should not affect the form of the signal, which has to remain pulse like. The subsampling is processed in such a way that the input pulses are only shifted towards the nearest sampling point. Two objectives have to be considered here:

(a)   The energy of the input signal must not change, i.e. the number of pulses in the sub-sampled signal must be the same as in the original pulse train.
(b)   The pulses must not be shifted too far away from their original position.

If objective (a) is considered when enlarging the sampling period, the pulses are shifted further away from their original time position. This will violate objective (b) if the sampling period becomes too large. For this reason, an upper limit of $T \leq 10$ ms has been introduced for the sampling of the input pulses.

Sampling of the output signal

For a linear dynamic system, first information about the dominating time constants and the dynamic order can be extracted from its step response. Based on that, estimates for the optimal sampling period and the model structure are possible.

Due to the physiological properties of muscle stimulation, its input signal must be pulse-like, and hence the system's *step* response cannot be measured directly. Using the experimental data we can only extract the *impulse* response, $g(t)$, from which the step response $h(t)$ can be derived by integration.

By averaging a number of experimental impulse responses, a mean response of the system has been derived, as shown in the top of Figure 3.3. The corresponding step response is depicted at the bottom of Figure 3.3.

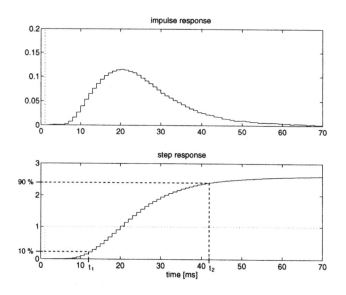

Figure 3.3   Impulse and step response.

From the step response the rise time of the system can be determined to be

$$t_r = t_2 - t_1 \approx 30 \text{ ms} . \tag{3.15}$$

Here, t_1 and t_2 are the times corresponding to 10% and 90% of the steady state.

For linear systems a simple heuristic relationship between the rise time and the necessary sampling period can be applied (Åström and Wittenmark 1990),

$$\frac{t_r}{T} \approx 4 \dots 10 . \tag{3.16}$$

With equation (3.15), the necessary sampling period is in the range of $T \approx 3 \dots 7.5$ ms for our application.

From the shape of the step response it can be surmised that the system may be of second order with a dead-time of $t_d \approx 7$ ms.

3.2.3   Linear structure

The experiments for the linear parameter identification have two objectives:

- to find a suitable sampling period T^* for the data;
- to find the optimal dynamical order n^* and the optimal delay k^* of an ARX model structure.

Thus, the three design parameters T, n and k have to be optimised.

Identification experiments

The operating point is chosen to be $\underline{0}$, i.e. $y^o = 0$ and $u^o = 0$, because this regime is known as the idle regime of the muscle. In the following parts of this section the $\tilde{\ }$ is omitted to mark the deviation from the operating regime, since equation (3.10) becomes

$$y(t) = \underline{\psi}^T(t)\underline{\Theta} + e(t) \,. \tag{3.17}$$

For each structure defined by a set of design parameters (T, n, k) the corresponding optimal parameter vectors $\underline{\Theta}^*(T, n, k)$ can be determined using standard linear regression techniques. As the optimisation criterion, we use the squared output error criterion (Press et al. 1992),

$$J(\underline{\hat{\Theta}}) = \frac{1}{L} \sum_{t=l_0+1}^{l_0+L} \left[y(t) - \hat{y}(t, \underline{\hat{\Theta}}) \right]^2 \,, \tag{3.18}$$

where $y(t)$ is the target output and

$$\hat{y}(t) = \underline{\psi}^T(t)\underline{\hat{\Theta}} \tag{3.19}$$

is the estimated output at sample instant t. The optimal parameter vector $\underline{\hat{\Theta}}^*$ can be estimated by minimising (3.18). A necessary condition for this is that the derivative with respect to the parameter vector vanishes,

$$\frac{\partial J(\underline{\hat{\Theta}})}{\partial \underline{\hat{\Theta}}} = 0 = \frac{2}{L} \sum_{t=l_0+1}^{l_0+L} \left\{ \underline{\psi}(t) \left[y(t) - \underline{\psi}^T(t) \underline{\hat{\Theta}}^* \right] \right\} \,. \tag{3.20}$$

From there the optimal parameter vector can be calculated as

$$\underline{\hat{\Theta}}^* = \left[\sum_{t=l_0+1}^{l_0+L} \underline{\psi}(t) \underline{\psi}^T(t) \right]^{-1} \sum_{t=l_0+1}^{l_0+L} \underline{\psi}(t) \, y(t) \,. \tag{3.21}$$

In this work, the optimal design parameters have been determined using the following two-step algorithm:

1   The conditioning of covariance matrices

$$R_{k.n} = \Psi_{k.n}^T \Psi_{k.n} \,, \tag{3.22}$$

where

$$\Psi_{k.n} = \begin{bmatrix} \underline{\psi}_{k.n}^T(l_0 + 1) \\ \vdots \\ \underline{\psi}_{k.n}^T(l_0 + L) \end{bmatrix} \quad \in \mathbb{R}^{L \times (2n+1)} \,, \tag{3.23}$$

are compared for various design parameters. Three structures with the covariance matrix having the best conditioning are selected (Ljung 1987).

2   The structures selected in the previous step are validated using a 5-fold cross–validation test (Weiss and Kulikowski 1991). The structure with the lowest error is selected.

Results

After applying the described algorithm to the data set introduced in section 3.1.2, the following set of design parameters has been found to be optimal:

$$
\begin{aligned}
\text{order:} \quad & n^* = 2\,; \\
\text{delay:} \quad & k^* = 2\,; \\
\text{sampling period:} \quad & T^* = 5\,\text{ms}\,.
\end{aligned}
\tag{3.24}
$$

After optimising the parameter vector of the selected linear model structure using equation (3.21), the linear model can be described in transfer function form as

$$
\hat{y}(t) = \frac{q^{-k}\,B(q^{-1})}{A(q^{-1})}\,u(t) = \frac{q^{-2}\,\left(0.0629 + 0.0385q^{-1}\right)}{1 - 1.4689q^{-1} + 0.5565q^{-2}}\,u(t)\,.
\tag{3.25}
$$

This model has been tested on a number of new data sets which have not been used for training. In Figure 3.4 the model response to some typical stimulation patterns is compared to the response of the real muscle. It is obvious that the linear model cannot predict the correct muscle output for all operating regimes. For low muscle activation, the output of the model is too high (e.g. top right), and for high activation of the muscle the model output is too low (e.g. top left). The muscle behaviour during a tetanic contraction cannot be correctly predicted either (bottom right). However, the simple linear model does provide a qualitatively correct response which *on average* matches the measured input–output data over the range of operation.

3.3   NONLINEAR MODELLING

The results obtained using linear system identification, section 3.2, show that the muscle behaviour is significantly nonlinear. In this section we apply some nonlinear modelling approaches, where information from the linear experiments about the local model structure is used.

3.3.1   Local model networks

Architecture

We consider the generalised form of basis function networks as depicted in Figure 3.5.
It can be described as

$$
\hat{y}(t) = \sum_{i=1}^{n_A} \rho_i(\underline{\phi}(t))\,\hat{f}_i(\underline{\psi}(t))\,,
\tag{3.26}
$$

where $\hat{y}(t) \in \mathbb{R}$ is the model output at time t, ρ_i are *validity functions* (or *basis functions*) which depend on some scheduling vector $\underline{\phi}(t) \in \mathbb{R}^{n_\phi}$, and $\hat{f}_i$ are *local models* which depend on the input vector $\underline{\psi}(t) \in \mathbb{R}^{n_\psi}$. The number of validity functions is given by n_A. This *Local Model Network*, first introduced by (Johansen and Foss 1992, Johansen and Foss 1993), is a generalised form of basis function networks (as compared to the Radial Basis Function (RBF) network described in (Moody and Darken 1989)) in the sense that the simple weights associated with the basis functions are replaced by more complex *local models*. Keeping this in mind the classical RBF network is treated as a special case of the wider class of Local Model Network.

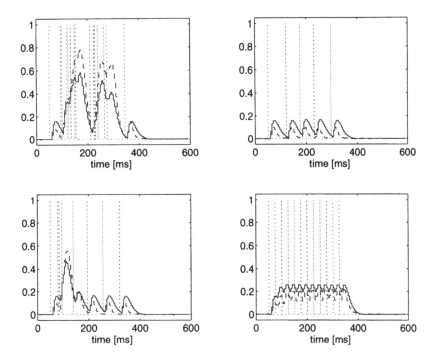

Figure 3.4   Comparison of linear model output [——] and target output [- - -] for typical input patterns [···].

Validity functions   Although any function with a locally limited activation might be applied as a validity function, Gaussian bells are used most widely. Other popular validity functions are B–splines, e.g. Kavli's ASMOD structures (Kavli 1993) and Friedmann's MARS (Friedmann 1991), and Kernel functions, e.g. Hlaváčková's KBFs (Hlaváčková 1995).

We restrict ourselves to the radial form of Gaussian bells, which can be described as

$$\tilde{\rho}_i(\underline{\phi}) = \exp\left[-\frac{(\underline{\phi} - \underline{c}_i)^T (\underline{\phi} - \underline{c}_i)}{s_i^2}\right] . \tag{3.27}$$

Here, $\underline{c}_i \in \mathbb{R}^{n_\phi}$ are the centre vectors and s_i determine the width of the bell.

To achieve a *partition of unity* of the input space, the basis functions have to be normalised (Shorten and Murray-Smith 1994), i.e.

$$\rho_i(\underline{\phi}) = \frac{\tilde{\rho}_i(\underline{\phi})}{\sum_{j=1}^{n_A} \tilde{\rho}_j(\underline{\phi})} . \tag{3.28}$$

Note that this can change the shape of the validity functions in a significant way (this is discussed in detail in Chapter 8).

Local models   In general, the local models $\hat{f}_i$ in equation (3.26) can be of any form; they can be nonlinear or linear, have a state space or input–output description, or be discrete or

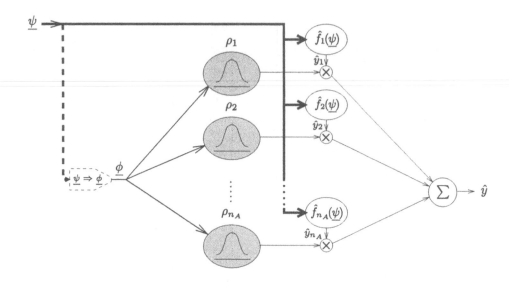

Figure 3.5   General architecture of a Local Model Network.

continuous time. They can be of different character, using physical models of the system for operating conditions where they are available, and parametric models for conditions where there is no physical description available.

For input–output modelling of dynamic systems, we choose to work with a model structure in the form of the discrete time NARX model as described by equation (3.3) with data vector (3.4) in section 3.1.1. The models $\hat{f}_i$ in equation (3.26) become local (N)ARX approximations of the overall system f. The contribution of the ith local model is defined by the activation of the corresponding validity function $\rho_i(\underline{\phi})$. The scheduling vector can then be chosen as a part of the entire data vector,

$$\underline{\phi}(t) \in \mathbb{R}^{n_\phi} \subseteq \mathbb{R}^{n_\psi} , \qquad n_\phi \leq n_\psi , \tag{3.29}$$

With local (N)ARX structures, equation (3.26) becomes

$$\hat{y}(t) = \hat{f}(\underline{\psi}(t)) = \sum_{i=1}^{n_A} \rho_i(\underline{\phi}(t)) \, \hat{f}_i(\underline{\psi}(t), \underline{\Theta}_i) , \tag{3.30}$$

where $\underline{\Theta}_i$ are local parameter vectors.

In section 3.3.2 we use the simplest form of local models – scalar weights; in section 3.3.2 linear ARX structures are applied.

Learning

The learning process in Local Model Networks can be divided into two tasks:

1   find the optimal number, position and shape of the validity functions, i.e. define the *structure* of the network.

2 find the optimal set of parameters for the local models, i.e. define the *parameters* of the network.

Approaches for performing these two tasks are discussed in this section.

Parameter optimisation   A common approach for the optimisation of the parameters is to use a least squares output error criterion. For use with LMNs, the following two variants are described in (Murray-Smith and Johansen 1995):

- **Local optimisation:** The parameters of each local model are estimated independently. A weighted least squares optimisation criterion can be defined for each local model, where the weighting factor is the current activation of the corresponding validity function,

$$J(\hat{\underline{\Theta}}_i) = \frac{1}{L} \sum_{t=l_0+1}^{l_0+L} \rho_i(\underline{\phi}(t)) \left[y(t) - \hat{y}_i(t, \hat{\underline{\Theta}}_i) \right]^2, \qquad i = 1, \ldots, n_A. \quad (3.31)$$

- **Global optimisation:** The parameters of all local models are estimated at the same time. The standard least squares error criterion can be used,

$$J(\hat{\underline{\Theta}}_{all}) = \frac{1}{L} \sum_{t=l_0+1}^{l_0+L} \left[y(t) - \hat{y}(t, \hat{\underline{\Theta}}_{all}) \right]^2, \quad (3.32)$$

where the global parameter vector $\underline{\Theta}_{all}$ consists of the parameter vector of all local models,

$$\underline{\Theta}_{all} = \left[\underline{\Theta}_1^T, \ldots, \underline{\Theta}_{n_A}^T \right]^T. \quad (3.33)$$

Considering the case where local scalar weights or linear local models are used the parameters in $\underline{\Theta}_1, \ldots, \underline{\Theta}_{n_A}$ are *linear* relative to the network output $\hat{y}$. Thus, linear regression methods, as described in section 3.2.3, can be applied to optimise the parameter vectors.

A discussion of local vs. global optimisation in Local Model Networks can be found in Chapter 7.

Structure optimisation   Structure optimisation in Local Model Networks is a more challenging task than the parameter optimisation, and most of the effort of a typical learning process is spent here. The aim is to adapt the number, position and shape of the validity functions to the complexity of the system. The following approaches for the optimisation of the structure of nonlinear dynamic systems are described in (Haber and Unbehauen 1990) and can be used for Local Model Networks:

(a) *Forward regression*: The model structure grows according to the complexity of the system.

(b) *Backward regression (pruning)*: A complex model is used as an initial structure for a model reduction algorithm, where it is attempted to extract the essence by pruning useless parameters (Reed 1994, Jutten and Fambon 1995).

If Gaussian bells are used as validity functions, their width can be adjusted according to the mean distance to the n-nearest-neighbours. For the radial Gaussian bells described by equation (3.27) the radius s_i can be set as

$$s_i = k_s \frac{1}{n} \sum_{j=1}^{n} |\underline{c}_i - \underline{c}_{i,j}|, \quad (3.34)$$

where $\underline{c}_i$ is the current centre and $\underline{c}_{i,j}$ is the jth nearest neighbour centre to $\underline{c}_i$. The scaling factor k_s defines the degree of overlap between the validity functions and has the same value for all centres.

3.3.2 Modelling experiments

The structure of local models used for the nonlinear modelling experiments is based on the results of the linear identification reported in section 3.2.3. The sampling period, the dynamic order and the delay are chosen according to relation (3.24). Thus, the input data vector becomes

$$\underline{\psi}(t) = [u(t-2),\ u(t-3),\ y(t-1),\ y(t-2)]^T \ . \tag{3.35}$$

RBF model

We describe the use of a simple RBF network to model isometric contraction of a muscle as introduced in section 3.1.2 (Donaldson *et al.* 1995).

We consider scalar weights as the simplest form of local models, $\hat{f}_i = \theta_i$, which results in a Radial Basis Function (RBF) network (Moody and Darken 1989),

$$\hat{y}(t) = \sum_{i=1}^{n_A} \rho_i(\underline{\phi}(t))\theta_i \ . \tag{3.36}$$

As the 'local models' do not depend on the data vector, all dynamic information has to be presented in the scheduling vector,

$$\underline{\phi}(t) = \underline{\psi}(t) = [u(t-2),\ u(t-3),\ y(t-1),\ y(t-2)]^T \ . \tag{3.37}$$

The model output depends linearly on the parameters θ_i. Thus, standard linear regression techniques based on the error criterion (3.32) can be applied.

Structure selection   As mentioned above, the selection of the network structure is a more difficult task than the weight optimisation. In our case, the basis functions have to be placed in a 4-dimensional input space, $\underline{\phi} \in \mathbb{R}^4$. Forward regression seems appropriate for such a high dimensional optimisation space. A large number of clustering techniques has been proposed for use with RBF networks; we adopt the approach outlined in (Neumerkel *et al.* 1993).

The cluster consists of all data presented to the network at time step t, including the weighted target output,

$$\underline{\phi}'(t) = [u(t-2),\ u(t-3),\ y(t-1),\ y(t-2),\ \gamma y(t)]^T \ . \tag{3.38}$$

The sample $\underline{\phi}'(t)$ becomes a cluster (i.e. defines the position of a RBF unit) if the minimal distance to existing clusters $\underline{c}'_i$ exceeds some threshold ϵ,

$$\min_{i=1}^{n_A} |\underline{\phi}'(t) - \underline{c}'_i| > \epsilon \ . \tag{3.39}$$

When this relation holds, a new cluster is created at $\underline{\phi}'(t)$ and n_A is increased by one. Otherwise, $\underline{\phi}'(t)$ is considered to belong to the nearest existing cluster and no new cluster is created. The factor γ controls the significance of changes in the output.

This procedure is started with no clusters existing (which means of course that the first data sample forms a cluster) and continues until all data points have been considered. From clusters $\underline{c}_i' \in \mathbb{R}^{n_\phi+1}$ the centres $\underline{c}_i \in \mathbb{R}^{n_\phi}$ are built by removing the last element $\gamma y(t)$ from the cluster vectors.

We considered the two nearest neighbours heuristic by setting $n = 2$ in equation (3.34).

Uniform LMN model

In this section we consider the use of local linear models instead of simple weights, which generalises the RBF model to the LMN. Initial results with this approach have been reported in (Gollee *et al.* 1994), with further details in (Gollee and Hunt 1997).

If no physical information about the system to be modelled is available (i.e. in the case of black box modelling), it is straightforward to choose linear structures for the local models, as linearisation is a common approach for the local description of nonlinear systems. With local ARX structures equation (3.30) has the form

$$\hat{y}(t) = \sum_{i=1}^{n_A} \rho_i(\underline{\phi}(t)) \left[\underline{\psi}^T(t)\underline{\Theta}_i \right] , \qquad (3.40)$$

where $\underline{\Theta}_i \in \mathbb{R}^{n_\psi}$ are local parameter vectors, which correspond to a linearised description of the system in a part of the operating space.

The scheduling vector $\underline{\phi}(t)$, which is generally a subvector of $\underline{\psi}(t)$ (see equation (3.29)), needs to be chosen in such a way that the nonlinear properties of the system are in some way represented by it. It is straightforward to use the delayed system output for scheduling as the nonlinear behaviour surely depends on the activity of the muscle, which can be described by the force produced by it. Thus, $\underline{\phi}(t)$ has been chosen as

$$\underline{\phi}(t) = y(t - 1) . \qquad (3.41)$$

Having a scheduling vector with only one element reduces the input space for the validity functions to one dimension. Applying a network with uniformly distributed validity functions seems appropriate as the scheduling space can easily be filled.

The model output depends linearly on the parameter vectors $\underline{\Theta}_i$, and standard linear regression techniques can be applied. For reasons outlined in Chapter 7 we choose to work with the local optimisation criterion (3.31).

3.3.3 Results

RBF model

The network was taught with a total of 4130 input–output data pairs, i.e. 32 data sets were used for training. The clustering algorithm, equations (3.38, 3.39), generated 56 units, with the parameters chosen as $\epsilon = 0.25$ and $\gamma = 4$. For width selection the values $n = 2$ and $k_s = 2$ were used in equation (3.34).

The performance of the trained network was tested on a number of new data sets which were not used in the training procedure. A comparison of the model response to the real muscle output is shown in Figure 3.6 for a number of data sets.

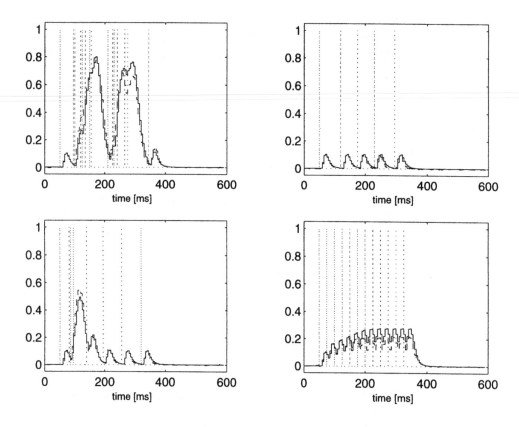

Figure 3.6 Comparison of RBF network output [—] and target output [- - -] for typical input patterns [···].

Table 3.1   Local model parameters.

Model no.	transfer function	gain
1	$\frac{q^{-2}(0.0609+0.0484\,q^{-1})}{1-1.2341\,q^{-1}+0.3782\,q^{-2}}$	0.7581
2	$\frac{q^{-2}(0.0832+0.0491\,q^{-1})}{1-1.3098\,q^{-1}+0.4522\,q^{-2}}$	0.9298
3	$\frac{q^{-2}(0.0954+0.0556\,q^{-1})}{1-1.4436\,q^{-1}+0.5607\,q^{-2}}$	1.2901
4	$\frac{q^{-2}(0.0959+0.0508\,q^{-1})}{1-1.5310\,q^{-1}+0.6315\,q^{-2}}$	1.4597
5	$\frac{q^{-2}(0.0892+0.0430\,q^{-1})}{1-1.6018\,q^{-1}+0.6835\,q^{-2}}$	1.6182
6	$\frac{q^{-2}(0.0621+0.0313\,q^{-1})}{1-1.7123\,q^{-1}+0.7749\,q^{-2}}$	1.4927
7	$\frac{q^{-2}(0.0533+0.0175\,q^{-1})}{1-1.8454\,q^{-1}+0.8943\,q^{-2}}$	1.4478
8	$\frac{q^{-2}(0.0480+0.0177\,q^{-1})}{1-1.9901\,q^{-1}+1.0314\,q^{-2}}$	1.5920

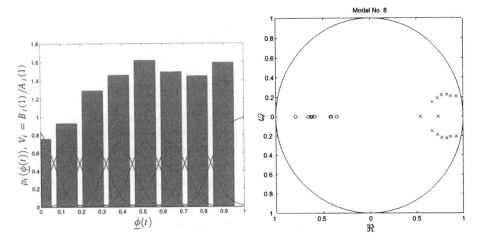

(a) Shape of the validity functions ρ_i for the chosen model structure (solid lines) and gains V_i (shaded bars) of the corresponding local model.

(b) Pole–zero configuration of the chosen model structure in the $\mathbb{Z}$–domain. o marks the location of the zeros, x marks the poles.

Figure 3.7 Model gains, pole–zero configuration and basis functions.

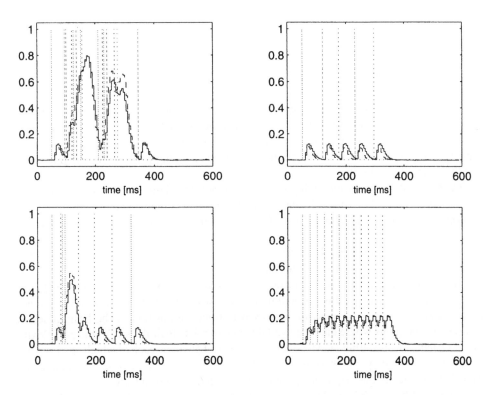

Figure 3.8 Comparison of LMN model output [—] and target output [- - -] for typical input patterns [$\cdots$].

Local Model Network model

We distributed the centres uniformly between 0 and 0.9, which are the minimal and maximal output values. The network parameters are trained in a similar manner as for the RBF model. Experimentation with the number and width of validity functions showed that the best values (in the sense of a minimal error on unknown data) were achieved with a network with eight validity functions, where the width is $s = 0.5$ for all units.

To give a feeling for the differences between the local linear models in the LMN, its transfer functions and gains are reported in Table 3.1. The shape of the validity functions and the corresponding gains of the chosen network structure are depicted in Figure 3.7(a). The locations of the poles and zeros of the local linear models are shown in Figure 3.7(b). The performance of the network, tested on a number of new data sets, is shown in Figure 3.8.

Table 3.2   Modelling results.

Model type	$rmse = \sqrt{\frac{1}{L} \sum_{t=l_0+1}^{l_0+L} [y(t) - \hat{y}(t)]^2}$
linear	0.054
RBF	0.029
LMN	0.027

3.4   CONCLUSIONS

We employed linear identification techniques as a first step in the investigation of a significantly nonlinear modelling problem. Linear identification theory is well-established and provides a large toolbox of analysis techniques. Using linear algorithms we determined 'optimal' structure parameters, namely the necessary sampling period, the dynamic order and the delay of the system.

The results of the linear modelling approach support the claim that the muscle contraction shows significant nonlinear features. The linear model obtained could only provide a description which is on average correct. Thus, the need for nonlinear modelling techniques was emphasised.

Two types of Local Model Networks were employed as nonlinear model structures, the Radial Basis Function network and the LMN with local linear ARX models.

For the RBF model, all dynamic information needs to be represented in the scheduling vector, which results in a 4-dimensional input space which has to be filled with basis functions. A simple clustering heuristic was used to limit the position of the basis function centres to the regions where data are available and to relate their density to the complexity of the system. The resulting model structure consists of 56 units, each of which has 1 weight parameter, 4 position parameters and 1 width factor. This gives a total of 336 parameters. The clustering technique requires 2 parameters which have to be adjusted.

When local linear ARX models are used instead of the simple scalar weights, the Local Model Network structure simplifies significantly. The dynamic information can be presented to the local dynamic models, thus the dynamics are modelled locally. The scheduling vector only needs to represent a value which indicates the nonlinear characteristics of the system. In our case we used the delayed output, i.e. the tension produced by the muscle, which indicates the muscle activation and thus the regime it is operating in. This reduces the scheduling space to one dimension. Such a limited space can be filled uniformly, without applying structure optimisation. We found a structure with 8 validity functions to be sufficient. Here, each validity function has one position parameter, one width factor and 4 local model parameters, which gives a total of 48 parameters. The network structure is more transparent than the RBF network, as its parameters can be interpreted in terms of linear transfer functions and local gains, as illustrated in Figure 3.7(a) and Table 3.1.

In order to compare the modelling results for the different structures described here, we give the root mean squared error, *rmse*, for each structure in Table 3.2. The models have been validated, on extra test data not used to adjust their parameters, simulating with an infinite prediction horizon.

Both basis function networks perform similarly, whereas the linear modelling results are significantly worse. Comparing performance on some typical data sets for the RBF network, depicted in Figure 3.6, and for the LMN network, shown in Figure 3.8, both approaches are about equally good for most operating conditions. However, the RBF network cannot model the tetanic muscle contraction as well as the LMN, as shown in the bottom right in the figures. The LMN has been used as the basis of a nonlinear muscle control structure. Details can be found in (Gollee and Hunt 1997).

REFERENCES

ÅSTRÖM, K. J. AND B. WITTENMARK (1990) *Computer-Controlled Systems. Theory and Design.* 2nd edn. Prentice Hall, Inc.. Englewood Cliffs, NJ.

DONALDSON, N. D., H. GOLLEE, K. J. HUNT, J. C. JARVIS AND M. K. N. KWENDE (1995) 'A radial basis function model of muscle stimulated with irregular inter-pulse intervals'. *Medical Engineering & Physics* **17**(6), 431–441.

FRIEDMANN, J. H. (1991) 'Multivariate adaptive regression splines'. *The Annals of Statistics* **19**, 1–141.

GOLLEE, H. AND K. J. HUNT (1997) 'Nonlinear modelling and control of electrically stimulated muscle: a local model network approach', submitted for publication.

GOLLEE, H., K. J. HUNT, N. DE N. DONALDSON, J. C. JARVIS AND M. K. N. KWENDE (1994) A mathematical analogue of electrically stimulated muscle using local model networks. In 'Proc. 33rd IEEE Conf. on Decision and Control'. Lake Buena Vista, Florida.

HABER, R. AND H. UNBEHAUEN (1990) 'Structure identification of nonlinear dynamic systems — a survey on input/output approaches'. *Automatica* **26**(4), 651–677.

HAMBRECHT, F. T. (1992) A brief history of neural prostheses for motor control of paralyzed extremities. In R. B. Stein, P. H. Peckham and D. P. Popović (Eds.). 'Neural Prostheses. Replacing Motor Function After Disease or Disability'. Oxford University Press. pp. 3–14.

HILL, A. V. (1938) 'The heat of shortening and the dynamic constants of muscle'. *Proc. R. Soc. Lond. [Biol.]* **126**, 136–195.

HLAVÁČKOVÁ, K. (1995) An upper estimate of the error of approximation of continuous multivariable functions by KBF networks. In 'Proc. 3rd European Symposium on Artificial Neural Networks'. Brussels. pp. 333–340.

HUXLEY, A. F. (1957) 'Muscle structure and theories of contraction'. *Prog. Biophys. and Biophys. Chem.* **7**, 257–318.

JARVIS, J. C. (1993) 'Power production and working capacity of rabbit tibialis anterior muscles after chronic electrical stimulation at 10 Hz'. *J. Physiol.* **470**, 157–169.

JOHANSEN, T. A. AND B. A. FOSS (1992) 'A NARMAX model representation for adaptive control based on local models'. *Modelling, Identification and Control* **13**(1), 25–39.

JOHANSEN, T. A. AND B. A. FOSS (1993) 'Constructing NARMAX models using ARMAX models'. *Int. J. Control* **58**, 1125–1153.

JUTTEN, C. AND O. FAMBON (1995) Pruning methods: A review. In 'Proc. 3rd European Symposium on Artificial Neural Networks'. Brussels. pp. 129–140.

KAVLI, T. (1993) 'ASMOD – an algorithm for adaptive spline modelling of observation data'. *International Journal of Control* **58**(4), 947–967.

LEONTARITIS, I. J. AND S. A. BILLINGS (1985) 'Input–output parametric models for nonlinear systems. Part I: Deterministic non-linear systems'. *International Journal of Control* **41**, 303–328.

LJUNG, L. (1987) *System Identification. Theory for the User*. Prentice Hall. Englewood Cliffs, NJ.

MOODY, J. AND C. DARKEN (1989) 'Fast-learning in networks of locally-tuned processing units'. *Neural Computation* **1**, 281–294.

MURRAY-SMITH, R. AND T. A. JOHANSEN (1995) Local learning in local model networks. In '4th IEE Intern. Conf. on Artificial Neural Networks'. pp. 40–46.

NEUMERKEL, D., R. MURRAY-SMITH AND H. GOLLEE (1993) Modelling dynamic processes with clustered time-delay neurons. In 'Proc. International Joint Conference on Neural Networks, Nagoya, Japan'.

PRESS, W. H., S. A. TEUKOLSKY, W. T. VETTERLING AND B. P. FLANNERY (1992) *Numerical Recipes (C): The Art of Scientific Computing*. 2nd edn. Cambridge University Press. UK.

REED, R. (1994) 'Pruning algorithms: A survey'. *IEEE Trans. on Neural Networks* **4**(5), 740–747.

SALMONS, S. AND J. C. JARVIS (1992) 'Cardiac assistance from skeletal muscle: a critical appraisal of the various approaches'. *Br Heart J* **68**, 333–338.

SHORTEN, R. AND R. MURRAY-SMITH (1994) On normalising basis function networks. In '4th Irish Neural Network Conf.'. Univ. College Dublin.

SILBERNAGL, S. AND A. DESPOPOULOS (1991) *Taschenatlas der Physiologie*. 4th edn. Georg Thieme Verlag. Stuttgart, New York.

WEISS, S. M. AND C. A. KULIKOWSKI (1991) *Computer Systems that Learn*. Morgan Kaufmann. San Mateo, CA.

ZAHALAK, G. I. (1992) An overview of muscle modeling. In R. B. Stein, P. H. Peckham and D. P. Popović (Eds.). 'Neural Prostheses. Replacing Motor Function After Disease or Disability'. Oxford University Press. pp. 17–57.

CHAPTER FOUR

Process Modelling Using the Functional State Approach

AARNE HALME, ARTO VISALA and XIA-CHANG ZHANG

Process modelling using the functional state concept is an approach where models are built using a two-level hierarchy. In the upper level, the process is treated like a finite state automaton, where the states are so called functional states. The lower level models the process locally using conventional approaches, such as state or input–output models. The functional state models are designed using process knowledge either using knowledge of the process structure or following an expert system approach. In addition to the functional state structure, the transition controls must be defined. Normally there are no uniquely defined transition controls between functional states, which makes a great difference between this and the conventional automaton approach. A basic task of this approach is to identify on-line transitions between functional states. This chapter provides an introduction to the functional state concept, and illustrates it with a simple 'wire model', where the methods used to detect changes in process dynamics and transitions between states are discussed. A neural network classifier using the Wiener model and Laguerre representation of the input signals is introduced. An example of the use of a Kalman filter for the representation of multiple dynamics is given. The methods are applied to a practical biotechnical process control task.

4.1 RECONSIDERING THE "STATE" CONCEPT

In systems theory the basic definition of the state of a process is that it parameterises the relation formed by all possible input–output signal pairs generated by the process. At a certain time, knowing the state at that moment and a future input the corresponding output is in principle uniquely determined. In practice, we know that a practical process, like a bioreactor used as a case-example later on, does not respond that way even if we know all its usual state variables (temperature, pH, biomass, substrate, nutrients concentrations etc.) and rerun the process from exactly the same values as before. Explaining the differences in terms of stochastic disturbances is not a relevant approach because in practice a bioreactor is usually a very stable and quite undisturbed environment. The most probable explanation is that the state of the process was actually not known, only some of its components.

A natural way to enlarge the state concept of the bioreactor would be to consider the "biological state" of the cell population in the reactor. This would mean, however, modelling of complicated metabolite pathways and things like cell age or mRNA distribution in the population. Most of those components are not measurable using today's technology. This does not necessarily mean that such an enlargement cannot be used in practice. For example, age and protein distributions have been successfully used to predict certain behaviours of the population that were difficult to explain in any other way. Generally speaking, however, much progress must be made before such a biological state can be properly defined and applied in everyday modelling work. Meanwhile, one possibility could be to characterise the overall behaviour of the process by symbolic and logical descriptions which define a certain number of structures, within which the detailed behaviour of the process is similar.

The approach which has been developed further in this chapter is the following. Especially in cases where the "inner structure" of the process, i.e. the cell's physiological or biochemical state in our bioreactor, has an important role in the process dynamics, it may be better to enlarge the (conventional) state with a symbolic functional state, i.e. to use variable structure modelling, than to increase the degree of the model by increasing the number of conventional state variables. The concept of the *functional state* is described in more detail later on. Generally, in a certain functional state the process behaves in a certain definite way representing a certain "structure". When the functional state changes, a structural change happens and the behaviour of the system changes accordingly. With even more practical terms the process is partitioned into several "macrostates", called functional states, which are identified separately. Inside each functional state the process is modelled by using a conventional model specifically valid in these conditions. The idea is that the "sub-models", local models, are simpler but valid only in the conditions defined by the functional states. The concept of the functional states should not be confused with the usual operating point of a process. Functional states represent certain possible nonlinear dynamics of the process in a discrete way, whereas the operating point varies continuously.

One of the fundamental questions related to the practical use of the functional state concept is the automatic recognition of the states from measurement information available. Only in this way can the concept be utilized effectively in control, estimation and optimisation. As illustrated later on, functional states make a process behave like an automaton, however, with the difference that the state transitions cannot be decided in a simple way from the controls. In fact, depending on the method used to determine the functional states, there are many approaches to transition recognition. In (Halme 1988) a knowledge-based system approach to recognise the functional states was introduced. Later on in this chapter algorithmic methods are introduced that can be used to recognise the transitions between the functional states. By combining these methods with the logical representations of the control policies, a state machine can be constructed which controls the transitions between the functional states. This in turn makes it further possible to automatically select models, methods or rules that are valid in a certain functional state and thus to operate with them in detailed analysis.

4.2 HOW THE FUNCTIONAL STATE MODELLING APPROACH WORKS

The enlarged state concept simplifies process modelling because specific state and measurement equations can be used within a certain functional state. Considering the functional states, an analogy may be found in finite automata theory. However, because the number of functional states is quite low, there is no need to use the complicated formalism of automata theory in this context. To be a useful concept the functional state should have two basic properties. The first one is that it is well defined in the sense that the process is always in one and only one functional state at a time. The second one is that it should be possible to model the transition dynamics between the functional states. The transition between the functional states is defined by a control policy that causes the transition. The policy is not defined uniquely by numerical control signals but rather by a symbolic or logical presentation which defines how, for example, the bioreactor is operated to run it from a given functional state to another. The dynamics may be defined in its simplest form by the expected transition time.

Within a certain functional state, the process dynamics can be modelled by using a usual numeric state model, "local model". Switching from one model to another is done when the functional state changes. The functional state concept is illustrated in Figure 4.1. The numerical state vector x containing, for example, reactor concentrations of biomass, substrates and products, temperature etc., can be common to all functional states even if not all of its components are used in every functional state M_i. The case of not all components being needed when modelling the state dynamics in a certain functional state can be easily taken into account by a proper definition of the state function $f(\cdot, M_i)$. Correspondingly, the information vector y, the components of which represent all measurement information available from the process (obtained by *in situ* probes or by sampling), can be considered common to all functional states. In case some components are not measured, they can be just ignored from the information vector by a proper definition of the measurement function $g(\cdot, M_i)$. Thus, in the deep level, transition from a certain functional state to another means just switching from a certain state dynamics – measurement equation pair to another (note that formally, if f and g are defined in the enlarged state space the functions can be considered unchangeable). The numeric state can be made continuous by making the initial conditions in the functional state M_i correspond to the last value in the state M_{i-1}. When knowing the functional state automata, i.e. a machine that controls the state changes, one can fairly easily use the whole apparatus, e.g. for estimation or other similar purposes, principally in the same way as if it was a complex high order classical state representation.

4.3 DETECTING TRANSITIONS BETWEEN FUNCTIONAL STATES

The main difficulty when using the functional state concept is how to automatically detect the state transitions (remember that only control policies for state transitions are supposed to be known). If this could be done with a moderate reliability and without too much delay, the partial models describing the process in different functional states could be used with no difficulty. The problem is how to construct the functional state machine without explicitly constructing the state transition function. The "wire model" below shows that it is perhaps easier to detect the transition events between the functional states than to recognise the presence in them. According to the assumption that the functional states are well defined, leaving the present state is the same event as entering the next one. Because the initial functional state is usually known it can be assumed that by knowing the transition control policy and detecting the transition event to the next state we can construct the state machine that controls the functional state transitions.

The idea adopted in the following for detection of transition events is to recognise

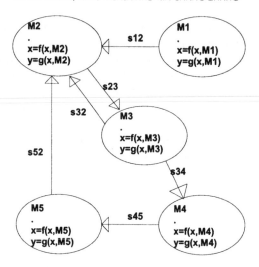

Figure 4.1   Diagram illustrating the functional state modelling approach.

changes in the input–output behaviour of the process. The behaviour of the system is regarded as remaining constant as long as similar inputs are mapped to similar outputs. If the system is supposed to be time-invariant it is sufficient to consider it, at any time instant t, as a mapping of the input time histories $u_{[t-T,t)}$ to the outputs $y(t)$. In practice, it is sufficient to consider only a certain length of time history $u_{[t-T,t)}$ because most systems have the so-called finite settling time property (finite time memory). Thus, we suppose that in a certain functional state the system is characterised by a certain input–output mapping of the form

$$y(t) = G[u_{[t-T,t)}]. \tag{4.1}$$

This mapping is generally between vector signal spaces and represents the input–output relation of the system from a certain fixed initial state (usually zero). We call it in the following the G-functional. It specifies a hyperplane in a feature space characterising the input signals and the corresponding instantaneous outputs. The mapping is normally not known analytically. This is true even if we are able to construct a state model valid in the functional state considered (e.g. a Monod-type of model in our bioreactor case). Only in the linear case is the question quite straightforward. However, the G-functional can be generally represented by a convolution form. In the linear case the convolution takes the well-known form:

$$y(t) = \int_0^T h_1(\tau)x(t-\tau)d\tau, \quad T \to \infty. \tag{4.2}$$

In the nonlinear case there is also a representation available. This is known as a Volterra series:

$$
\begin{aligned}
y(t) &= \int_0^T h_1(\tau_1)x(t-\tau_1)d\tau_1 + \int_0^T \int_0^T h_2(\tau_1,\tau_2)x(t-\tau_1)x(t-\tau_2)d\tau_1 d\tau_2 + \dots \\
&\quad \int_0^T \dots \int_0^T h_n(\tau_1,\tau_2,\dots\tau_n)x(t-\tau_1)x(t-\tau_2)\dots x(t-\tau_n)d\tau_1 d\tau_2 \dots d\tau_n \\
&\quad + \dots,
\end{aligned}
\tag{4.3}
$$

where the kernel functions $h_2, h_3, ...h_n$ are the so-called generalised impulse responses (Schetzen 1980). The Volterra series representation is indeed a very general one. It corresponds to the power series representation of normal analytical functions, but is valid with more general conditions. In practice, a finite truncation of the series must be used. Also the inputs are sampled and the discrete time version is used in computations. It is easy to see that the following representation is valid in the discrete case

$$y(t) = \sum_{i=1}^{N} h_1(i)x(k - i) + \sum_{i_1=1}^{N}\sum_{i_2=1}^{N} h_2(i_1, i_2)x(k - i_1)x(k - i_2) + ...$$

$$\sum_{i_1=1}^{N}\sum_{i_2=1}^{N} ... \sum_{i_n=1}^{N} h_n(i_1, i_2...i_n)x(k - i_1)x(k - i_2)...x(k - i_n). \qquad (4.4)$$

This is the single-input–single-output version. The representation can be extended in a natural way to multi-input–multi-output version (in fact even to more general forms (Halme 1972)).

In the case where the system's initial state varies and we are interested in its contribution to the outputs, one can also generalise the linear case. As is well-known, in the linear case the output includes two terms, one as defined by equation (4.1) and the other which depends on the impulse response h_1 and the initial state. This transient usually dies out after the settling time of the system. The same is generally valid for the nonlinear systems, but the formal representation is more complex. In practice, if one likes to include the initial state to the G-functional it can be thought of as an additional pulse-type input to the multi-input version of equation (4.4).

The G-functional can also be approximated by heuristic means. Use of a neural network (e.g. a multilayer perceptron) is one possibility and the GMDH method (Group Method of Data Handling, see for example (Hecht-Nielsen 1990)) another. It is worth noting their relation to the Volterra series representation. We suppose that the input data array is the same sampled signal history that is used in the convolution representation. GMDH uses polynomials in the intermediate layer functions. It is quite easy to see that the final network represents a similar polynomial type function to the Volterra series of the same order. The screening policy applied in the method leaves, however, many of the terms out and thus the network presentation includes usually far fewer terms than a full polynomial function of type equation (4.4). In a neural network, suppose that the threshold functions in multi-layer perceptron are sigmoids. According to Weierstrass's theorem these functions can be approximated on a finite close interval with a given accuracy by a polynomial function. It is easy to see that if the polynomial approximations are taken instead of the sigmoid functions, the network has polynomial type representation similar to equation (4.4) (including also fewer terms than the full representation).

Suppose then that we have modelled the input–output mapping with one of the representations given above. It can be identified or "trained" on the input-output data of a given functional state by just using some of the standard least squares or gradient type methods. After training, the parameters are frozen, or only updated at a low rate. The model is run in parallel with the process and the outputs are compared. If there is a great enough difference between them, the decision that the process has arrived at a new functional state is made. This can be done again by using different techniques. The method adopted here is based on two sliding time data windows, as shown in Figure 4.2, and applying an appropriate statistical criteria. Window 1 is called the detecting window and window 2 the reference window. The reference window is always longer than the detection window. Statistical

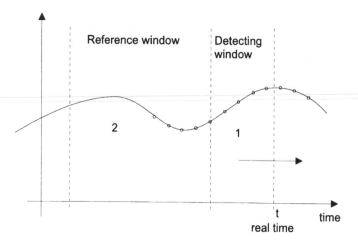

Figure 4.2   Illustration of the detecting and reference data windows.

properties of the error signal (the error is obtained by subtracting the real and model output) are calculated in both windows and compared to judge whether the data in the windows are realisations from the same or different stochastic processes. Different properties mean that the nature of the error signal has changed and accordingly the process must behave differently than before. This approach has been successfully used before in fault detection problems (Halme and Selkinaho 1984). The judgement can be based simply, for example, on comparing the variances, when supposed that the error signal after training is a zero mean Gaussian signal. The H_0-assumption is that there is no change in the variances. The following detector signal is used

$$z(k) = \frac{f_D(e(k)^2)}{f_R(e(k)^2)},\tag{4.5}$$

where f_D and f_R denote operators continuously forming averages from the error signal $e(k)$ variances in the detecting and reference windows respectively. Instead of taking a straightforward average, one can use RC-filtered "forgetting" averaging operators defined by

$$f_D(e(k)^2) = f_D(e(k-1)^2) + \frac{1}{L_D}[e(k)^2 - f_D(e(k-1)^2)]\tag{4.6}$$

$$f_R(e(k)^2) = f_R(e(k-1)^2) + \frac{1}{L_R}[e(k)^2 - f_R(e(k-1)^2)]\tag{4.7}$$

where L_R and L_D are the length of the reference and detecting windows respectively. The test variable $z(k)$ formed by this way continuously follows an F-distribution with the degrees of freedom defined by the length of the windows. A rare value of this signal means that the H_0-assumption is false with a high probability and we can conclude a change in the process behaviour. Typical values used in our tests are $L_R = 50$ and $L_D = 4$. In theory, according to the F-tables the threshold level can be set so that the desired probability (e.g. 99%) for false H_0-assumption can be attained.

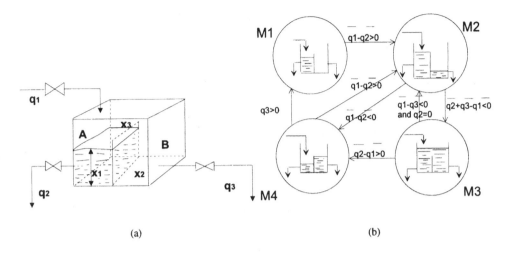

(a)   (b)

Figure 4.3 The "wire model" used to test the functional state concept. In the right hand figure the flows in the conditions are moving time averages.

4.4 A "WIRE MODEL" ILLUSTRATING THE CONCEPT

It is necessary to test the ideas by beginning with a simple model. The model must be simple enough so that its behaviour is completely understood. It must, however, be complex enough to demonstrate different functional states, which in turn requires some kind of structural changes in the process. We have chosen the process illustrated in Figure 4.3 – a flow tank divided into two compartments by a wall. The tank has one inflow which goes to compartment A only, and two outflows, one from each compartment. The outflows, which are free flows, are taken from the bottom of the compartments. Both are controlled by an on–off valve. The inflow comes "through the cover" and is controlled continuously. The process has four clearly detectable different functional states which are also illustrated in Figure 4.3. In all functional states the system dynamics can be described by a simple nonlinear state equation. It is also easy to determine the control policies which conduct the transitions between the functional states. The state equations are given in equations (4.9)-(4.11) and some feasible control policies are shown in Figure 4.3. The control policies are described by simple logic conditions on the moving averages of the inflow and outflows (determining the temporal mass balance of the system). Note that the conditions do not determine the input and output signals uniquely. It is, however, easy to plan such input signals that satisfy certain conditions and thus cause the corresponding functional state transitions. The functional states denoted by $M_1, ..., M_4$ are clearly well defined in the sense that the process can be in only one functional state at a time.

Suppose in the following that the physical structure of the system is not known by the observer and the level heights in the tank cannot be measured. For simulations and demonstrations the level heights are defined as:

■ x_1 = the level height in the left compartment A
■ x_2 = the level height in the right compartment B
■ x_3 = the common level height, when the level heights in both compartments are above the wall height h between the compartments, then $x_3 = x_2 = x_1$.

The only information available for the observer is in the input and output signals:

- u_1 = inflow q_1
- u_2 = valve controlling outflow q_2
- u_3 = valve controlling outflow q_3
- y_1 = outflow q_2
- y_2 = outflow q_3.

The input u_1 varies continuously. The inputs u_2 and u_3 are binary signals having values 0 (valve closed) and 1 (valve open). The process has been simulated by using the SIMNON-package. Figure 4.4 illustrates the general behaviour of the system under certain control excitation after starting from the functional state M_1.

$$M_1: \quad \dot{x}_1 = \frac{1}{A_1}(u_1 - c_1(u_2)\sqrt{x_1})$$
$$y_1 = c_1(u_2)\sqrt{x_1} \tag{4.8}$$
$$y_2 = 0$$

$$M_2: \quad \dot{x}_2 = \frac{1}{A_2}(u_1 - c_1(u_2)\sqrt{h} - c_2(u_3)\sqrt{x_2})$$
$$y_1 = c_1(u_2)\sqrt{h} \tag{4.9}$$
$$y_2 = c_2(u_3)\sqrt{x_2}$$

$$M_3: \quad \dot{x}_2 = \frac{1}{A_1 + A_2}(u_1 - c_1(u_2)\sqrt{x_3} - c_2(u_3)\sqrt{x_3})$$
$$y_1 = c_1(u_2)\sqrt{x_3} \tag{4.10}$$
$$y_2 = c_2(u_3)\sqrt{x_3}$$

$$M_4: \quad \dot{x}_1 = \frac{1}{A_1}(u_1 - c_1(u_2)\sqrt{x_1})$$
$$\dot{x}_2 = \frac{1}{A_2}(u_1 - c_2(u_3)\sqrt{x_2})$$
$$y_1 = c_1(u_2)\sqrt{x_1} \tag{4.11}$$
$$y_2 = c_2(u_3)\sqrt{x_2}$$

4.5 SIMULATION DEMONSTRATIONS

4.5.1 Recognition of the functional states

The "wire model" has been simulated and studied in different conditions in order to test and illustrate the ideas presented above for detection of the functional state transitions. No information about level heights is assumed to be available for functional state detection.

In some cases the change of functional state can be detected easily by a pure logic type of criterion. For example, supposing that the present functional state is M_1 and the control policy chosen is (u_1=sufficiently large, u_2=open, u_3=open) the u_3=open) the change to M_2 is simply detected by observing the transition in q_3 from zero to non-zero. However, this criterion cannot be used if another policy is chosen, for example (u_1=sufficiently large,

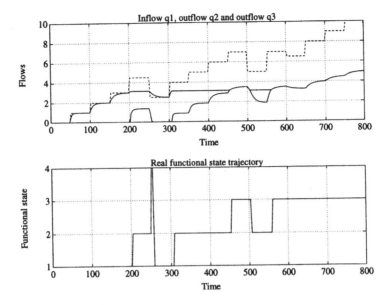

Figure 4.4 The behaviour of the system in the test run. The signals obtained from the "wire model" simulator. The test run starts in functional state M_1. In the upper figure q_1 is marked with "- - -", q_2 is the upper solid line and q_3 the lower solid line.

u_2=open, u_3=closed). Changes between M_3 and M_4 cannot be detected by such simple logic criteria, because in both functional states the outputs are non-zero continuous signals.

For the detection of functional state transitions on the basis of prediction errors two sets of black-box models were identified, Volterra-series models and multilayer perceptron (MLP) neural network models. Although there are four separate functional states in the process, only three different "dynamics" were identified. The transition from M_4 to M_1 happens when y_2 (outflow q_3) becomes zero.

- "Left dynamics", input u_1 (=inflow q_1), output y_1 (=outflow q_2), valid in states M_1 and M_4.
- "Right dynamics", input u_1 (=inflow q_1), output y_2 (=outflow q_3), valid in state M_2.
- "Common dynamics", input u_1 (=inflow q_1), output y_2 (=outflow q_3), valid in state M_3.

The "left dynamics" describes thus the behaviour of the left compartment only and the "right dynamics" the right compartment when the left one overflows.

In the following, both Volterra- and neural network models have as input 4 values from the signal memory of q_1, i.e. the input to the models consists of a sliding data window of 4 elements, which are separated by 12 time units. The window length is thus 36 time units. The sampling period is 3 time units. Signals were scaled by the constant 10 before processing with the models. In the following figures signals are, however, rescaled so that comparisons can be made with original signals.

The identification results relative to identification or "training" data in the case of "left dynamics" are shown in Figure 4.5, the Volterra model, and Figure 4.6, the neural network model.

The neural network used is a multilayer perceptron with one hidden layer which has 6 processing units. The model has 37 "weights" or parameters; the hidden layer has 30 and the

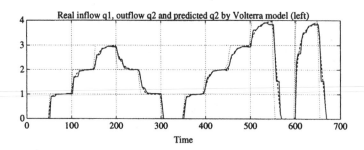

Figure 4.5   The response of the identified Volterra series model "—" and the response of the process "- - -" to the training pattern of the input signal u_1 "..." in the functional state M_1, i.e. "left dynamics".

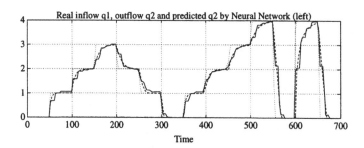

Figure 4.6   The response of the trained neural network "—" and the response of the process "- - -" to the training pattern of the input signal u_1 "..." in the functional state M_1, i.e. "left dynamics".

output layer 7 parameters, which were estimated or "trained" using the back-propagation algorithm (Rumelhart *et al.* 1986). Weights were initiated to small random numbers. The back-propagation algorithm was started with learning rate 0.75 and the learning rate was decreased several times during training when performance improvement seemed to slow down. Typically 200–500 iterations of the whole training set were needed before the network optimum was found. It should be noted that training with the Levenberg–Marquardt method gives much faster convergence and better results than the back-propagation algorithm used here.

The Volterra series presentation is linear in parameters. Each parameter corresponds to one different product permutation of less than or equal to the selected degree (in our case 3) of the input window elements (in our case 4). The discrete Volterra model used is of third degree and has 34 parameters. (In the case of 3 (5) inputs there are 19 (55) parameters.) The model parameters or "Volterra kernels" were estimated by using standard least squares estimation algorithm for linear models. When the identification data contains p time instants and the Volterra model has n parameters, the batch estimation problem can be written in SISO case as

$$\underline{y} = \underline{U}\underline{h}, \tag{4.12}$$

where $p \times n$ matrix $\underline{U}$ contains in rows all different product permutations of the input

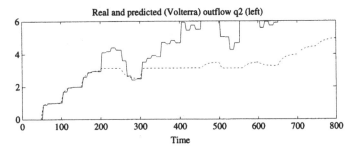

Figure 4.7   The response of trained "left dynamics" – model "—", the real response "- - -".

window elements calculated for observation times (one row for each time instant), $p \times 1$ vector $\underline{y}$ contains the corresponding outputs observed and $n \times 1$ vector $\underline{h}$ is the kernel or parameter vector to be estimated.

The estimation of Volterra kernels is straightforward and takes much less time than the estimation of neural network weights. The fitting seems to be good in both cases. The responses given by neural networks seem to be smoother in some cases. This is obviously due to saturation effects in sigmoid functions used in the neural network.

In general, it must be noted that nonlinear black-box models, like Volterra series or neural networks, do not contain any *a priori* information about the modelled process. There must be enough identification or training data, which represents the whole operating area and dynamics well enough. The nonlinear black-box models do not usually give reasonable responses if used outside the scope of identification data, a fact which is sometimes forgotten. Defining a training set that trains the model for a large set of control signals is difficult. Instead, knowing a typical reference control signal simplifies the task of fitting the model to be accurate around this signal.

The functional state transition detection is demonstrated by using the same test run as shown already in Figure 4.4. In the following figures the results are shown when the Volterra models are used. The neural network approach gives the same kind of results. Each model is used for functional state transition detection from the state in which it is valid to the next state specified by the control policy (in Figure 4.4). Figure 4.7 shows the response of the model identified to "left dynamics", Figure 4.8 the response of the model identified to "right dynamics" and Figure 4.9 shows the response of the model identified to "common dynamics" in the connection of the test run. Also real variable trajectories are shown. In Figure 4.10 the detected functional state and the real functional state are represented. It can clearly be seen that the model responses are similar, with real outputs only in those functional states in which the models were identified to be valid. When transition happens in the real process, the monitored error increases and the transition can be detected.

In the functional state detection sliding window parameters were $L_D = 5$ and $L_R = 50$ and the test signal threshold was 2. The test signals $z(k)$ for all three separate models were calculated in parallel. On transition, all filtered $f_D(e(k)^2)$ variables were initiated to zero and all $f_R(e(k)^2)$ variables were initiated to 0.04, which is a typical square error value when the transition transients have died out.

There is always some detection delay due to sliding window techniques used when the test signal is formed. In principle, the more reliable detection is wanted, the longer detection and reference windows are needed and the longer becomes the detection delay. High

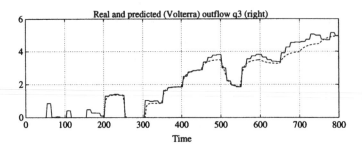

Figure 4.8   The response of trained "right dynamics" – model "—", the real response "- - -".

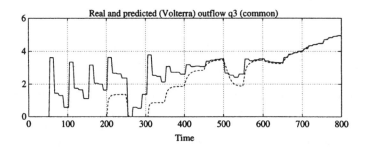

Figure 4.9   The response of trained "common dynamics" – model "—", the real response "- - -".

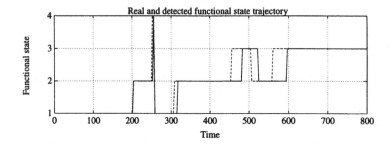

Figure 4.10   The functional state detection in the case of test run presented in fig 4.4. The detected "—" and the real "- - -" functional state.

reliability means long delays, which can cause problems as can be seen in the following state estimation example.

4.5.2   Application to state estimation

The use of "local" models is next illustrated in the connection with state estimation. Local models are simple numerical models of the process, which are valid only in certain functional states. Until now it has been supposed that no information about the level heights is available. However, suppose now that noisy level measurements are available and they are

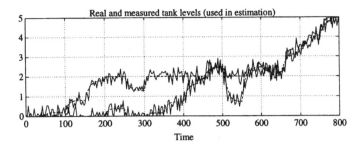

Figure 4.11   The noisy level measurements and the real levels in the test run in fig 4.4

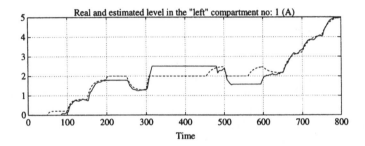

Figure 4.12   The estimated "—" and the real level "- - -" in the "left" compartment A.

improved by using the extended Kalman filter. Note that the state dynamics of the system does not allow the presentation with one equation only.

On the basis of the functional state detection the state estimation configuration is changed. The Kalman filter with "left dynamics" is used in functional states M_1 and M_4 for estimation of x_1 the level in the left compartment. The Kalman filter with "right dynamics" is used in functional state M_2 for estimation of x_2 the level in the right compartment. The Kalman filter with "common dynamics" is used in functional state M_3 for estimation of both x_1 and x_2. Filters were programmed so that the continuity of state estimates and covariances is guaranteed.

The noisy level measurements are presented in Figure 4.11. The estimated and real level in left compartment x_1 are shown in Figure 4.12. The estimated and real level in the right compartment x_2 are shown in Figure 4.13. It can be seen that when the filter used is valid (detected functional state is correct) the estimates converge rather quickly. During the detection delay, however, the estimates can drift to faulty values, because the Kalman filter is not valid, i.e. wrong state dynamics are used in state estimation. This can be seen clearly in Figure 4.12, because x_1 is not estimated in state M_2.

The noise covariance 0.25 was used when noisy measurements were generated. All three Kalman filters had the same parameters: 0.25 as the measurement noise covariance, and 0.05 as system noise covariance. The parameters of the local state models in each Kalman filter were supposed to have been identified correctly. It is quite clear that the Kalman filter works in this case only when connected to the functional state detection.

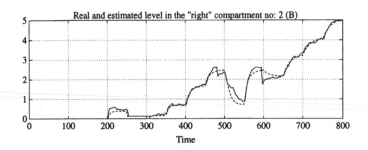

Figure 4.13 The estimated "—" and the real "- - -" level in "right" compartment B.

4.6 FUNCTIONAL STATE RECOGNITION ON THE BASIS OF A LAGUERRE REPRESENTATION

The obvious problem when applying the functional state concept detection presented in the previous chapters is that consideration of the history of the input signals is based on the sampled values. This may mean enormous number of inputs in practical cases. This problem can be circumvented by representing the input signals in orthogonal bases (Laguerre). This was originally derived by Norbert Wiener in the 1940s, his major contributions in this field are recorded in (Wiener 1958). The Wiener model is briefly described in the following on the basis of (Schetzen 1980). The Laguerre representation of the input signals is then discussed. Laguerre representations are finally used as the feature vector for the MLP-classifier, which detects changes in functional states.

4.6.1 Orthogonal development of Wiener G-functionals and the Wiener model

The SISO Volterra series model above in equation (4.3) can be expressed in the operator form as

$$y(t) = \sum_{n=0}^{\infty} H_n[x(t)],$$ (4.13)

where each *homogeneous* Volterra functional H_n corresponds to an integral term of order n in the series with Volterra kernel function $h_n(\tau_1, ..., \tau_n)$.

A pth-degree *nonhomogeneous* Volterra functional is in general

$$g_p(h_p, h_{p-1(p)}, ..., h_{0(p)}; x(t)) = \sum_{n-0}^{p} H_{n(p)}[x(t)]$$

$$= h_{0(p)} + \sum_{n=1}^{p} \int_0^{\infty} ... \int_0^{\infty} h_{n(p)}(\tau_1, \tau_2, ..., \tau_n) ...$$

$$x(t - \tau_1)x(t - \tau_2)...x(t - \tau_n)d\tau_1 d\tau_2 ... d\tau_n.$$

(4.14)

(Originally he developed this theory partly in connection with communication theory, which partly explains the approach and formalism in the following.) Wiener G-functionals form a set of nonhomogeneous Volterra functionals

$$g_n(k_n, k_{n-1(n)}, \ldots, k_{0(n)}; x(t)),$$

for which

$$\overline{H_m[x(t)]g_n(k_n, k_{n-1(n)}, \ldots, k_{0(n)}; x(t))} = 0 \text{ for } m < n, \tag{4.15}$$

when $x(t)$ is a white Gaussian input. $\overline{f(t)}$ is the average value of $f(t)$ in the interval $-T < t < T$, $T \to \infty$. It means that the nonhomogeneous functional $g_n(k_n, k_{n-1(n)}, \ldots, k_{0(n)}; x(t))$ is *orthogonal* to any homogeneous Volterra functional of degree less than n for the white Gaussian input $x(t)$. The G-functionals so derived are denoted by $G_n[k_n; x(t)]$. Only the nth order Wiener kernel k_n is shown because the derived Wiener kernels $k_{n-1(n)}, \ldots, k_{0(n)}$ are determined uniquely by k_n.

The Wiener representation of the system is

$$y(t) = \sum_{n=0}^{\infty} G_n[k_n; x(t)], \tag{4.16}$$

where k_n are the Wiener kernels of the system. The given system's Volterra kernels can be obtained uniquely from the system's Wiener kernels. In practice, when the system is modelled with Wiener G-functionals, probably fewer terms and parameters are needed for a certain approximation accuracy than with Volterra models.

Each Wiener kernel k_p of any physical nonlinear system of the Wiener class can be expanded to the p-dimensional Laguerre series

$$k_p(\tau_1, \ldots, \tau_p) = \sum_{n_1=0}^{\infty} \cdots \sum_{n_2=0}^{\infty} c_{n_1 \ldots n_p} l_{n_1}(\tau_1) \ldots l_{n_p}(\tau_p), \tag{4.17}$$

in which

$$l_n = \sqrt{2p} \sum_{k=0}^{n} \frac{(-1)^k n!}{k![(n-k)!]^2} (2pt)^{n-k} e^{-pt} u_{-1}(t), \tag{4.18}$$

and

$$c_{n_1 \ldots n_p} = \int_0^{\infty} \cdots \int_0^{\infty} k_p(\tau_1, \ldots, \tau_p) l_{n_1}(\tau_1) \ldots l_{n_p}(\tau_p) d\tau_1 \ldots d\tau_p. \tag{4.19}$$

Also the Volterra kernels $h_n(\tau_1 \ldots \tau_n)$ can be expanded in this way. With this presentation of k_p the Wiener G-functional can be presented in the form

$$G_p[k_p; x(t)] = G_p \left[\sum_{n_1=0}^{\infty} \cdots \sum_{n_p=0}^{\infty} c_{n_1 \ldots n_p} l_{n_1}(\tau_1) \ldots l_{n_p}(\tau_p); x(t) \right], \tag{4.20}$$

$$= \sum_{n_1=0}^{\infty} \cdots \sum_{n_p=0}^{\infty} c_{n_1 \ldots n_p} G_p[l_{n_1} \ldots l_{n_p}; x(t)]. \tag{4.21}$$

A G-functional G_p can be completely specified in terms of the coefficients $c_{n_1 \ldots n_p}$ of its Wiener kernel expansion.

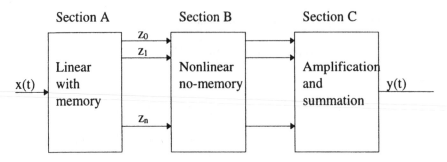

Figure 4.14 The general Wiener model.

It can be shown that

$$G_p[k_p; x(t)] \;=\; \sum_{n_1=0}^{\infty} \cdots \sum_{n_p=0}^{\infty} c_{n_1 \dots n_p} G_p[l_{n_1} \dots l_{n_p}; x(t)] \tag{4.22}$$

$$=\; \sum_{n_1=0}^{\infty} \cdots \sum_{n_p=0}^{\infty} c_{n_1 \dots n_p} \prod_{i=1}^{N} G_{k_i}[l_{m_i}^{(k_i)}; x(t)], \tag{4.23}$$

where $l_{n_1} \dots l_{n_p} = \prod_{i=1}^{p} l_{n_i} = \prod_{j=1}^{N} l_{m_j}^{(k_j)}$ when $\sum_{j=1}^{N} k_j = p$.

It can be further shown that

$$G_p[l_i^{(p)}; x(t)] = \sum_{m=0}^{\frac{p}{2}} \frac{(-1)^m p!}{m!(p-2m)!} \left(\frac{A}{m}\right) \left[\int_0^{\infty} l_i(\tau) x(t-\tau) d\tau \right]^{p-2m} = H_p[z_i(t)],$$

$$\tag{4.24}$$

where $H_p[z_i(t)]$ is the pth degree orthogonal *Hermite* polynomial and $z_i(t)$ is the response of a linear system, with impulse response $l_i(t)$, to input $x(t)$. The transfer functions $L_i(s)$ of the linear system with impulse response $l_i(t)$ is its Laplace transformation

$$L_i(s) \;=\; \sqrt{2p} \frac{(p-s)^i}{(p+s)^{i+1}} \tag{4.25}$$

$$z_i(t) \;=\; \mathcal{L}^{-1} \{ L_i(s) \mathcal{L}\{x(t)\} \}. \tag{4.26}$$

The corresponding discrete transfer function (z-transformation) has been derived by Tustin approximation.

Thus

$$G_p[k_p; x(t)] \;=\; \sum_{n_1=0}^{\infty} \cdots \sum_{n_p=0}^{\infty} c_{n_1 \dots n_p} \prod_{i=1}^{N} G_{k_i}[l_{m_1}^{(k_i)}; x(t)] \tag{4.27}$$

$$=\; \sum_{n_1=0}^{\infty} \cdots \sum_{n_p=0}^{\infty} c_{n_1 \dots n_p} \prod_{i=1}^{N} H_{k_i}[z_{m_i}(t)]. \tag{4.28}$$

The general Wiener model consists of three parts and is shown in Figure 4.14. Section A is a linear SIMO (single-input–multi-output) dynamic Laguerre system, section B is a nonlinear static system and section C a MISO amplification and summation system. The static

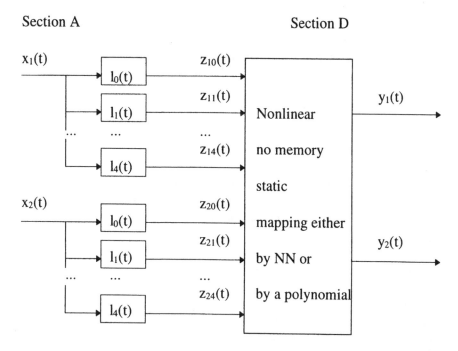

Figure 4.15 Reduced form of the Wiener model.

mapping in section B is accomplished with orthogonal Hermite polynomials in the detailed Wiener model (Schetzen 1980).

The Wiener model was derived by *expanding the system kernels* with an orthogonal Laguerre representation (also other orthogonal expansions can be used). However, in the practical implementation the coefficients in the Laguerre *expansion of the past of the input* are calculated in section A. Present values of the Laguerre system output are thus the co-efficients of the time series representing what the system can imperfectly remember of its past inputs.

The SISO Wiener model above can be generalised to the MIMO case. Then the outputs of section A are the separate responses $z_{j(i)}(t)$ of the Laguerre systems for each input $x_i(t)$. The sections A and B (Hermite polynomials) are common for all outputs.

4.6.2 A neural network-classifier for the functional states inspired by the Wiener model

If the static nonlinear mapping in section B and C of Figure 4.14 is approximated with a neural network, the model structure shown in Figure 4.15 results. The dynamic linear part is based on Laguerre systems for all inputs. The static no memory part is marked with section D.

The system structure in Figure 4.15 can be interpreted as a NN-approximation with an input signal pretreatment. The present values of the Laguerre system outputs $z_{j(i)}(t)$ contain information about the dynamic behaviour of the inputs in the near past. Although derived for the input–output relationship, the same structure can also be used in modelling of entirely or partly autonomous systems such as batch or fed-batch processes. Some of the outputs or components of the state vector are then fed back to the inputs. In the classifier case below there is no explicit feedback, but the on-line measurements reflect the real state variables implicitly. The outputs are redefined as classifier outputs for the functional states. Laguerre system outputs $z_{j(i)}(t)$ are then used as a feature vector to a classifier for the functional states. Section D in this case is an MLP (Multilayer Perceptron), which maps the "past behaviour" to a unique binary code for each functional state.

4.7 CASE STUDY: A BIOPROCESS

In the following case study, a genetically modified string of *B. Subtilis* is fermented in batches to produce alpha-amylase. This is a batch fermentation which goes through a number of "phases". The following five phases can be identified:

1 Lag phase.
2 I growth phase.
3 Intermediate phase.
4 II growth phase.
5 Declining phase.

These phases, which represent rough biological states of the biomass, can be identified by an experienced person by just looking at the CO_2 (in exhaust gas), DO (dissolved oxygen), oxygen uptake rate and pH-control acid/base addition curves. It is also possible to automate the recognition by using relatively simple rules that compare those curves to certain standard type of behaviour, as shown in (Halme 1988). A typical run is shown in Figure 4.16a. It is not easy to identify these functional states by using the conventional bioreactor state variables at certain time instants only, i.e. biomass, substrate, DO, and product concentrations in this case. This is because of the intermediate phase where the growth stops and then starts again. For example, suppose that the situation is the following:

■ biomass medium level
■ substrate medium level
■ DO low level
■ product low level

and suppose further that the history of the run is not known. It is difficult to conclude whether the process is running in the middle of the I growth phase or the beginning of the intermediate phase. Thus it would be also difficult to predict the further behaviour of the process by knowing only this information. This in fact means that the conventional state vector is not sufficient to describe the process state. What happens in practice in the intermediate phase is that the glucose initially available in the growth medium is finished and more glucose is being hydrolysed. The growth continues when enough glucose is again available. Now, one may think that the problem can be solved easily by increasing the state dimension by adding one or two suitable state variables which represent some intermediate product concentrations or mRNA for example, and which would explain the behaviour of the process. This is partly true, but after doing this it can be noted that only slightly better explanation of product (alpha-amylase) formation has been obtained. The problem is still the same: it is not easy to separate the process phases by just using the state variables.

The fact is that to fully describe the state of the process a very complicated structured model would be probably needed (do not confuse the terms "structured model" and "variable structure system"). Such models, which may be called *deep models*, are of course interesting and worth developing especially because they explain and clarify the behaviour

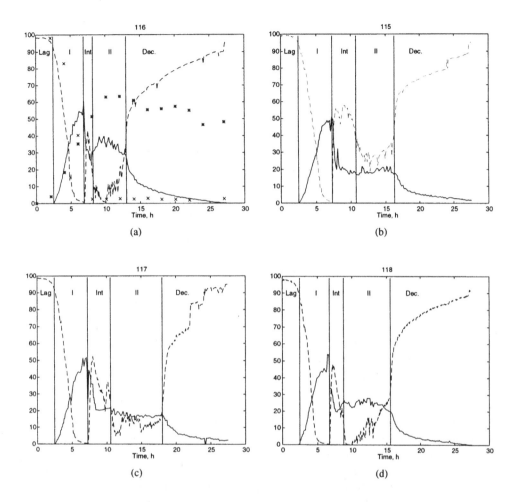

Figure 4.16 The data from four *B. Subtilis* fermentations. For all fermentations CO_2 ($25\times$ % Vol) and DO (% Sat) are shown with "—" and "- - -" respectively. The analysed biomass with "*" ($10\times$ mg/ml) and substrate with "×" ($2\times$ mMol/l) concentrations are also shown for the fermentation 116. The division into functional states is also shown by vertical lines and labels.

of the process. From the process control point of view such models probably are, however, too complicated to use in practice. Another possible approach is to use a *shallow model* and describe the "additional" part of the state by a symbolic or logic expression based on expert knowledge of the process, i.e. use the functional state concept. Another good example of the fermentation where different functional states can be clearly recognised is the classical baker's yeast fermentation (batch, fed-batch or continuous). These have been analysed e.g. in (Bellgardt *et al.* 1983) and (Zhang *et al.* 1994).

Because the above bioreactor process is a batch process, it is not practical to use an input–output mapping of type (1) for detection of the functional states. Instead, the mapping can be as well the state equation of the system. Its exact form is not known of course, but in its structure the state variable histories are used to predict theries are used to predict their own

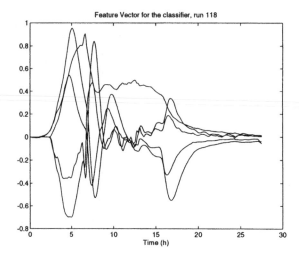

Figure 4.17   The feature vector from run 118.

values at the real time or in the future. All the process variables mentioned above are state variables or are closely related to such variables in the reactor.

The MLP classifier with a Laguerre system for temporal feature extraction, as described above, has been applied to the *B. subtilis* batch fermentation to recognise the functional states to give some insight into the process behaviour in individual runs. Four batch runs have been made from identical initial conditions using the same inoculation. The runs are presented in Figures 4.16a to 4.16d. The runs are not particularly close to each other, but the same functional states can be found in each run and separated with vertical lines.

Two on-line measurements, CO_2 in exhaust gas and dissolved oxygen in the process broth, were processed with the discrete Laguerre system: sample time T was 0.0833h, Laguerre parameter p for DO Laguerre system was 2 and for CO_2 Laguerre system p was 3.4. At first the first three Laguerre outputs $[z_0 \quad z_1 \quad z_2]$ from both systems were fed to the classifier. The results were not good enough. In Figures 4.16a-d it can be seen that the measurement trajectories are quite different and noisy. Particularly the DO level in the Int,- and II-functional states differs in different runs. However, the DO changes are important for classification. The following feature vector which emphasizes the change in the measurements turned out to be the most suitable: from the CO_2 Laguerre system three terms were calculated $[z_0 \quad z_1 - z_0 \quad z_2 - z_0]$ and from the DO Laguerre system two terms were calculated $[z_1 - z_0 \quad z_2 - z_0]$. For example, the feature vector in run 118 is shown in Figure 4.17.

The MLP had five inputs described above, one hidden layer had four neurons and the output layer included five neurons, one for each functional state. The target output was defined so that only the output corresponding to the current functional state had value 1 and the other outputs 0. The classifier was trained with data from three runs using the Levenberg–Marquardt method in Matlab^{TM} and validated with one run. In Figure 4.18 the results are shown, for training with runs 115, 116 and 117 and validating with run 118. In Figure 4.19 the training set consisted of runs 115, 116 and 118 and the classifier was validated with run 117. The results are not perfect, but it should be noted that runs are quite different, particularly in the Intermediate and the II growth states.

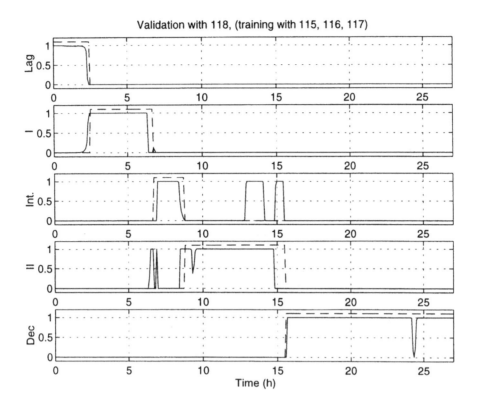

Figure 4.18 The classifier output "—" and the defined functional state "- - -" for run 118 when the classifier was trained with runs 115, 116 and 117.

Clearly when this kind of classification is used on line there should be some kind of monitoring intelligence to interpret and filter the classifier output. In general, when trying to recognise the functional states all the information available should be utilised and combined in order to get as reliable results as possible. The fact which can be utilised especially in this case is the order in which the functional states (phases) appear.

4.8 CONCLUSIONS

This chapter introduced the concept of the functional state and presented two different ways to detect or recognise such states automatically. The model based approach used input–output models for each functional state. The classifier method used the Laguerre systems to extract information about the process dynamics from the on-line measurements. Laguerre presentation of the signals and systems can be utilised in many other ways in control and monitoring, too. The two methods presented are generally applicable and they seem to work relatively well and be reliable. There is a natural delay always present when detection

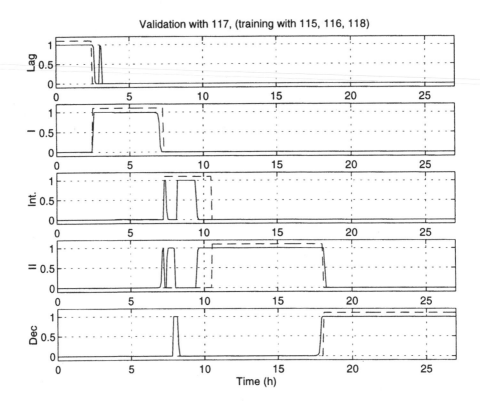

Figure 4.19   The classifier output "—" and the defined functional state "- - -" for run 117 when the classifier was trained with the runs 115, 116 and 118.

is done after the functional state has changed. It should be noted, however, that many examples exist where the functional states can be recognised from a single variable. For example the functional states in an aerobic yeast growth process and their use in monitoring and control have been reported in (Zhang *et al.* 1994a), (Zhang *et al.* 1994b) and (Zhang 1995). In this case a single variable called respiratory quotient can be used to separate the functional states. The situation is the same in many mechanical systems.

REFERENCES

BELLGARDT, K.-H., W. KUHLMANN AND H.-D. MEYER (1983) Deterministic growth model of *saccaromyces cerevisiae*, parameter identification and simulation. In 'Proc. 1st IFAC Workshop on Modelling and Control of Biotechnical Processes, Helsinki, Finland 1982'. Pergamon Press, Oxford.

HALME, A. (1972) Generalized polynomial operators for nonlinear systems analysis. *Acta Polytechnica Scandinavica*, Ma 24, Helsinki.

HALME, A. (1988) Expert system approach to recognize the state of fermentation and to diagnose faults in bioreactor. In 'Proc. SCI/IFAC 4th International Congress on Computer Applications in Fermentation Technology'. Ellis Horwood Ltd, Chichester.

HALME, A. AND J. SELKINAHO (1984) Instrument fault detection using an adaptive filtering method. In 'Proceedings of IFAC 9th World Congress'. Pergamon Press, Oxford.

HECHT-NIELSEN, R. (1990) *Neurocomputing*. Addison-Wesley.

RUMELHART, D. E. AND J. L. MCCLELLAND (1986) *Parallel Distributed Processing: Explorations in the Microstructures of Cognition,* Vol. 1: *Foundations.* MIT Press. Cambridge, MA.

SCHETZEN, M. (1980) *The Volterra and Wiener Theories of Nonlinear Systems.* John Wiley, New York.

WIENER, N. (1958) *Nonlinear Problems in Random Theory.* The Technology Press, MIT, and John Wiley, New York.

ZHANG, X.-C. (1995) Aspects of Modelling and Control of Bioprocesses. Doctoral thesis. Automation Technology Laboratory, HUT, Finland. Research reports No. 13.

ZHANG, X.-C., A. VISALA, A. HALME AND P. LINKO (1994a) 'Functional state modeling and fuzzy control of fed-batch aerobic baker's yeast'. *Journal of Biotechnology* **37**, 1–10.

ZHANG, X.-C., A. VISALA, A. HALME AND P. LINKO (1994b) 'Functional state modelling approach for bioprocesses: local models for aerobic yeast growth processes'. *Journal of Process Control* **4**(3).

CHAPTER FIVE

Markov Mixtures of Experts

MARINA MEILĂ and MICHAEL I. JORDAN

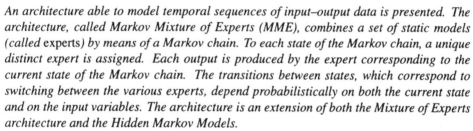

An architecture able to model temporal sequences of input–output data is presented. The architecture, called Markov Mixture of Experts (MME), combines a set of static models (called experts*) by means of a Markov chain. To each state of the Markov chain, a unique distinct expert is assigned. Each output is produced by the expert corresponding to the current state of the Markov chain. The transitions between states, which correspond to switching between the various experts, depend probabilistically on both the current state and on the input variables. The architecture is an extension of both the Mixture of Experts architecture and the Hidden Markov Models.*

It is shown that the parameters of the experts and the transition probabilities can be simultaneously estimated from input–output data only. The algorithm presented is an iterative algorithm based on the Estimation–Maximisation procedure; as a consequence, the unobserved states are estimated as well.

The algorithm is used in a fine motion task for a robot arm. Such a task is performed in stages; during each stage, the arm is required to move maintaining contact with a given surface of the surrounding objects. Due to multiple sources of uncertainty, the state of contact (i.e. which are the surfaces in contact with the arm) must be estimated from measurements and previous state information. This is achieved by a MME with the state of contact as hidden state variable.

5.1 INTRODUCTION

The present chapter introduces an architecture able to model temporal sequences of input–output data by dynamically switching between local static models. It is extending to the domain of dynamical systems a number of connectionist approaches which have been successful in modelling complex nonlinear static relationships and have contributed to the increased interest in nonparametric techniques in general.

More specifically, the problem to be addressed is the following: Assume that in a sequence of output data each point is generated by one of m fixed input–output mappings called *experts*. The experts are known to belong to a given class of functions parameterised by a *parameter vector* θ but their actual parameter values have to be estimated. Time is discrete. At each time step, the expert that will process the current input is chosen randomly, by sampling from a probability distribution which depends on the expert chosen at the previous time step and on the current input. Thus a *Markov chain* (Rabiner and Juang 1986) with m states and input dependent transition probabilities is associated to the process. The transition probability matrix A depends on a set of parameters denoted by W. The input and the output are measured, but the state of the Markov chain is hidden to the observer. This model is called *Markov Mixture of Experts* (MME).

One application where such models occur is in capital markets where price evolution is described by one of a few different models, with shifts between the models at random times. Another example is an agent acting in different environments; the outputs depend on both its actions (the inputs) and on the unknown environment; also, as a consequence of its actions, the agent can move to another environment. A machine having several distinct operating regimes can be viewed in such a way too; switching from one regime to another is controlled by both known factors (the inputs) and unknown ones (which motivate the probabilistic representation). Finally, an example which will be discusssed further in this paper: a robot arm performing an assembly task. During such a task, the arm needs to move maintaining contact with given surfaces, but avoiding contact (collision) with others. Due to noise in the measurements, the state of contact cannot be known directly; therefore it plays the role of a hidden variable and it needs to be estimated from the response of the robot arm to the input commands.

This paper will solve the problem of estimating the unknown parameters of the MME model from the given input–output observations within the *Maximum Likelihood* (Söderström and Stoica 1989) framework. After a more rigorous problem statement an iterative algorithm, based on the *Estimation–Maximisation* (Dempster *et al.* 1977) method, to learn the solution from data will be presented. The application of the algorithm is exemplified on two tasks.

5.2 THE MARKOV MIXTURE OF EXPERTS MODEL

This section will introduce the Markov Mixture of Experts Model. To help the intuition behind the formal definitions, we shall begin with an example.[1]

5.2.1 Parameter and state estimation in fine motion tasks

For a large class of robotics tasks, such as assembly tasks or manipulation of relatively light-weight objects, under appropriate damping of the manipulator the dynamics of the objects can be neglected. For these tasks the main difficulty is in having the robot achieve its goal despite uncertainty in its position relative to the surrounding objects. Uncertainty is due to inaccurate knowledge of the geometric shapes and positions of the objects, of their physical properties (surface friction coefficients), or to positioning errors in the manipulator. The basic method to deal with this problem is *controlled compliance* (Mason 1981). Under compliant motion, the task is performed in stages; in each stage the robot arm maintains permanent contact with a selected surface or feature of the environment; the stage ends when contact with the feature corresponding to the next stage is made. The goal itself is usually defined as satisfaction of a set of contact constraints.

---

[1]This application is examined by (Meilă and Jordan 1996).

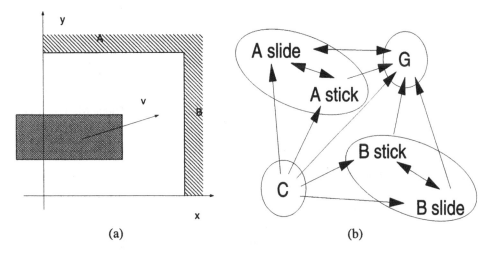

Figure 5.1 Restricted movement in a 2D space (a) and the corresponding reachability graph (b). The nodes represent movement models: C is the free space, A and B are surfaces with static and dynamic friction, G represents jamming in the corner. The velocity v has positive components.

Decomposing the given task into subtasks and specifying each goal or subgoal in terms of contact constraints has proven to be a particularly fertile idea, from which a fair number of approaches have evolved. One of its consequences is the introduction of a new variable, the *state of contact*, that takes *discrete* values within the set of possible states of contact between the robot arm and the surrounding surfaces. Each of the state of contact based approaches to control has to face and solve the problem of estimating it (i.e. checking if the contact with the correct surface is achieved) from noisy measurements.

To each state of contact we associate a *movement model*; that is: a relationship between positions, nominal and actual[2] velocities that holds over a domain of the position-nominal velocity space. From this point of view, the states of contact represent the domains of validity of each movement model. Because trajectories are continuous, a point can move from a state of contact only to one of the neighbouring states of contact. This can be depicted by a directed graph with vertices representing states of contact and arcs for the possible transitions between them, called the *reachability graph*. An example of a 2D space and its reachability graph is shown in Figure 5.1. Ideally, in the absence of noise, the states of contact and every transition through the graph can be perfectly observed. To deal with the uncertainty in the measurements, we will attach probabilities to the arcs of the graph in the following way: Let us denote by Q_i the set of configurations corresponding to state of contact i and let the movement of a point x with uniform nominal velocity v for a time ΔT be given by $x(t + \Delta T) = f^*(x, v, \Delta T)$; x and v are vectors of the same dimension. Now, let x', v' be the noisy measurements of the true values x, v; $x \in Q_j$ and $P[x, v|x', v', j]$ the posterior probability distribution of x, v given the measurements and the state of contact. Then, the probability of transition to a state $q' = i$ from a given state $q = j$ in time T_s can

---

[2]The nominal velocity is the velocity of the arm if it was moving in free space under the same active force. It is proportional in magnitude and identical in orientation to the force vector exerted by the actuators (the *active* force). When the object is subject to reactive forces from the surfaces it is in contact with, the actual velocity does not coincide with the nominal velocity.

be expressed as:

$$P[i|x', v', j] = \int_{\{x,v|x \in Q_j, f^*(x,v,T_s) \in Q_i\}} P[x, v|x', v', j]dx\, dv = a_{ij}(x', v') \tag{5.1}$$

Defining the transition probability matrix $A = [a_{ji}]_{i,j=1}^m$ and assuming a probability density $p[x|i, x_{true}]$, $x_{true} \in Q_i$ for the measurement noise leads to a Hidden Markov Model (HMM) with output x having a continuous emission probability distribution and where the state of contact plays the role of a hidden state variable.

5.2.2 Defining the MME model and the learning problem

Now, let us introduce the Markov Mixture of Experts model in its general form.

Suppose that a discrete time process is modelled by a Markov chain with state $q(t)$ taking values in a finite set $\{1, \ldots, m\}$. At each time step, the transition probabilities of the chain, defined by

$$a_{ij} = Pr[q(t) = i|q(t-1) = j] \quad t = 0, 1, \ldots$$

with

$$\sum_i a_{ij} = 1 \quad j = 1, \ldots, m \tag{5.2}$$

depend on an input signal u, possibly a vector, and on a set of parameters, W:

$$a_{ij} = a_{ij}(u(t), W).$$

The same input u is also processed in parallel by a number of m static modules called *experts*. Each expert outputs a value $f(u(t), \theta_i)$, $i = 1, 2, \ldots, m$ which is a scalar or a vector of fixed dimension n_y. The output of expert i is then corrupted by additive Gaussian noise with 0 mean and covariance matrix Σ_i. In the following, it will be assumed for simplicity that the noise covariance is of the form

$$\Sigma_i = \sigma_i^2 I_{n_y}$$

(with I_n being the unit matrix of order n) and the vector $[\sigma_1 \sigma_2 \ldots \sigma_m]$ will be denoted by σ.

The output y of the system is chosen from among the outputs of the experts, using the state of the Markov chain as an indicator variable. Thus, if the state $q(t) = i$, then the output

$$y(t) = f(u(t), \theta_i) + v_i(t)$$

where

$$p(v_i) = \frac{1}{(2\pi\sigma_i^2)^{n_y/2}} \exp(-\frac{\|v_i\|^2}{2\sigma_i^2}) \quad i = 1, 2, \ldots, m$$

and $v_i(t)$ is independent of everything else.

Only the input u and the output y are accessible to measurement; the state q of the Markov chain is hidden to the exterior observer.

The model is schematically represented in Figure 5.2. Taking the functions a_{ij} and f to be constant functions (independent of the input), it is easy to see that the MME model contains the hidden Markov model as a special case; in fact, the former can be viewed as a time-variant, continuous output distribution HMM. It is also a generalisation of the mixture

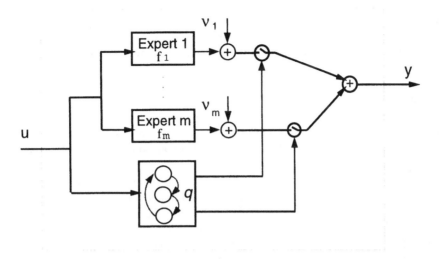

Figure 5.2 The Markov Mixture of Experts architecture.

of experts architecture (Jordan and Jacobs 1994), to which it reduces when the columns of the transition probability matrix are all equal.

$$a_{ij} = a_{ij'} \quad \forall i$$

It becomes the model of (Bengio and Frasconi 1995) when A and f are neural networks.

Suppose that for a MME model we have observed a sequence of outputs $y_{\overline{0.T}}$ corresponding to the sequence of inputs $u_{\overline{0.T}}$, also known. (Here $s_{\overline{t_1.t_2}} = [\, s(t_1), s(t_1 + 1), \ldots, s(t_2)]$ denotes the sequence of values of the variable s over the time interval $t_1, \ldots, t_2$). In the following the set $\{(u(t), y(t)), t = 0, 1, \ldots, T\}$ will be referred to as the *data set*. Our main goal is to estimate the values of the parameters θ and W from this data set.

The next section will show a way to achieve this. Meanwhile, note that the relationship between u and y depends on the values of the parameters as well as on the sequence of states of the Markov chain $q_{\overline{0.T}}$ and the noise covariances σ. In general, it is assumed that their values are not known; as it will be seen, the algorithm developed in section 5.3 will compute as intermediate results estimates for these quantities as well. This property of the algorithm has two consequences: the first is that estimates of the hidden state sequence and noise level can be obtained with little or no overhead calculations; the second is that if any of these values are known, they can be used in a straightforward manner to improve the parameter estimates' accuracy.

5.3 ESTIMATING THE PARAMETERS OF THE MME MODEL

5.3.1 The maximum likelihood principle

As stated before, the problem to be solved is the estimation of the unknown parameters θ and W of the MME model from a data set consisting of $u_{\overline{0.T}}$, $y_{\overline{0.T}}$. The estimation will be done within the Maximum Likelihood (ML) framework.

ML is a principle which states that the estimate of an unknown parameter from some data is chosen from among its admissible values, such that the *likelihood* of the data is maximised w.r.t. that parameter. The likelihood is the probability of the data set, as function of the parameters.

It should be pointed out that ML is not the only criterion used in parameter estimation. A more comprehensive method, which allows the incorporation of prior information about the parameters, is *Bayesian estimation* (see for example (MacKay 1992)). Discussing this and other criteria like the Akaike Information Criterion (AIC) (Söderström and Stoica 1989), the description length (Rissanen 1978) or the simple Mean Squared Error (MSE) is beyond the scope of this work.

Once the criterion is formulated, the problem becomes an optimisation problem: find the values of the parameters that maximise the likelihood function L. Since it is rarely the case that the maximum of L can be found analytically, here we will resort to an iterative but general method of maximisation, called the EM algorithm.

We will assume for the moment that the data set is extended to include the state sequence $q_{\overline{0,T}}$ besides the input and the output sequences. With this *complete* data set, a new likelihood function, the *complete likelihood* L_c is defined. In the following we will use its logarithm, called *log-likelihood*:

$$
\begin{aligned}
l_c(\theta,\,W,\,\sigma) &= \ln L_c(\theta,\,W,\,\sigma) \\
&= \ln P\left(y_{\overline{0,T}},\,q_{\overline{0,T}}\,|\,u_{\overline{0,T}},\,W,\,\theta,\,\sigma\right) \\
&= \ln\left\{\,Pr[q_{\overline{0,T}}\,|\,u_{\overline{0,T}},\,W,\,\theta,\,\sigma]\cdot\prod_{t=0}^{T}p(y(t)|q(t),\,u(t),\,\theta,\,\sigma)\right\}.
\end{aligned}
\tag{5.3}
$$

Since the noise is assumed to be zero mean Gaussian, the probability density of output $y(t)$ given the state $q(t) = k$ is given by:

$$
p(y(t)|k,\,u(t),\,\theta,\,\sigma) = \frac{1}{(2\pi\sigma_k^2)^{n_y/2}}\exp\left[-\frac{\|y(t)-f(u(t),\theta_k)\|^2}{2\sigma_k^2}\right].
\tag{5.4}
$$

Thus, the complete likelihood l_c becomes

$$
\begin{aligned}
l_c(W,\,\theta,\,\sigma) &= \sum_{k=1}^{m}z_{q(0)}^{k}\ln\pi_k(0) + \sum_{t=0}^{T-1}\sum_{i,j=1}^{m}z_{q(t)}^{j}z_{q(t+1)}^{i}\ln a_{ij}(u(t),\,W) \\
&\quad + \sum_{t=0}^{T}\sum_{k=1}^{m}z_{q(t)}^{k}\left[-\frac{\|y(t)-f(u(t),\theta_k)\|^2}{2\sigma_k^2}-\frac{n_y}{2}\ln(2\pi\sigma_k^2)\right],
\end{aligned}
$$

where $\pi(0)$ denotes the initial probability distribution for the Markov chain state

$$
\pi_i(0) = P[q(0) = i]
\tag{5.5}
$$

and

$$
z_{q(t)}^{k} = \begin{cases} 1 & q(t) = k. \\ 0 & \text{otherwise.} \end{cases}
\tag{5.6}
$$

The above quantity being a likelihood function, maximising it would provide us with estimates of the parameters. However, the state sequence is not known and neither is l_c which depends on it; what we can and will compute is its average given the observed data.

$$J(\theta, W, \sigma) = E[l_c \mid y_{\overline{0.T}}, u_{\overline{0.T}}, W, \theta, \sigma] \qquad (5.7)$$

$$= -\frac{1}{2} \sum_{k=1}^{m} \frac{1}{\sigma_k^2} \sum_{t=0}^{T} \gamma_k(t) \|y(t) - f(u(t), \theta_k)\|^2 - \frac{n_y}{2} \sum_{k=1}^{m} \ln(\sigma_k^2)$$

$$+ \sum_{t=0}^{T-1} \sum_{i,j=1}^{m} \xi_{ij}(t) \ln a_{ij}(u(t), W) + C,$$

where C is a constant and

$$\gamma_k(t) = E[z_{q(t)}^k \mid y_{\overline{0.T}}, u_{\overline{0.T}}, W, \theta, \sigma] = P[q(t) = k \mid u_{\overline{0.T}}, y_{\overline{0.T}}, W, \theta, \sigma] \qquad (5.8)$$

$$\xi_{ij}(t) = P[q(t) = j, q(t+1) = i \mid u_{\overline{0.T}}, y_{\overline{0.T}}, W, \theta, \sigma] \qquad (5.9)$$

$$= E[z_{q(t)}^j z_{q(t+1)}^i \mid u_{\overline{0.T}}, y_{\overline{0.T}}, W, \theta, \sigma].$$

The last two quantities are well known in the HMM literature and can be computed efficiently by means of a procedure which parallels the *forward-backward* algorithm (Baum *et al.* 1970). They still depend on the parameters; to obtain a numerical value for them we will use some initial estimates of W, θ, σ.

5.3.2 The maximisation step

Equations (5.7–5.9) implement the *Expectation* (E) step of the EM procedure. Its result is the expected value of the complete log-likelihood as a function of the unknown parameters. In computing this expected value, we made use of an initial estimate of the parameters. Now we will obtain new estimates by maximising J w.r.t. W, θ and σ. This represents the second step, *Maximisation* (M), of EM:

$$W^{\text{new}}, \theta^{\text{new}}, \sigma^{\text{new}} = \underset{W, \theta, \sigma}{\text{argmax}} \, J(W, \theta, \sigma). \qquad (5.10)$$

Since each parameter appears in only one term of J the maximisation is equivalent to:

$$\theta_k^{\text{new}} = \underset{\theta_k}{\text{argmin}} \sum_{t=0}^{T} \gamma_k(t) \|y(t) - f(u(t), \theta_k)\|^2, \qquad (5.11)$$

$$W^{\text{new}} = \underset{W}{\text{argmax}} \sum_{t=0}^{T-1} \sum_{i,j=1}^{m} \xi_{ij}(t) \ln a_{ij}(u(t), W), \qquad (5.12)$$

$$\sigma_k^{2\,\text{new}} = \frac{\sum_{t=0}^{T} \gamma_k(t) \|y(t) - f(u(t), \theta_k)\|^2}{n_y \sum_{t=0}^{T} \gamma_k(t)}. \qquad (5.13)$$

By solving (5.11)–(5.13) a new set of values for the parameters is found and the current iteration of the EM procedure is completed. Now the newly computed estimates replace the old estimates and a new iteration can start. It is provable that the EM algorithm converges to a local maximum of the likelihood (Dempster *et al.* 1977), (Neal and Hinton 1993).

Moreover, it can be proven that, if the M step is replaced by one where the new values of the parameters are chosen so that $J(W, \theta, \sigma)$ is not maximised, but merely *increased*, the resulting iterative algorithm still converges asymptotically to a local maximum of the likelihood. This version is called *Generalised EM* (GEM). It is expected that GEM should converge slower than EM on a given problem; however, given that only the maximisation w.r.t. σ can be performed exactly[3] in all cases, one may often have to resort to GEM.

In the end of this section, let us look closer to equations (5.11) and (5.12). None of these expressions can be maximised analytically in the general case. Equation (5.11) is of the form of a weighted least squares problem. This problem has a closed form solution when the function f is linear in θ_k, which is the usual linear weighted least squares solution. In another special case, when f is a nonlinear function of a linear combination of the components of θ_k, an iterative procedure called *Iterative Reweighted Least Squares* (IRLS) can be used. For more details on this technique see (McCullagh and Nelder 1983). In the other cases, general optimisation methods, like gradient ascent, have to be used.

Equation (5.12) has a closed form solution when a_{ij} do not depend on the input. In this case

$$a_{ij}^{new} = \frac{\sum_{t=0}^{T-1} \xi_{ij}(t)}{\sum_{t=0}^{T-1} \gamma_j(t)} \quad i, j = 1, \ldots, m \tag{5.14}$$

which coincides with the Baum–Welch (Baum *et al.* 1970) solution for classical HMMs. This should not be a surprise because in this case the MME becomes a HMM.

In other cases, solving for a_{ij} may require an iterative process. But one can note the following: if there exists a $W^\star$ to satisfy simultaneously

$$a_{ij}(u(t), W^\star) = \frac{\xi_{ij}(t)}{\gamma_j(t)} \tag{5.15}$$

for all $i, j = 1, \ldots, m;\ t = 0, \ldots, T-1$ then $W^\star$ will be the solution of (5.12) under the constraint (5.2). The derivation of (5.15) is based on the use of the Lagrange multipliers and is omitted here. Even if such a $W^\star$ does not exist or cannot be found, changing the W in a way that moves the a_{ij} towards the "targets" $\xi_{ij}(t)/\gamma_j(t)$ will cause an increase in J. In this case, we will be performing a GEM (Generalised EM) step.

5.4 AN EXAMPLE: LINEAR EXPERTS AND CONSTANT A

In this section we will examine, by means of simulation, the learning performance of the MME architecture in the special case where the matrix A does not depend on the input and the experts are linear in the parameters.[4] In this case, the data generation mechanism can be summarised as follows:

$$y(t) = \theta_{q(t)}^T u(t) + \sigma_{q(t)} v(t) \tag{5.16}$$
$$q(t+1) \sim a_{.q(t)}(u(t), W)$$

with $y(t) \in R$, $\theta, u(t) \in R^{n_u}$ and $v(t)$ being a white noise with zero mean and identity covariance matrix.

The maximisation step for the vectors θ_k reduces in this case to a weighted linear regression which can be performed analytically.

---

[3]Remember, however, that the ML estimate for the variance is *biased*. After convergence, it should be corrected as, for instance, in (Söderström and Stoica 1989).

[4]This example is taken from (Meilă *et al.* 1994).

$$\theta_k = \left[\sum_{t=0}^{T} \gamma_k(t)u(t)u^T(t)\right]^{-1} \sum_{t=0}^{T} \gamma_k(t)u(t)y(t). \tag{5.17}$$

For comparison we will use a version of the Mixture of Experts, having the same experts structure as the corresponding HMM model. The difference is that, in the former case, at time t, the index of the selected expert, $q(t)$, is an i.i.d. discrete uniformly distributed random variable and the input–output dependency is a linear regression as in (5.16).

The above structure of the model, corresponding to a totally noninformative HMM, leads to a particular case of the algorithm: deciding the index of the expert will depend only on the information at time instant t by the residuals

$$\epsilon_k(t) = \|y(t) - \theta_k^T u(t)\|. \tag{5.18}$$

As a consequence, an EM iteration of the learning algorithm for this simplified model will consist of:

E: for $t = 0, \ldots, T$ choose the current expert index to be $k = \text{argmax}_{k'} \epsilon_{k'}(t)$.

M: estimate the parameters θ_k and σ_k for each expert.

This procedure will be called the *static classification* iterative estimation (SCIE) algorithm (in contrast with the dynamic classification accomplished by the MME). The MME and SCIE were compared w.r.t. three performance criteria:

- *Parameter estimation accuracy* represented by the value of the sum of squared errors for the parameter estimates J_r
- *Explaining the data generating mechanism.* In this application, the information required is the sequence of expert indices, explaining what is the dynamics of the un-observed state, and the performance criterion is N_c the number of times when

$$\hat{k}(t) \overset{\Delta}{=} \text{argmax}_k \gamma_k(t) = k^*(t) \overset{\Delta}{=} q(t).$$

- Further use of the estimated data generating mechanism for *optimal prediction or filtering*. When using batch learning methods, the effect of replacing the observed system output for any time index with a smoothed value (computed based on the the observed input vector at the same time index and on the whole system history) can be used for filtering the data (which is supposed to be entirely available at the processing time). The performance criterion for this case is the mean squared error of the filtered signal

$$\sigma_f^2 = \frac{1}{T} \sum_{t=0}^{T} \left(y(t) - \hat{y}_{\hat{k}}(t)\right)^2$$

with $\hat{y}_{\hat{k}}(t)$ being the output of the $\hat{k}(t)$th expert at time t.

The data were generated by the model in (5.16) with $m = 3$ and the following parameter values:

Regression parameters	Transition probability matrix
$\theta_1 = [0.7 \quad -0.1 \quad -0.3 \quad 0.2]^T$ $\theta_2 = [0.3 \quad 0.8 \quad 0.1 \quad 0.4]^T$ $\theta_1 = [-0.05 \quad -0.9 \quad 0.2 \quad 0.3]^T$	$a^T = \begin{bmatrix} 0.9 & 0.05 & 0.05 \\ 0.02 & 0.96 & 0.02 \\ 0.06 & 0.06 & 0.88 \end{bmatrix}$ $\sigma = [0.1; 0.4; 0.05]$

Table 5.1 Results of the *MME* and *SCIE* procedures in 80 convergent simulations.

T	Regression Estimate Accuracy $J_r = \|\hat{\theta} - \theta^*\|^2$ $\frac{mean}{std.dev.}$		Filtering Performance σ_f^2 $\frac{mean}{std.dev.}$		Recognition Performance[%] $N_c = \text{Card}\{\hat{k}(t) = k^*(t)\}_{t=\overline{0.T}}$ $\frac{mean}{std.dev.}$	
	MME	SCIE	MME	SCIE	MME	SCIE
2000	$\frac{0.0007}{0.0004}$	$\frac{0.0956}{0.2586}$	$\frac{0.094}{0.008}$	$\frac{0.111}{0.07}$	$\frac{99.4}{0.19}$	$\frac{88.51}{8.07}$

Table 5.2 Results of the *MME* and *SCIE* procedures (including the nonconvergent realisations) for various levels of the additive output noise σ.

σ	Regression Estimate Accuracy $J_r = \|\hat{\theta} - \theta^*\|^2$ $\frac{mean}{std.dev.}$		Filtering Performance σ_f^2 $\frac{mean}{std.dev.}$		Recognition Performance[%] $N_C = \text{Card}\{\hat{k}(t) = k^*(t)\}_{t=\overline{0.T}}$ $\frac{mean}{std.dev.}$	
	MME	SCIE	MME	SCIE	MME	SCIE
0.5	$\frac{0.025}{0.04}$	$\frac{0.67}{0.37}$	$\frac{0.024}{0.009}$	$\frac{0.012}{0.0017}$	$\frac{96.8}{0.88}$	$\frac{59.6}{5.7}$
0.1	$\frac{0.032}{0.1}$	$\frac{0.15}{0.24}$	$\frac{0.01}{0.0004}$	$\frac{0.01}{0.01}$	$\frac{99.86}{0.1}$	$\frac{87.6}{11.7}$
0.05	$\frac{0.016}{0.44}$	$\frac{0.188}{0.36}$	$\frac{0.0031}{0.0049}$	$\frac{0.0069}{0.01}$	$\frac{99.7}{1.65}$	$\frac{90.2}{13.8}$
0.02	$\frac{0.00002}{0.00002}$	$\frac{0.112}{0.235}$	$\frac{0.0004}{0.00001}$	$\frac{0.007}{0.013}$	$\frac{99.9}{0.7}$	$\frac{91}{15}$
0.01	$\frac{0.12}{0.45}$	$\frac{0.21}{0.42}$	$\frac{0.00027}{0.0012}$	$\frac{0.0044}{0.009}$	$\frac{99.9}{0.58}$	$\frac{90.9}{15}$

The results obtained with the MME procedure are well clustered into two classes: in one class, all parameter estimates $(\hat{\theta}, \hat{\sigma}^2, \hat{a}_{ij})$ converge to values close to the true parameters. In the other class, the estimated parameters converge as well, but far from the true values, as a result of the local type of convergence of EM procedures.

Experiment 1   In order to accurately estimate the performance obtainable with the MME procedure, 200 runs of the procedure were performed, with different random initial values of the parameters θ, σ, A.

In almost half of the simulations (80 trials), the parameters converge to values close to the true parameters. It was found that selecting as a threshold $J_r = 0.1$, the two types of convergence behaviour can be easily detected. Since in practice the true parameters are not known, σ_f can be used as a global convergence detector, performing equally well as J_r.

The results obtained in the convergent simulation cases are compared with the results of running an SCIE procedure, starting from the same initial conditions as MME procedure.

Table 5.1 presents the statistics of the three important performance indices: J_r, σ_f and n_K. For all these performance measures, the MME procedure clearly outperforms the SCIE procedure.

Experiment 2   In the second experiment, five groups of 50 simulation were performed. Each group consists of 50 runs of MME and SCIE procedures, on a given training set generated with the same σ for all three experts, but different for each of the five groups, in order to illustrate the influence of the noise level on the resulting performance criteria. The results are summarised in Table 5.2. Now all the simulations were considered in computing the statistics in the table, leading to results emphasising the worst cases in the simulations.

One conclusion can be drawn from the experimental results: the usefulness of the system identification structure based on the HMM modelling, which allows a better regression parameter estimation and a better recognition accuracy.

5.5 LEARNING FINE MOTION BY MME

An implementation example and further simulation results of the fine motion problem introduced in section 5.2 will be presented now.

5.5.1 Problem and implementation

The problem is to learn a predictive model for the movement of a point in the 2-dimensional space shown in Figure 5.1a. The inputs were 4-dimensional vectors of positions (x, y) and nominal velocities (v_x, v_y) and the output was the predicted position (x_{out}, y_{out}). The coordinate range was $[0, 10]$ and the admissible velocities were confined to the upper right quadrant $(v_x, v_y \geq V_{min} > 0)$ and had magnitudes (after multiplication by T_s) between 0 and 4. The restriction in direction guaranteed that the trajectories remained in the coordinate domain; it was also reflected in the topology of the reachability graph, which is not complete (has no transition to the free space from another state, for example). The magnitude range was chosen so that all the transitions in the graph have a non-negligible chance of being observed. A detailed description of the physical model is given in the Appendix, section 5.8.

We implemented this model by a MME. The state $q(t)$ represents the expert used up to time t and $a_{ij}(t)$ the probability of using expert i between t and $t + 1$ given that $q(t) = j$.

Implementation of the experts f: The experts $f(u, \theta_k)$ were chosen to be linear in the parameters, corresponding to the piecewise linearity of the true model (see the Appendix). Linearity is achieved by introducing the additional input variables v_x/v_y, v_y/v_x, $x\, v_y/v_x$, $y\, v_x/v_y$, each of them having a new parameter as coefficient. But each additional variable affects only one of the six experts (see the Appendix). The increase in the number of parameters is insignificant. To avoid infinite input values the values of v_x, v_y were lower-bounded by the small constant $V_{min} = 1/50$.

Implementation of the transition probability matrix: To implement the transition matrix A we used a bank of *gating networks*, one for each column of A. Examining Figure 5.1 it is easy to see that there exist experts that share the same final state of contact (for instance, **A stick** and **A slide** both represent movements whose final position is on surface **A**). Since transition probabilities depend only on the final position the columns of the matrix A corresponding to these experts are equal. This brings the number of distinct gating networks to 4.

The boundaries of the decision regions are curved surfaces, so that to implement each of them we used a two layer perceptron with softmax[5] output, as presented in Figure 5.3. The number of hidden units in each gating net was chosen considering the geometry of the decision regions to be learned.

---

[5] See section 5.9 for the definition of the softmax function.

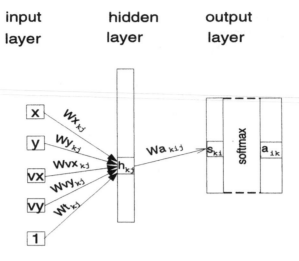

Figure 5.3   Gating network computing the transition probabilities a_{ik} for state of contact k.

5.5.2   Training and testing criteria

The training set consists of $N = 5000$ data points, in sequences of length $T \leq 6$, all starting in the free space. The starting position of the sequence and the nominal velocities at each step were picked randomly. It was found that for effective learning it is necessary that the state frequencies in the training set are roughly equal. The distribution over velocities and the sequence length T were chosen so as to meet this requirement. The (x, y) values obtained by simulation were corrupted with additive Gaussian noise with standard deviation σ.

In the M step, the parameters of the gating networks were updated by gradient ascent, with a fixed number of epochs for each M step. Section 5.10 shows how the gradient was computed. For the movement models least squares estimation was used. To ensure that models and gates are correctly coupled, initial values for θ are chosen around the true values. In the present case, this is not an unrealistic assumption. W was initialised with small random values. Because of the long time to convergence, only a small number of runs have been performed. The observed variance of the results over test sets was extremely small, so that the values presented here can be considered as typical.

Three criteria were used to measure the performance of the learning algorithm: experts' parameters deviation from the true values, square root of prediction MSE and hidden state misclassificaton. Because training was performed on a distribution that is not expected to appear in practice, all the models were tested on both the training distribution and on a distribution which is uniform over (v_x, v_y) (and therefore highly non-uniform over the states of contact), subsequently called the "uniform V distribution".

Performance comparisons were made with a ME architecture having identical experts but only one gating net. For the number of hidden units in it, various values have been tried; the results presented here represent the performance of the best model obtained. The performance of a k-nearest neighbour (k-NN) model is also shown.

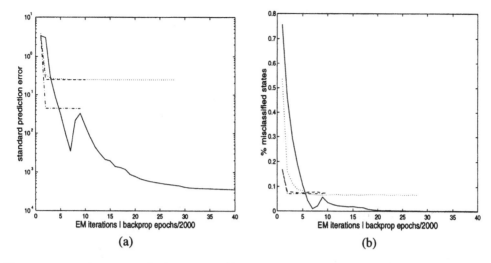

Figure 5.4 Prediction standard error (MSE$^{1/2}$) (a) and percentage of wrong expert choices (misclassified states) (b) on the training set during learning for the MME and ME models. The abscissa was scaled according to the number of back-propagation epochs for the two models. — MME, no noise; $\cdots$ MME, noise 0.2; $-\cdot-$ ME, no noise; $--$ ME, noise 0.2.

Table 5.3 Performance (MSE$^{1/2}$) for the k-NN method. Memorised (training) set: size = 11,000; noise = σ_m. Test set: size = 11,000; noise = σ_t.

σ_m	k	$\sigma_t = 0$	$\sigma_t = 0.2$
0	100	0.333762	0.425333
0.2	200	0.341495	0.455458

5.5.3 Results

Learning curves. Figure 5.4 presents the learning curves for two MME and two ME models. The horizontal axes show the number of EM epochs. Since the number of back-propagation epochs for each M step is 2000, the number of training epochs for the ME models was scaled down accordingly. It is visible that the ME models converge faster than the MME models. The difference may be due to the somehow arbitrary choice of the number of epochs in the M step; the MME model also contains more parameters in the gating nets (220 vs ME's 180 parameters). The experts contain 64 parameters.

The only model which achieves 0 misclassification error is the MME trained without noise. For training in noise, the final prediction error on the training set is the same for both architectures, but the results on test sets will show a difference favouring the MME model.

Comparison with k-NN. The k-NN method performed much worse than the other two algorithms. Therefore here (Table 5.3) only a summary of the best results obtained is presented.

Test set performance. Figure 5.5 presents the test set performance of several trained models. Each test set contained $\geq$ 50,000 datapoints. The input noise added to (x, y) was

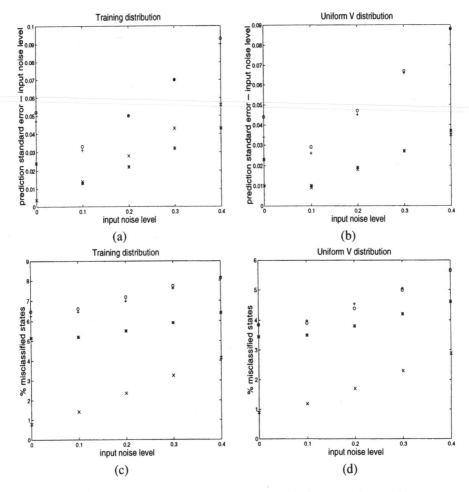

Figure 5.5 Test set performance of the MME and ME models for various levels of the input noise σ_{in}. Prediction $MSE^{1/2} - \sigma_{in}$ for the uniform V distribution (a) and for the training distribution; the percentage of misclassified states for the uniform V (c) and training distribution (d). × – MME, no noise; * – MME noise 0.2; ○ – ME, no noise; + – ME, noise 0.2.

Gaussian with variance between 0 and $(0.4)^2$. It can be seen that the ME models perform similarly, w.r.t. both prediction error and state classification. The two MME models perform better than the ME models, with a significant difference between the model trained with noise and the model trained with ideal data. The results are consistent over distributions, noise levels and performance criteria.

Parameter error. In the present problem the true parameters for the experts can be computed exactly. The values of the MSE of the learned parameters in various models are shown in Table 5.4 (where σ represents the noise level in the training set):

Prediction along a trajectory. To illustrate the behaviour of the algorithm in closed loop, Figure 5.6 presents a sample trajectory and the predictions of the MME model in open and closed loop mode. In the latter, the predicted and measured positions were averaged (with ratios 1/2) to provide the models' (x, y) input for the next time step.

Table 5.4 MSE of experts parameters for various ME and MME models. σ represents the noise level in the training set.

Model	σ noise	$\mathrm{MSE}_\theta^{1/2}$
MME	0	0
MME	0.05	0.0474
MME	0.2	0.0817
ME	0	0.1889
ME	0.2	0.1889

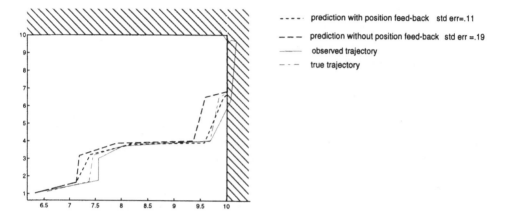

Figure 5.6 Sample trajectory in the (x, y) space, together with the open and closed loop predictions of the MME model. The noise level is 0.2.

The simulations show that, although input noise is not explicitly taken into account by the model, the MME architecture is tolerant to it and is able to achieve both learning and good prediction performance in noisy conditions.

The comparison with k-NN confirms that the problem is not simple and that well tailored algorithms are required to solve it. MME outperforms ME in all situations, with a small additional computational effort during training, showing the advantage yielded by taking into account the time-dependencies in the data.

5.6 DISCUSSION

The learning algorithm and the architecture presented here allow us to model a wide range of dynamic systems. Theoretically any finite state discrete time system can be represented in the form of a MME, and continuous state spaces can be often conveniently discretised. But the aim of using a MME is to take advantage of the natural modularity of the modelled process and to divide it into subprocesses of significantly lower complexity.

In the special case when the MME reduces to a mixture of experts, the probability of a state depends only on the input u, and it is easy to see (consult also (Jordan and Jacobs 1994)) that the gating network partitions the input space, assigning to each expert a subset of it. As the domains of the experts are generally contiguous, the mixture of experts can be viewed as a combination of local models, each being accurate in a confined region of the input space.

From the same point of view, the MME is dividing the data between *operating regimes* which are locally accurate models of dynamic processes. The partition performed is twofold: the dynamic component is confined to the level of the Markov chain and, at the same level, the set of input–output pairs is partitioned into subsets which are assigned to the static experts. The former is ensured by the structure of the model, the latter has to be learned during the E step of the EM algorithm.

This suggests that a critical condition for accurate learning is the correct assignment of the data points to the experts responsible for them. This is basically a clustering problem. Therefore, although the learner is provided with "input–output" pairs, it is essentially performing an alternation of supervised and unsupervised learning tasks.

5.6.1 Computational aspects

Local maxima. The EM algorithm is guaranteed to converge to a local maximum of the likelihood. This may not be sufficient when the value of the likelihood at the local maximum is much smaller than the value at the global maximum. Another issue is that in some cases we may be in search of parameters with physical meaning. In these cases, we need to find the (local) maximum that is closest to the parameters of the data generating process.

The first issue can be addressed by performing several runs of the learning algorithm, each starting from a different initial point in the parameter space, and by choosing at the end the most likely solution. To address the second issue, besides validating the values of the parameters after convergence, prior knowledge about the process can be used to find a good initial estimate.

Influence of f and a_{ij}. We have imposed no conditions on the form of f and a_{ij} so far. They influence the success of learning in several ways:

1. They determine the difficulty of the Maximisation step, as discussed in 5.3.2.

2. f influences the difficulty of the clustering problem; a linear f reduces it to Gaussian clustering, whereas if the class of allowed I/O functions becomes too rich or if the mapping $\theta \to f(\cdot, \theta)$ is not one-to-one the assignment problem can become ambiguous.

3. The form of f and a_{ij} determines the number of parameters[6] and thereby the number of data points necessary for accurate learning.

4. By 2. and 3. the shape of the likelihood function is influenced and subsequently the number and distribution of the local maxima.

Number of states. So far we have assumed that m, the number of experts, is known. This is not always the case in practice. Simulations suggest that starting with a large enough m could be a satisfactory strategy. Sometimes the superfluous clusters are automatically "voided" of data points in the process of learning. This can be attributed to the fact that, for any fixed model structure, maximising likelihood is equivalent to minimising the description length of the data (Rissanen 1978).

On the other hand, the number of parameters in A increases proportionally to m^2, increasing both the computational burden and the *data complexity*. The clustering becomes also harder for a large number of alternatives. Therefore it is important to know m or to have a good upper bound on it.

Data complexity. The data complexity of a model is, loosely speaking, the number of data points required to attain a certain level of the prediction error. In the case where we

---

[6]More precisely, the *complexity* of the model.

can talk about true parameters, the prediction error is reflected by the accuracy of the parameters. For a more rigorous and detailed discussion of data complexity consult (Geman *et al.* 1992) and its references. For the purposes of this paper it is sufficient to state that for a given class of models, the data complexity increases with the number of parameters. As a consequence, an increase in m will induce a quadratic increase in the number of parameters and in the training set as well.

Simulations have shown that training this architecture requires large amounts of data. Therefore it is necessary to use prior knowledge to reduce the number of parameters. This and other uses of knowledge outside the data set for constraining the model are the topics of the next subsection.

5.6.2 Incorporating prior knowledge

Prior knowledge is extremely valuable in any practical problem. Although the ML paradigm doesn't provide a systematic way of incorporating it, prior knowledge can and should be used at several levels of the model.

First, it will help in finding the appropriate model structure. An important structural parameter is the number of states of the Markov chain. For instance, knowing the number of the movement models in the fine motion task has allowed us to pair each expert with a movement model, thus having simple linear experts whose parameters have a physical meaning. For systems that are not intrinsically discrete-time, prior knowledge should guide the choice for an appropriate discretisation step. A too long discretisation step may miss transitions between states leading thus to poor modelling, whereas a too short one will cause unnecessarily low transition probabilities between different states.

Closely related to the choice of the number of states is the choice of the functional form of the experts. This has been shown in the modelling of the robot arm motion. Prior knowledge should help find the class of functions f with the suitable complexity. This issue, as well as the issue of the best representation for the inputs and outputs, arise in any modelling problem.

A similar selection concerns the form of the functions in A. In view of condition (5.2) it is often reasonable to group the transition probabilities into gating networks as we have done in section 5.5 thus having a vector of parameters W_j for each set of functions $\{a_{ij} | i = 1, \ldots, m\}$. If it is known *a priori* that certain transitions cannot occur, then their probabilities can be set to 0, thereby reducing both the complexity of the gating network and speeding up the learning process by reducing uncertainty in the state estimation. For instance, in the fine motion problem, when the state space grows by adding new objects to the environment, the number of possible transitions from a state of contact (also called the *branching factor*) will be limited by the number of the states of contact which are within a certain distance and can be reached by a simple motion, which will grow much more slowly than m. Therefore, the number of nonzero elements in A will be approximately proportional to m. This is generally the case when transitions are local, providing another reason why locality is useful.

Finally, when the experts have physical meaning, prior knowledge should give "good" initial estimates for the experts parameters. This is important not only to decrease the learning time, but also to avoid convergence to spurious maxima of the likelihood.

Relearning. Certain aspects of relearning have been discussed at the end of the previous section. (By relearning we mean learning a set of parameters, under the assumption that another learning problem has already been solved and that the old model is identical in structure to the new one. Sometimes it is also assumed that the old and new parameter values are "close".) Since EM is a batch learning algorithm, it cannot automatically adapt to changes in the parameters of the data generating process. But knowledge exterior to the data set (that was called prior knowledge in the above paragraphs) can still be used to make relearning easier than learning from scratch. This process can be viewed as the reverse of incorporating prior knowledge: we must now (selectively) discard prior knowledge before learning from the new data begins. Obviously, the model structure remains unchanged; the question is whether to keep some of the parameter values unchanged as well. The answer is yes in the case when we know that only some of the experts or gating nets have suffered changes, in other words when the change was local. In this case the learning algorithm can be easily adapted to *locally relearn*, i.e. to re-estimate only a subset of the parameters.

5.7 CONCLUSION

To conclude, the MME architecture is a generalisation of both HMMs and the Mixture of Experts architecture. From the latter it inherits the ability to take advantage of the natural decomposability of the processes to model, when this is the case, or to construct complex models by putting together simple local models otherwise. The embedded Markov chain allows it to account for simple dynamics in the data generation process.

The parameters of the MME can be estimated by an iterative algorithm which is a special case of the EM algorithm. As a consequence, the *a posteriori* probabilities of the hidden states are also computed. Multiple aspects pertaining to the implementation of the algorithm have been discussed. The trained model can be used as a state estimator or for prediction as the forward model in a recursive estimation algorithm.

5.8 APPENDIX: EXACT MOVEMENT MODEL FOR THE FINE MOTION EXAMPLE

Here the equations describing the exact movement model of the point in the 2D space shown in Figure 5.1 are given. First, some notation. The position of the point at the current time is denoted by (x, y) and its position after moving for a time T_s with constant velocity vector (v_x, v_y) by (x^{new}, y^{new}).

The static and dynamic friction coefficients on surfaces A and B are μ_A^s, μ_A^d, μ_B^s, μ_B^d; we will consider that $x \in [0, X]$, $y \in [0, Y]$. With the movement model notation introduced in Figure 5.1 the movement equations are:

C:

$$x^{new} = x + v_x T_s \tag{5.19}$$
$$y^{new} = y + v_y T_s. \tag{5.20}$$

A slide:

$$x^{new} = x + \text{sgn}(v_x)\text{sgn}(v_y)\mu_A^d(Y - y) + T_s(v_x - \mu_A^d v_y) \tag{5.21}$$
$$y^{new} = Y. \tag{5.22}$$

A stick:

$$x^{new} = x + v_x \frac{Y - y}{v_y} \tag{5.23}$$
$$y^{new} = Y. \tag{5.24}$$

B slide:

$$x^{new} = X \tag{5.25}$$

$$y^{new} = y + \text{sgn}(v_x)\text{sgn}(v_y)\mu_B^d(X - x) + T_s(v_y - \mu_B^d v_x). \tag{5.26}$$

B stick:

$$x^{new} = X \tag{5.27}$$

$$y^{new} = y + v_y\frac{X - x}{v_x}. \tag{5.28}$$

G:

$$x^{new} = X \tag{5.29}$$

$$y^{new} = Y. \tag{5.30}$$

Equations (5.21–5.28) are slightly more general than needed for the present case. For velocities restricted to v_x, $v_y \geq V_{min} > 0$ the sgn functions are 1 and can be dropped, yielding all equations but (5.23) and (5.28) linear.

These can also be brought to a linear form by introducing the input variables v_x/v_y, v_y/v_x, $(x\,v_y)/v_x$, $(y\,v_x)/v_y$. The coefficients of the extended set of variables are denoted by $\theta_x^C, \ldots, \theta_{v_y/v_x}^{\text{Bstick}}, \ldots \theta_{v_y}^G$ and represent the parameters of the model to be learned.

The domain of each movement model is the region in the 4-dimensional space where the given model is valid. This region is defined in terms of a set of inequalities which must be satisfied by all its points. For example, the (preimage of) free space C is the set of points for which $0 < x^{new} < X$, $0 < y^{new} < Y$. If we define the functions:

$$d1(x, y, v_x, v_y) = x + v_x T_s \tag{5.31}$$

$$d2(x, y, v_x, v_y) = y + v_y T_s$$

the following is an equivalent definition for the domain of C :

$$0 < d1 < X$$
$$0 < d2 < Y$$

but now the functions $d1$ and $d2$ can also be used in inequalities defining the domains of neighbouring movement models. The complete list of functions which define the domain boundaries is:

$$d1 = x + v_x T_s$$
$$d2 = y + v_y T_s$$
$$d3 = v_x - \mu_A^s v_y$$
$$d4 = v_y - \mu_B^s v_x$$
$$d5 = x + \frac{Y-y}{v_y} v_x$$
$$d6 = y + \frac{X-x}{v_x} v_y$$
$$d7 = x + \text{sgn}(v_x)\,\text{sgn}(v_y)\,\mu_A^d(Y - y) + T_s(v_x - \mu_A^d v_y)$$
$$d8 = y + \text{sgn}(v_x)\,\text{sgn}(v_y)\,\mu_B^d(X - x) + T_s(v_y - \mu_B^d v_x)$$

In addition, for all the movement models hold the inequalities:

$$0 \leq x \leq X$$
$$0 \leq y \leq Y$$
$$V_{min} \leq v_x$$
$$V_{min} \leq v_y$$
$$|v| \leq V_{max}$$

It can be seen that the model boundaries are not linear in the inputs. Even if made linear by augmenting the set of inputs as above, some domains are not convex sets and others, although convex cannot be represented as the result of a tessellation with softmaxed hyperplanes.

5.9   THE SIGMOID AND SOFTMAX FUNCTIONS

The definition of the *sigmoid* function used in the present example is

$$\text{sigmoid}(x) \equiv h(x) = \tanh(x) \equiv \frac{e^x - e^{-x}}{e^x + e^{-x}}, \tag{5.32}$$

where x is a real scalar. The sigmoid function maps the real line into the interval (-1,1). Its derivative w.r.t. the variable x can be expressed as

$$h'(x) = 1 - h^2(x).$$

The *softmax function* is the multivariate analogue of the sigmoid. It is defined on $\mathcal{R}^n \to \mathcal{R}^m$ by:

$$\text{softmax}_i(x, W) \equiv g_i(x, W) = \frac{\exp(W_i^T x)}{\sum_j \exp(W_j^T x)}, \quad i = 1, ..m$$

with W, x vectors of the same dimension.) If we make the notation $s_i = W_i^T x$, then the partial derivatives of the softmax function are given by

$$\frac{\partial g_i}{\partial s_j} = g_i(\delta_{ij} - g_j) \ \forall i, j, \tag{5.33}$$

where δ_{ij} represents Kronecker's symbol.

5.10   COMPUTING THE GRADIENT

Here we show how to compute the derivatives used in the generalised M-step of the EM algorithm.

To obtain the new iterates of the parameters of the gating networks, one has to maximise the function given by (5.12) w.r.t. to W

$$J_W = \sum_{t=0}^{T-1} \sum_{ij} \xi_{ij}(t) \ln \left(a_{ij}(u(t), W) \right).$$

In the present implementation, the functions a_{ij} are computed by a two layer perceptron with softmax output. The complete calculation is given below, and the notation is explained by Figure 5.3.

$$a_{ik}(u(t), W) \equiv a_{ik}(t) = \text{softmax}_i(h_k, \text{Wa}_k)$$

$$h_{kj}(t) = \text{sigmoid}[\text{Wx}_{kj} x(t) + \text{Wy}_{kj} y(t) + \text{Wvx}_{kj} v_x(t) + \text{Wvy}_{kj} v_y(t) + \text{Wt}_{kj}],$$

where $k = 1, ..., m$, $i = 1, ..., m'_k$, $j = 1, ..., n_k+1$ and m, m'_k, n_k represent the number of states, of outputs and of hidden units for gate k, respectively. The parameter matrices and vectors Wx, Wy, Wvx, Wvy, Wt and Wa have the appropriate dimensions. The last unit in each hidden layer is fixed to 1 to produce the effect of a threshold in s_k.

Replacing the derivatives of the softmax and using the fact that

$$\sum_i \xi_{ij}(t) = \gamma_j(t) \quad \forall t, j$$

we obtain the following formulas for the partial derivatives of J_W (where $j \to k$ means that the final state of contact of expert j is k):

$$\frac{\partial J_W}{\partial \mathrm{Wa}_{kij}} = \sum_t \sum_l \frac{\partial J_W}{\partial a_{lk}(t)} \frac{\partial a_{lk}(t)}{\partial \mathrm{Wa}_{kij}}$$

$$= \sum_t \sum_l \frac{\sum_{j' \to k} \xi_{lj'}(t)}{a_{lk}(t)} \left[-a_{lk}(t)(a_{ik}(t) - \delta_{il})h_{kj}(t) \right]$$

$$= \sum_t h_{kj}(t) \left[\sum_{j' \to k} \xi_{ij'}(t) - a_{ik}(t) \sum_{j' \to k} \gamma_{j'}(t) \right]$$

$$\frac{\partial J_W}{\partial \mathrm{Wx}_{kj}} = \sum_t \sum_l \frac{\partial J_W}{\partial a_{lk}(t)} \frac{\partial a_{lk}(t)}{\partial h_{kj}(t)} \frac{\partial h_{kj}(t)}{\partial \mathrm{Wx}_{kj}}$$

$$= \sum_t h'_{kj}(t)x(t) \sum_i \mathrm{Wa}_{kij} \left[\sum_{j' \to k} \xi_{ij'}(t) - a_{ik}(t) \sum_{j' \in k} \gamma_{j'}(t) \right].$$

The experts f are linear in the parameters, therefore their derivatives computation is straightforward.

REFERENCES

BAUM, L. E., T. PETRIE, G. SOULES AND N. WEISS (1970) 'A maximization technique occurring in the statistical analysis of probabilistic functions of Markov chains'. *Ann. Math. Stat.* **41**, 164–171.

BENGIO, Y. AND P. FRASCONI (1995) An input output HMM architecture. In J. D. Covan, G. Tesauro and J. Alspector (Eds.). 'Neural Information Processing Systems - 7'. MIT press. Cambridge, MA. pp. 427–434.

DEMPSTER, A. P., N. M. LAIRD AND D. B. RUBIN (1977) 'Maximum likelihood from incomplete data via the EM algorithm'. *Journal of the Royal Statistical Society, B* **39**, 1–38.

GEMAN, S., ELIE BIENENSTOCK AND RENE DOURSAT (1992) 'Neural networks and the bias/variance dilemma'. *Neural Computation* **4**, 1–58.

JORDAN, M. I. AND R. A. JACOBS (1994) 'Hierarchical mixtures of experts and the EM algorithm'. *Neural Computation* **6**, 181–214.

MACKAY, D. J. (1992) 'Bayesian interpolation'. *Neural Computation* **4**(3), 415–447.

MASON, M. T. (1981) 'Compliance and force control for computer controlled manipulation'. *IEEE Trans. on Systems, Man and Cybernetics.*

MCCULLAGH, P. AND J. A. NELDER (1983) *Generalized Linear Models.* Chapman & Hall, London.

MEILĂ, M. AND M. I. JORDAN (1996) Learning fine motion by Markov mixtures of experts. In D. Touretzky, M. Mozer and M. Hasselmo (Eds.). 'Advances in Neural Information Processing Systems 8'.

MEILĂ, M., I. TĂBUŞ AND M. I. JORDAN (1994) System identification using hidden Markov models with auxiliary input. In 'Proceedings of the Control and Computers Conference'. Timişoara. pp. 121–128.

NEAL, R. M. AND G. E. HINTON (1993) 'A new view of the EM algorithm which justifies incremental and other variants'. submitted to *Biometrika*.

RABINER, R. L. AND B. H. JUANG (1986) 'An introduction to hidden Markov models'. *ASSP Magazine* **3**(1), 4–16.

RISSANEN, J. (1978) 'Modeling by shortest data description'. *Automatica* **14**, 465–471.

SÖDERSTRÖM, T. AND P. STOICA (1989) *System Identification*. Prentice Hall, Englewood Cliffs, NJ.

CHAPTER SIX

Active Learning with Mixture Models

DAVID A. COHN, ZOUBIN GHAHRAMANI and MICHAEL I. JORDAN

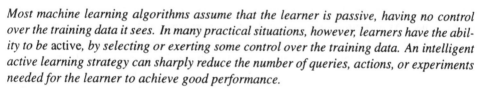

Most machine learning algorithms assume that the learner is passive, having no control over the training data it sees. In many practical situations, however, learners have the ability to be active, *by selecting or exerting some control over the training data. An intelligent active learning strategy can sharply reduce the number of queries, actions, or experiments needed for the learner to achieve good performance.*

Recent years have seen a great number of heuristic approaches to active learning, each with the goal of providing an efficient means for gathering training data. In this chapter, we show that for statistically based machine learning algorithms, one can compute the "optimal" way to select training data. We describe one such model, the mixture of Gaussians, in which the data is modelled locally by Gaussians distributions. We then show how to efficiently select queries or actions which minimise the expected variance of the model's estimate, resulting in a learner which significantly outperforms learners which gather data randomly or heuristically.

6.1 ACTIVE LEARNING – BACKGROUND

The goal of machine learning is to create systems that can improve their performance at some task as they acquire experience or data. In most machine learning research, the learner is a passive recipient of this data. This approach ignores the fact that, in many situations, the learner's most powerful tool is its ability to act, gather data, and influence the world it is trying to understand. For example, the learner may be able to select joint angles or torques to learn the kinematics or dynamics of a robot arm, select the locations for sensor measurements to identify and locate buried hazardous wastes, or query a human expert to classify an unknown word in a natural language understanding problem. Research in active learning attempts to determine how this selection can be used most effectively.

Formally, active learning is a closed-loop phenomenon in which the learner selects actions or makes queries which influence what data are added to its training set. When actions/queries are selected properly, the data requirements for some problems decrease drastically, and some NP-complete learning problems become polynomial in computation time (Anguin 1988, Baum and Lang 1991). In practice, active learning offers its greatest rewards in situations where data are expensive or difficult to obtain, or when the environment is complex or dangerous. In industrial settings, each training point may take days to gather and cost thousands of dollars; a method for optimally selecting these points could offer enormous savings in time and money.

6.1.1   Goals for the active learning system

There are several goals that an active learning system may wish to achieve. One form of active learning falls under the rubric of *optimisation*, in which the goal of the system is to select inputs to maximise or minimise some response variable. For example, the task of the learner may be to select drill sites to maximise petroleum yield. There is an extensive literature on optimisation, examining both cases where the learner knows the functional form of the response variable (e.g. that it is some polynomial function of its inputs), and cases in which it has no such knowledge. The latter case is covered by the *stochastic optimisation* literature.

A closely related form of active learning is adaptive control, where the goal is to learn a control policy for an unknown system. Control problems present the added complication that the consequence of a specific action may not be known until many time steps after it is taken. Therefore, to select actions optimally the learner must account for its expected future performance. As in optimisation, one is also often concerned with performing well *during* the learning task, and must trade off exploitation of the current policy for exploration which may improve it. The subfield of dual control (Fe'ldbaum 1965) is specifically concerned with finding an optimal balance of exploration and control while learning.

In this chapter, we will restrict ourselves to examining the problem of *supervised* active learning. The goal is to learn a mapping $X \rightarrow Y$ based on a set of (potentially noisy) training examples $\mathcal{D} = \{(x_i, y_i)\}_{i=1}^m$, where $x_i \in X$ and $y_i \in Y$. For the robot control problem, this mapping may be $state \times action \rightarrow new\ state$; for the hazard location problem, it may be $sensor\ reading \rightarrow target\ position$. Supervised active learning has been examined extensively in the classical statistics literature as the problem of *optimal experiment design* (Fedorov 1972).

6.1.2   Selecting actions

Within the framework of supervised active learning, there are several ways in which the learner may be able to select the data it obtains. Under some circumstances, the selection process is *unconstrained*: the learner can completely specify arbitrary inputs to be queried. Often, however, the selection process may be *constrained* by the environment, i.e., the set of possible queries may depend on the state of the environment. For example, in the robot learning problem, $state \times action \rightarrow new\ state$, the learner may be free to select the next *action* component of its input, but the *state* component is determined by its previous *state* and *action*. In these cases, the learner must traverse a trajectory through the input domain to learn the input–output mapping.

In some cases, the input selection process takes the form of *query filtering* (Cohn *et al.* 1994, Freund *et al.* 1993, Lewis and Catlett 1994). The learner is presented with a set or stream of allowable queries, and must choose some subset of them. An example of this would be monitoring a stream of unlabelled speech and deciding what utterances to ask a human expert to classify – rather than being able to construct interesting examples itself, the learner must wait for them to appear in the input stream. Finally, one branch of active learning is concerned with obtaining a concise training set by filtering an existing data set, under the assumption that the learner has both the inputs and outputs available to it at the time of selection (Plutowski and White 1993).

In this chapter, we will assume that the learner can iteratively select a new input $\tilde{x}$ (possibly from a constrained set), observe the resulting output $\tilde{y}$, and incorporate the new example $(\tilde{x}, \tilde{y})$ into its training set. Depending on the research field and problem domain, the act of selecting an $\tilde{x}$ to be labelled may be thought of as "making a query", "performing an experiment", or "taking an action". The question we will be concerned with is how to choose which $\tilde{x}$ to try next.

There are many heuristics the learner may use for choosing $\tilde{x}$, including choosing places where it doesn't have data (Whitehead 1991), performs poorly (Linden and Weber 1993), has low confidence (Thrun and Möller 1992), expects a change in its model (Cohn *et al.* 1990, Cohn *et al.* 1994), or has previously found data that resulted in learning (Schmidhuber and Storck 1993). In this chapter we will consider how one may select $\tilde{x}$ in a statistically "optimal" manner for some classes of machine learning algorithm. In the following section we introduce the approach and briefly review how it has been applied to neural networks (MacKay 1992, Cohn 1996). In Section 6.3 we consider an alternative, statistically based learning architecture: the mixture of Gaussians. Section 6.4 presents the empirical results of applying statistically based active learning to this architecture. While optimal data selection for a neural network is computationally expensive and approximate, we find that optimal data selection for the mixture of Gaussians is both efficient and accurate.

6.2 ACTIVE LEARNING – A STATISTICAL APPROACH

We begin by defining $P(x, y)$ to be the unknown joint distribution over x and y, and $P(x)$ to be the known marginal distribution of x (commonly called the *input distribution*). Given input x and training set $\mathcal{D}$, we denote the learner's output $\hat{y}(x; \mathcal{D})$.[1] We can then write the expected error of the learner as follows:

$$\int_x E_T\left[\left(\hat{y}(x; \mathcal{D}) - y(x)\right)^2 | x\right] P(x) dx, \tag{6.1}$$

where $E_T[\cdot]$ denotes expectation over $P(y|x)$ and over training sets $\mathcal{D}$. The expectation inside the integral may be decomposed as follows (Geman *et al.* 1992):

$$
\begin{aligned}
E_T\left[\left(\hat{y}(x; \mathcal{D}) - y(x)\right)^2 | x\right] &= E\left[(y(x) - E[y|x])^2\right] \\
&\quad + \left(E_{\mathcal{D}}\left[\hat{y}(x; \mathcal{D})\right] - E[y|x]\right)^2 \\
&\quad + E_{\mathcal{D}}\left[\left(\hat{y}(x; \mathcal{D}) - E_{\mathcal{D}}[\hat{y}(x; \mathcal{D})]\right)^2\right],
\end{aligned}
\tag{6.2}
$$

where $E_{\mathcal{D}}[\cdot]$ denotes the expectation over training sets $\mathcal{D}$ and the remaining expectations on the right-hand side are expectations with respect to the conditional density $P(y|x)$. It is important to remember here that in the case of active learning, the distribution of $\mathcal{D}$ may differ substantially from the joint distribution $P(x, y)$.

The first term in equation (6.2) is the variance of y given x – it is the *noise* in the distribution, and does not depend on the learner or on the training data. The second term is the learner's *squared bias*, and the third is its *variance*; these last two terms comprise the

---

[1] In this section, we present our equations in the univariate setting for clarity. In the following sections, we use multivariate notation for completeness.

mean squared error of the learner with respect to the underlying regression function $E[y|x]$. When the second term of equation (6.2) is zero, we say that the learner is *unbiased*. We shall assume that the learners considered in this chapter are approximately unbiased; that is, that their squared bias is negligible when compared with their overall mean squared error.[2] Thus we focus on algorithms that minimise the learner's error by minimising its variance:

$$\sigma_{\hat{y}}^2 \equiv E_{\mathcal{D}}\left[\left(\hat{y}(x; \mathcal{D}) - E_{\mathcal{D}}[\hat{y}(x; \mathcal{D})]\right)^2\right].\tag{6.3}$$

(For readability, we will drop the explicit dependence on x and $\mathcal{D}$ in the remainder of this chapter – unless denoted otherwise, $\hat{y}$ and $\sigma_{\hat{y}}^2$ are functions of x and $\mathcal{D}$.)

It is worth emphasising that equation (6.3) is the learner's variance at a given x. To compute the learner's average variance over the domain, we must integrate over X. In practice, we will compute a Monte Carlo approximation of this integral, evaluating $\left\langle \sigma_{\hat{y}}^2 \right\rangle$ at a number of unlabelled *reference points* drawn according to $P(x)$.

6.2.1   Selecting data to minimise learner variance

In an active learning setting, the learner chooses the x component of the data; the variance therefore depends on the variability in y given the chosen x component of our training set $\mathcal{D}$. We indicate this by rewriting equation (6.3) as

$$\sigma_{\hat{y}}^2 = \left\langle \left(\hat{y} - \langle\hat{y}\rangle\right)^2 \right\rangle,$$

where $\langle\cdot\rangle$ denotes $E_{\mathcal{D}}[\cdot]$ given a fixed x-component of $\mathcal{D}$. When a new input $\tilde{x}$ is selected and queried, and the resulting $(\tilde{x}, \tilde{y})$ added to the training set, $\sigma_{\hat{y}}^2$ should change. We will denote the expectation (over values of $\tilde{y}$) of the learner's new variance as

$$\left\langle \tilde{\sigma}_{\hat{y}}^2 \right\rangle = E_{\mathcal{D}\cup(\tilde{x},\tilde{y})}\left[\sigma_{\hat{y}}^2 | \tilde{x}\right].\tag{6.4}$$

In the multivariate setting, x and y are replaced by the vectors $\mathbf{x}$ and $\mathbf{y}$, and $\sigma_{\hat{y}}^2$ is replaced by the estimator's output covariance matrix $\Sigma_{\hat{y}}$. Then, we write equation (6.4) as

$$\left\langle \tilde{\Sigma}_{\hat{y}} \right\rangle = E_{\mathcal{D}\cup(\tilde{\mathbf{x}},\tilde{\mathbf{y}})}\left[\Sigma_{\hat{y}} | \tilde{\mathbf{x}}\right].\tag{6.5}$$

We wish to select data to minimise the value of some function of equation (6.5), integrated over X. The most common functions are the determinant and matrix trace, though many other choices are possible (Atkinson and Donev 1992). We will attempt to minimise the trace of equation (6.5), which will minimise the expected value of $(\hat{\mathbf{y}} - \mathbf{y})^T (\hat{\mathbf{y}} - \mathbf{y})$.

Intuitively, the minimisation proceeds as follows: we assume that the learner has an estimate of $\Sigma_{\hat{y}}$, its output variance at $\mathbf{x}$. If, for some new input $\tilde{\mathbf{x}}$, we knew the conditional distributions $P(\mathbf{y}|\mathbf{x})$ and $P(\tilde{\mathbf{y}}|\tilde{\mathbf{x}})$, we could compute an estimate of the learner's new variance at $\mathbf{x}$ given the additional example at $\tilde{\mathbf{x}}$. While the true distributions of $\mathbf{y}$ and $\tilde{\mathbf{y}}$ are unknown, many learning architectures provide approximations to them. Using these estimated distributions, we can estimate $\left\langle \tilde{\Sigma}_{\hat{y}} \right\rangle$, the expected variance of the learner after querying at $\tilde{\mathbf{x}}$.

---

[2]In fact, only a weaker assumption is needed: we must only assume that our selection of training data does not adversely affect the learner's bias. This is one of the assumptions that is tested by the experiments described in section 6.4. One can also test this assumption using cross-validation on the training set to determine whether or not the bulk of the learner's error is due to variance.

Given the estimate of $\left\langle \tilde{\Sigma}_{\hat{y}} \right\rangle$, which applies to a given $\mathbf{x}$ and a given query $\tilde{\mathbf{x}}$, we must integrate $\mathbf{x}$ over the input distribution to compute the integrated average variance of the learner. Here, we will again compute a Monte Carlo approximation of this integral, evaluating $\left\langle \tilde{\Sigma}_{\hat{y}} \right\rangle$ at a number of reference points drawn according to $P(\mathbf{x})$. By querying an $\tilde{\mathbf{x}}$ that minimises the average expected variance over the reference points, we have a solid statistical basis for choosing new examples.

6.2.2 Example: active learning with a neural network

In this section we review the use of techniques from optimal experiment design (OED) to minimise the estimated variance of a neural network (Fedorov 1972, MacKay 1992, Cohn 1996). We will assume we have been given a learner $\tilde{\mathbf{y}} = f_{\hat{\mathbf{w}}}()$, a training set $\mathcal{D} = \{(\mathbf{x}_i, \mathbf{y}_i)\}_{i=1}^{m}$ and a parameter vector estimate $\hat{\mathbf{w}}$ that maximises some likelihood measure given $\mathcal{D}$. If, for example, one assumes that the data were produced by a process whose structure matches that of the network, and that noise in the process outputs is normal and independently identically distributed, then the negative log likelihood of $\hat{\mathbf{w}}$ given $\mathcal{D}$ is proportional to

$$S^2 = \frac{1}{m} \sum_{i=1}^{m} (\mathbf{y}_i - \hat{\mathbf{y}}(\mathbf{x}_i))^T (\mathbf{y}_i - \hat{\mathbf{y}}(\mathbf{x}_i)) .$$

The maximum likelihood estimate for $\hat{\mathbf{w}}$ minimises this measure S^2.

The estimated output variance of the network is related to S^2 by

$$\Sigma_{\hat{y}} \approx S^2 \left(\frac{\partial \hat{\mathbf{y}}(x)}{\partial \mathbf{w}} \right)^T \left(\frac{\partial^2 S^2}{\partial \mathbf{w}^2} \right)^{-1} \left(\frac{\partial \hat{\mathbf{y}}(x)}{\partial \mathbf{w}} \right) , \quad \text{(MacKay 1992)}$$

where the true variance has been approximated by a truncated Taylor series expansion around S^2. This estimate makes the assumption that $\partial \hat{\mathbf{y}} / \partial \mathbf{w}$ is locally linear. Combined with the assumption that $P(\mathbf{y}|\mathbf{x})$ is Gaussian with constant variance for all $\mathbf{x}$, one can derive a closed form expression for $\left\langle \tilde{\Sigma}_{\hat{y}} \right\rangle$. See (Cohn 1996) for details.

In practice, $\partial \hat{\mathbf{y}} / \partial \mathbf{w}$ may be highly nonlinear, and $P(\mathbf{y}|\mathbf{x})$ may be far from Gaussian; in spite of this, empirical results show that this criterion often works well (Cohn 1996). It has the advantage of being grounded in statistics, and is optimal given the assumptions. Furthermore, the criterion is differentiable with respect to $\tilde{\mathbf{x}}$. This makes it possible, for continuous domains with continuous action spaces, to search the input space by following the gradient of the criterion.

For neural networks, however, this approach has many disadvantages. In addition to relying on simplifications and assumptions which hold only approximately, the process is computationally expensive. Computing the variance estimate requires inversion of a $|\mathbf{w}| \times |\mathbf{w}|$ matrix for each new example, and incorporating new examples into the network requires expensive retraining. Paass and Kindermann (Paass and Kindermann 1995) discuss a Markov-chain based sampling approach which addresses some of these problems. In the rest of this chapter, we consider the mixture of Gaussians, an alternative machine learning architecture that is much more amenable to optimal data selection.

6.3 MIXTURES OF GAUSSIANS

The mixture of Gaussians model is a powerful density estimation technique with roots in the statistics literature (Titterington *et al.* 1985), which has, over the last few years, found increasing use in machine learning (Cheeseman *et al.* 1988, Nowlan 1991, Specht 1991, Ghahramani and Jordan 1994). It is closely related to the "mixtures of experts" architecture (Jacobs *et al.* 1991, Jordan and Jacobs 1994), which is a regression model with a modular structure similar to decision trees and adaptive spline models (Breiman *et al.* 1984, Friedman 1991). Also closely related are the local models examined by Murray-Smith (Murray-Smith 1994), Takagi and Sugeno (Takagi and Sugeno 1985) and others (see Chapters 2, 3 and 7 for examples).[3]

Like the mixture of experts and related models, the mixture of Gaussians model partitions the data into local regions and fits simple models to these regions. The mixture of Gaussians model, however, differs from strictly regression-based approaches, in that the learner models the joint input–output *density* of the data. Similar density estimation approaches have been discussed by Specht (1991) for nonparametric models, and Nowlan (1991) and Tresp *et al.* (1994) among others, for Gaussian mixture models.

One benefit of this density estimation approach to learning is that there is no *a priori* distinction between inputs and outputs – one may specify any subset of the input–output dimensions, and compute the conditional distribution on the remaining dimensions. Consider the problem of learning the dynamics of a robot arm, for example. If one has only learned an input–output mapping for the forward model, the inverse dynamics must be learned separately. By estimating the joint input–output density, a single model may be learned for both. By conditioning on the model inputs, one has a model of the arm's forward dynamics; by conditioning on the outputs, one has a model of the arm's inverse dynamics. The density-estimation approach can also be used to sample from the output or compute the mode of some variable, rather than its mean. This obviates some of the problems in learning inverse mappings, such as a robot arm's inverse kinematics or dynamics, where the conditional distribution of the output is multimodal and the standard assumptions of regression-based approaches do not hold (Ghahramani 1994) (see Figure 6.1).

6.3.1 Formal description of the mixture of Gaussians

The mixture model approach models the data $\mathcal{D} = \{(\mathbf{x}_i, \mathbf{y}_i)\}_{i=1}^m$ as being generated independently from a mixture density

$$P(\mathbf{x}_i, \mathbf{y}_i) = \sum_{j=1}^{N} P(\mathbf{x}_i, \mathbf{y}_i | j; \theta_j) P(j), \qquad (6.6)$$

where θ_j denotes the parameters of component j. In this chapter, we assume that the data were produced by a mixture of N multivariate Gaussians g_j, for $j = 1, ..., N$ (see Figure 6.2). Practically any well-defined distribution could be used as a component of the mixture, and depending on prior knowledge about the structure of the domain, some components may be more reasonable choices than others. For our purposes, however, where we assume little prior domain knowledge, the flexibility and computational simplicity of the Gaussian distribution makes it an appealing choice.

---

[3]The research in (Murray-Smith 1994) also has an active component: Murray-Smith's Gaussian basis functions are used to filter the existing data to provide a more concise training set.

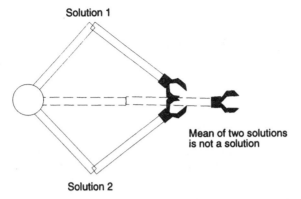

Figure 6.1 In non-convex problems, the average of two solutions may not itself be a solution. In the inverse kinematics problem above, for example, the robot has found two settings of joint angles that position its hand at the desired position. A regression-based approach would compute the mean of the solutions and select a set of joint angles (dashed lines) that did *not* reach the target. The density estimation approach allows one to compute the mode, in effect selecting one of the two actual solutions. See (Ghahramani 1994) for details.

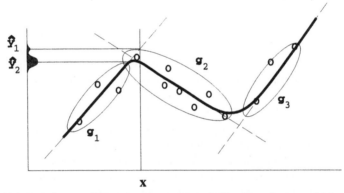

Figure 6.2 Using a mixture of Gaussians to compute $\hat{y}$. The Gaussians model the data density. Predictions are made by mixing the conditional expectations of each Gaussian given the input x.

Given the assumption that the data were produced by a mixture of Gaussians, the joint density $P(\mathbf{x}, \mathbf{y})$ is estimated by finding the parameters of the N Gaussians that are most likely to have produced the data. The EM algorithm can be used to efficiently estimate this maximum likelihood fit of the Gaussians to the data (Dempster *et al.* 1977). Given a particular input $\mathbf{x}$ then, the conditional density $P(\mathbf{y}|\mathbf{x})$ describes the input–output mapping estimated by the model. To estimate a single-valued vector function of the input $\mathbf{y} = f(\mathbf{x})$, rather than the full conditional density, one can evaluate $\hat{\mathbf{y}} = E(\mathbf{y}|\mathbf{x})$, the expectation of $\mathbf{y}$ given $\mathbf{x}$.

For each Gaussian g_i we will denote the input/output means as $\boldsymbol{\mu}_{\mathbf{x},i}$ and $\boldsymbol{\mu}_{\mathbf{y},i}$ and variances and covariances as $\Sigma_{\mathbf{x},i}$, $\Sigma_{\mathbf{y},i}$ and $\Sigma_{\mathbf{xy},i}$ respectively. We will denote the dimensions of $\mathbf{x}$ and $\mathbf{y}$ as d_x and d_y, and define $d = d_x + d_y$. The probability of data point $(\mathbf{x}, \mathbf{y})$, given

g_i, can then be expressed as

$$P(\mathbf{x}, \mathbf{y}|i) = (2\pi)^{-\frac{d}{2}} |\Sigma_i|^{-\frac{1}{2}} \exp\left[-\frac{1}{2}(\mathbf{v} - \mu_i)^T \Sigma_i^{-1}(\mathbf{v} - \mu_i)\right], \qquad (6.7)$$

where we have defined

$$\mathbf{v} = \begin{bmatrix} \mathbf{x} \\ \mathbf{y} \end{bmatrix}, \quad \mu_i = \begin{bmatrix} \mu_{\mathbf{x},i} \\ \mu_{\mathbf{y},i} \end{bmatrix}, \quad \Sigma_i = \begin{bmatrix} \Sigma_{\mathbf{x},i} & \Sigma_{\mathbf{xy},i} \\ \Sigma_{\mathbf{xy},i}^T & \Sigma_{\mathbf{y},i} \end{bmatrix}.$$

We first describe how the EM algorithm is used to estimate the parameters of the mixture of Gaussians. We then detail how the estimated model may be used to make predictions. Finally, we turn to the problem of active learning: how additional examples may be selected so as to minimise the variance in those predictions.

6.3.2 The EM algorithm for mixture models

From equation (6.6) and the independence assumption we see that the log likelihood of the parameters given the data set is

$$l(\theta|\mathcal{D}) = \sum_{i=1}^{m} \log \sum_{j=1}^{N} P(\mathbf{v}_i|j; \theta_j) P(j).$$

By the maximum likelihood principle, the best model of the data has parameters that maximise $l(\theta|\mathcal{D})$. This function cannot be maximised analytically because it involves a log of a sum. One approach to maximising it numerically is to iteratively change the parameters in the direction of the gradient of the log likelihood. A much more efficient and intuitive approach is given by the EM algorithm (Dempster *et al.* 1977).

Intuitively, there is a "credit-assignment" problem: it is not clear which component of the mixture generated a given data point and thus which parameters to adjust to fit that data point. The EM algorithm for mixture models is an iterative method for solving this credit-assignment problem. The intuition is that if one had access to a "hidden" random variable $\mathbf{z}$ indicating which data point was generated by which component of the mixture, then the overall maximisation problem would decouple into a set of simple maximisations. Using the binary indicator variables $\mathcal{Z} = \{\mathbf{z}_i\}_{i=1}^{m}$, defined such that $\mathbf{z}_i = (z_{i1}, \ldots, z_{iN})$ and $z_{ij} = 1$ iff $\mathbf{v}_i$ was generated by Gaussian j, a "complete-data" log likelihood function can be written

$$l_c(\theta|\mathcal{D}, \mathcal{Z}) = \sum_{i=1}^{m} \sum_{j=1}^{N} z_{ij} \log[P(\mathbf{v}_i|\mathbf{z}_i; \theta) P(\mathbf{z}_i; \theta)], \qquad (6.8)$$

which does not involve a log of a summation.

Since $\mathbf{z}$ is unknown, l_c cannot be utilised directly, so we instead work with its expectation, denoted by $Q(\theta|\theta_k)$, where k denotes the iteration number. As shown by (Dempster *et al.* 1977), $l(\theta|\mathcal{D})$ can be maximised by iterating the following two steps:

$$\begin{aligned} \textit{E-step:} \quad & Q(\theta|\theta_k) &=& \quad E[l_c(\theta|\mathcal{D}, \mathcal{Z})|\mathcal{D}, \theta_k] \\ \textit{M-step:} \quad & \theta_{k+1} &=& \quad \underset{\theta}{\arg\max}\ Q(\theta|\theta_k). \end{aligned} \qquad (6.9)$$

The Expectation or E-step computes the expected complete data log likelihood, and the Maximisation or M-step finds the parameters that maximise this likelihood. For densities from the exponential family the E-step reduces to computing the expectation over the missing data of the sufficient statistics required for the M-step.

With the mixture of Gaussians model, the E-step simplifies to computing $E[z_{ij}|\mathbf{v}_i, \theta_k]$, which, given the binary nature of z_{ij}, is the probability that generated data point i was generated by Gaussian j, $P(z_{ij} = 1|\mathbf{v}_i, \theta_k)$. We denote this by h_{ij}, which can be computed as

$$h_{ij} = \frac{P(j)|\hat{\Sigma}_j|^{-1/2} \exp\left[-\frac{1}{2}(\mathbf{v}_i - \hat{\boldsymbol{\mu}}_j)^T \hat{\Sigma}_j^{-1}(\mathbf{v}_i - \hat{\boldsymbol{\mu}}_j)\right]}{\sum_{l=1}^{m} P(l)|\hat{\Sigma}_l|^{-1/2} \exp\left[-\frac{1}{2}(\mathbf{v}_i - \hat{\boldsymbol{\mu}}_l)^T \hat{\Sigma}_l^{-1}(\mathbf{v}_i - \hat{\boldsymbol{\mu}}_l)\right]}. \tag{6.10}$$

The M-step re-estimates the means and covariances of the Gaussians[4] using the data set weighted by the h_{ij}:

$$\hat{\boldsymbol{\mu}}_j^{k+1} = \frac{\sum_{i=1}^{m} h_{ij}\mathbf{v}_i}{\sum_{i=1}^{m} h_{ij}},$$

$$\hat{\Sigma}_j^{k+1} = \frac{\sum_{i=1}^{m} h_{ij}(\mathbf{v}_i - \hat{\boldsymbol{\mu}}_j^{k+1})(\mathbf{v}_i - \hat{\boldsymbol{\mu}}_j^{k+1})^T}{\sum_{i=1}^{m} h_{ij}}.$$

6.3.3 Making predictions using the mixture of Gaussians

After some number of iterations of the EM algorithm, $\hat{\boldsymbol{\mu}}_j$ and $\hat{\Sigma}_j$ will converge to some locally optimal estimates of the Gaussian means and covariances, providing an estimate of the joint density $P(\mathbf{v}) = P(\mathbf{x}, \mathbf{y})$.[5] From this density we can estimate the conditional density of $\mathbf{y}$ given $\mathbf{x}$, and its expected value $\hat{\mathbf{y}}$.

For each Gaussian i, the conditional variance of $\mathbf{y}$ given $\mathbf{x}$ is

$$\Sigma_{\mathbf{y}|\mathbf{x},i} = \Sigma_{\mathbf{y},i} - \Sigma_{\mathbf{xy},i}^T \Sigma_{\mathbf{x},i}^{-1} \Sigma_{\mathbf{xy},i},$$

and the conditional expectation $\hat{\mathbf{y}}_i$ and variance of this expectation $\Sigma_{\hat{\mathbf{y}},i}$ given $\mathbf{x}$ are:

$$\hat{\mathbf{y}}_i = \boldsymbol{\mu}_{\mathbf{y},i} + \Sigma_{\mathbf{xy},i}\Sigma_{\mathbf{x},i}^{-1}(\mathbf{x} - \boldsymbol{\mu}_{\mathbf{x},i}) \tag{6.11}$$

$$\Sigma_{\hat{\mathbf{y}},i} = \frac{1}{n_i}\Sigma_{\mathbf{y}|\mathbf{x},i}\left(1 + (\mathbf{x} - \boldsymbol{\mu}_{\mathbf{x},i})^T \Sigma_{\mathbf{x},i}^{-1}(\mathbf{x} - \boldsymbol{\mu}_{\mathbf{x},i})\right). \tag{6.12}$$

Here, n_i is the amount of "support" for the Gaussian g_i in the training data. It can be computed as

$$n_i = \sum_{j=1}^{N} h_{ij}.$$

The expectations and variances in equation (6.12) are mixed according to the probability that g_i has of being responsible for $\mathbf{x}$, prior to observing $\mathbf{y}$:

$$h_i \equiv h_i(\mathbf{x}) = \frac{P(\mathbf{x}|i)}{\sum_{j=1}^{m} P(\mathbf{x}|j)},$$

where

$$P(\mathbf{x}|i) = (2\pi)^{-\frac{d_x}{2}} |\Sigma_{\mathbf{x},i}|^{-\frac{1}{2}} \exp\left[-\frac{1}{2}(\mathbf{x} - \boldsymbol{\mu}_{\mathbf{x},i})^T \Sigma_{\mathbf{x},i}^{-1}(\mathbf{x} - \boldsymbol{\mu}_{\mathbf{x},i})\right]. \tag{6.13}$$

---

[4]This derivation assumes equal priors for the Gaussians, which therefore cancel in the computation of h_{ij}. If the priors are viewed as mixing parameters they can also be learned in the maximisation step.

[5]Convergence is indicated when the likelihood in equation (6.9) stops increasing.

For input $\mathbf{x}$ then, the conditional expectation $\hat{\mathbf{y}}$ of the resulting mixture and its variance may be written:

$$\hat{\mathbf{y}} = \sum_{i=1}^{N} h_i \hat{\mathbf{y}}_i \qquad (6.14)$$

$$\Sigma_{\hat{\mathbf{y}}} = \sum_{i=1}^{N} \frac{h_i^2}{n_i} \Sigma_{\mathbf{y}|\mathbf{x},i} \left(1 + (\mathbf{x} - \boldsymbol{\mu}_{\mathbf{x},i})^T \Sigma_{\mathbf{x},i}^{-1} (\mathbf{x} - \boldsymbol{\mu}_{\mathbf{x},i})\right), \qquad (6.15)$$

where we have assumed that the $\hat{\mathbf{y}}_i$ are independent in calculating $\Sigma_{\hat{\mathbf{y}}}$. Both of these terms can be computed efficiently, in closed form. It is also worth noting that $\Sigma_{\hat{\mathbf{y}}}$ is one of several variance measures we might be interested in. If, for example, our mapping is multivalued (that is, if Gaussians with different $\hat{\mathbf{y}}_i$ overlap significantly in the $\mathbf{x}$ dimension), we may wish our prediction $\hat{\mathbf{y}}$ to reflect the most likely $\mathbf{y}$ value. In this case, $\hat{\mathbf{y}}$ would be the mode, and a preferable measure of uncertainty would be the (unmixed) variance of the individual Gaussians.

6.3.4   Active learning with a mixture of Gaussians

We consider how to select examples optimally for a mixture of Gaussians according to the error minimisation criterion (equation (6.1)), subject to the assumptions that the learner is unbiased, and that the input distribution $P(\mathbf{x})$ is known. With a mixture of Gaussians, one interpretation of the assumption that we know $P(\mathbf{x})$ is that we know $\boldsymbol{\mu}_{\mathbf{x},i}$ and $\Sigma_{\mathbf{x},i}$ for each Gaussian. In that case, EM would only estimate $\boldsymbol{\mu}_{\mathbf{y},i}$, $\Sigma_{\mathbf{y},i}$, and $\Sigma_{\mathbf{xy},i}$. Generally however, knowing the input distribution will not correspond to knowing the actual $\boldsymbol{\mu}_{\mathbf{x},i}$ and $\Sigma_{\mathbf{x},i}$ for each Gaussian. We may simply know, for example, that $P(\mathbf{x})$ is uniform, or can be approximated by some set of sampled inputs. In such cases, we must use EM to estimate $\boldsymbol{\mu}_{\mathbf{x},i}$ and $\Sigma_{\mathbf{x},i}$ in addition to the parameters involving $\mathbf{y}$. However, if we estimate these values from training data obtained by actively sampling $\tilde{\mathbf{x}}$, we will be estimating the joint distribution of $P(\tilde{\mathbf{x}}, \mathbf{y}|i)$ instead of $P(\mathbf{x}, \mathbf{y}|i)$. To obtain a proper estimate, we must correct equation (6.7) as follows:

$$P(\mathbf{x}, \mathbf{y}|i) = P(\tilde{\mathbf{x}}, \mathbf{y}|i) \frac{P(\mathbf{x}|i)}{P(\tilde{\mathbf{x}}|i)}. \qquad (6.16)$$

Here, $P(\tilde{\mathbf{x}}|i)$ is computed by applying equation (6.13) given the mean and $\mathbf{x}$ variance of the training data, and $P(\mathbf{x}|i)$ is computed by applying the same equation using the mean and $\mathbf{x}$ variance of a set of reference data drawn according to $P(\mathbf{x})$. Note also that $P(\mathbf{y}|\hat{\mathbf{x}}, i) = P(\tilde{\mathbf{x}}, \mathbf{y}|i)/P(\tilde{\mathbf{x}}|i)$ can alternatively be provided by an unnormalised mixture of experts architecture (Jacobs et al. 1991) and multiplied by the input density $P(\mathbf{x}|i)$ to yield the joint density.

If our goal in active learning is to minimise variance, we should select training examples $\tilde{\mathbf{x}}$ to minimise $\left\langle \tilde{\Sigma}_{\hat{\mathbf{y}}} \right\rangle$. With a mixture of Gaussians, we can compute $\left\langle \tilde{\Sigma}_{\hat{\mathbf{y}}} \right\rangle$ efficiently. The model's estimated distribution of $\tilde{\mathbf{y}}$ given $\tilde{\mathbf{x}}$ is explicit:

$$P(\tilde{\mathbf{y}}|\tilde{\mathbf{x}}) = \sum_{i=1}^{N} \tilde{h}_i P(\tilde{\mathbf{y}}|\tilde{\mathbf{x}}, i) = \sum_{i=1}^{N} \tilde{h}_i \mathcal{N}(\hat{\mathbf{y}}_i(\tilde{\mathbf{x}}), \Sigma_{\mathbf{y}|\mathbf{x},i}(\tilde{\mathbf{x}})),$$

where $\tilde{h}_i \equiv h_i(\tilde{\mathbf{x}})$, and $\mathcal{N}(\boldsymbol{\mu}, \Sigma)$ denotes the normal distribution with mean $\boldsymbol{\mu}$ and covariance matrix Σ. Given this, we can model the change in each g_i separately, calculating its expected variance given a new point sampled from $P(\tilde{\mathbf{y}}|\tilde{\mathbf{x}}, i)$ and weight this change by $\tilde{h}_i$. The new expectations combine to form the learner's new expected variance

$$\left\langle \tilde{\Sigma}_{\hat{\mathbf{y}}} \right\rangle = \sum_{i=1}^{N} \frac{h_i^2}{n_i + \tilde{h}_i} \left\langle \tilde{\Sigma}_{\mathbf{y}|\mathbf{x},i} \right\rangle \left(1 + (\mathbf{x} - \boldsymbol{\mu}_{\mathbf{x},i})^T \left(\Sigma_{\mathbf{x},i} \right)^{-1} (\mathbf{x} - \boldsymbol{\mu}_{\mathbf{x},i}) \right), \quad (6.17)$$

where the expectation can be computed exactly in closed form:

$$\left\langle \tilde{\Sigma}_{\mathbf{y}|\mathbf{x},i} \right\rangle = \left\langle \tilde{\Sigma}_{\mathbf{y},i} \right\rangle - \left\langle \tilde{\Sigma}_{\mathbf{xy},i}^T \right\rangle \left(\Sigma_{\mathbf{x},i} \right)^{-1} \left\langle \tilde{\Sigma}_{\mathbf{xy},i} \right\rangle$$

$$- \frac{n_i^2 \tilde{h}_i^2}{(n_i + \tilde{h}_i)^4} \Sigma_{\mathbf{y}|\tilde{\mathbf{x}},i} (\tilde{\mathbf{x}} - \boldsymbol{\mu}_{\mathbf{x},i}) \left(\Sigma_{\mathbf{x},i} \right)^{-1} (\tilde{\mathbf{x}} - \boldsymbol{\mu}_{\mathbf{x},i})^T$$

$$\left\langle \tilde{\Sigma}_{\mathbf{y},i} \right\rangle = \frac{n_i}{n_i + \tilde{h}_i} \Sigma_{\mathbf{y},i} + \frac{n_i \tilde{h}_i}{(n_i + \tilde{h}_i)^2} \left(\Sigma_{\mathbf{y}|\tilde{\mathbf{x}},i} + (\hat{\mathbf{y}}_i(\tilde{\mathbf{x}}) - \boldsymbol{\mu}_{\mathbf{y},i})(\hat{\mathbf{y}}_i(\tilde{\mathbf{x}}) - \boldsymbol{\mu}_{\mathbf{y},i})^T \right)$$

$$\left\langle \tilde{\Sigma}_{\mathbf{xy},i} \right\rangle = \frac{n_i}{n_i + \tilde{h}_i} \Sigma_{\mathbf{xy},i} + \frac{n_i \tilde{h}_i}{(n_i + \tilde{h}_i)^2} (\tilde{\mathbf{x}} - \boldsymbol{\mu}_{\mathbf{x},i})(\hat{\mathbf{y}}_i(\tilde{\mathbf{x}}) - \boldsymbol{\mu}_{\mathbf{y},i})^T.$$

If, as discussed earlier, we are also estimating $\boldsymbol{\mu}_{\mathbf{x},i}$ and $\Sigma_{\mathbf{x},i}$, we must take into account the effect of the new example on those estimates, and must replace $\boldsymbol{\mu}_{\mathbf{x},i}$ and $\Sigma_{\mathbf{x},i}$ in the above equations with

$$\tilde{\boldsymbol{\mu}}_{\mathbf{x},i} = \frac{n_i \boldsymbol{\mu}_{\mathbf{x},i} + \tilde{h}_i \tilde{\mathbf{x}}}{n_i + \tilde{h}_i}, \qquad \tilde{\Sigma}_{\mathbf{x},i} = \frac{n_i \Sigma_{\mathbf{x},i}}{n_i + \tilde{h}_i} + \frac{n_i \tilde{h}_i (\tilde{\mathbf{x}} - \boldsymbol{\mu}_{\mathbf{x},i})(\tilde{\mathbf{x}} - \boldsymbol{\mu}_{\mathbf{x},i})^T}{(n_i + \tilde{h}_i)^2}.$$

We can use equation (6.17) to guide active learning. By evaluating the expected new variance over a reference set given candidate $\tilde{\mathbf{x}}$, we can select the $\tilde{\mathbf{x}}$ giving the lowest expected model variance. Note that in high-dimensional spaces, it may be necessary to evaluate an excessive number of candidate points to get good coverage of the potential query space. In these cases, it is more efficient to differentiate equation (6.17) and hill-climb on $\partial \left\langle \tilde{\Sigma}_{\hat{\mathbf{y}}} \right\rangle / \partial \tilde{\mathbf{x}}$ to find a locally maximal $\tilde{\mathbf{x}}$. See, for example (Cohn 1996).

6.4 EXPERIMENTAL RESULTS

While the theory we have described here is appealing in its simplicity, it has still necessitated a number of assumptions which may not strictly hold. Among these are the assumption that the data truly were produced by a mixture of Gaussians, that the learner is approximately unbiased, and that the local maximum likelihood estimates from the EM algorithm are close to the true optima. When these assumptions all hold, the variance-minimising criterion is statistically optimal; how it performs when these assumptions are stretched or violated requires empirical testing.

For an experimental testbed, we used a task in which the learner's goal was to estimate the kinematics of a simulated 2-degree-of-freedom planar robot arm (described as the "Arm2D" problem in (Cohn 1996); see Figure 6.3). The inputs are joint angles (θ_1, θ_2), and the outputs are the Cartesian coordinates of the tip (x_1, x_2). One of the implicit assumptions of this model we have described here is that the noise is Gaussian in the output dimensions. To test the robustness of the algorithm to this assumption, we ran experiments using no noise, using additive Gaussian noise in the outputs, and using additive Gaussian noise in the inputs. The results of each were comparable; we report here the results using additive Gaussian noise in the inputs. Gaussian input noise corresponds to the case where the arm effectors or joint angle sensors are noisy, and results in non-Gaussian errors in the learner's outputs. The input distribution $P(\mathbf{x})$ is assumed to be uniform.

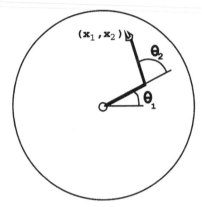

Figure 6.3　The arm kinematics problem. The learner attempts to predict tip position given a set of joint angles (θ_1, θ_2).

We compared the performance of the variance-minimising criterion by comparing the *learning curves* of a learner using the criterion with that of one learning from random samples. The learning curves plot the mean squared error and variance of the learner as its training set size increases. The curves are created by starting with an initial sample, measuring the learner's mean squared error or estimated variance on a set of reference points (independent of the training set), selecting and adding a new example to the training set, retraining the learner on the augmented set, and repeating.

On each step, the variance-minimising learner chose a set of 64 unlabelled reference points drawn from input distribution $P(\mathbf{x})$. It then selected a query $\tilde{\mathbf{x}} = (\theta_1, \theta_2)$ that it estimated would minimise $\langle \tilde{\Sigma}_{y|x} \rangle$ over the reference set. In the experiments reported here, the best $\tilde{\mathbf{x}}$ was selected from another set of 64 candidate points drawn at random from $P(\mathbf{x})$ on each iteration.[6]

There were three design parameters in the simulation that needed to be considered: the number of Gaussians, their initial placement, and the number of iterations of the EM algorithm. We set these parameters by optimising them on the learner using random examples, then used the same settings on the learner using the variance-minimisation criterion. Parameters were set as follows: models with fewer Gaussians have the obvious advantage of requiring less storage space and computation. Intuitively, a small model should also have the advantage of avoiding overfitting, which has been commonly observed in systems with extraneous parameters. Empirically, as we increased the number of Gaussians, generalisation improved monotonically with diminishing returns (for a fixed training set size and number of EM iterations). The test error of the larger models generally matched that of the smaller models on small training sets (where overfitting would be a concern), and continued

---

[6]As described earlier, we could also have selected queries by hill-climbing on $\partial \langle \tilde{\Sigma}_{y|x} \rangle /\partial \tilde{\mathbf{x}}$; in this low-dimensional problem it was more computationally efficient to consider a random candidate set.

to decrease on large training sets where the smaller networks "bottomed out". We therefore preferred the larger mixtures, and report here our results with mixtures of 60 Gaussians. We selected initial placement of the Gaussians randomly, chosen uniformly from the smallest hypercube containing all current training examples. We arbitrarily chose the identity matrix as an initial covariance matrix. The learner was surprisingly sensitive to the number of EM iterations. We examined a range of 5 to 40 iterations of the EM algorithm per step. Small numbers of iterations (5 – 10) appear insufficient to allow convergence with large training sets, while large numbers of iterations (30 – 40) degraded performance on small training sets. An ideal training regime would employ some form of regularisation, or would examine the degree of change between iterations to detect convergence; in our experiments, however, we settled on a fixed regime of 20 iterations per step.

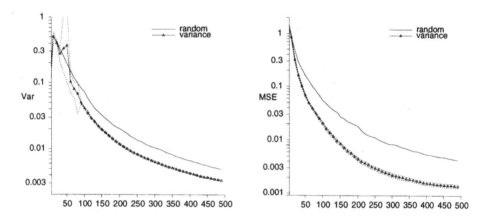

Figure 6.4   Variance and MSE learning curves plotted on a log scale for mixture of 60 Gaussians trained on the kinematics problem. Dotted lines denote standard error for average of 10 runs, each started with one initial random example.

Figure 6.4 shows the variance and MSE learning curves for a mixture of 60 Gaussians trained on the kinematics problem with 1% input noise added. The estimated model variance using the variance-minimising criterion is significantly better than that of the learner selecting data at random. The mean squared error, however, exhibits even greater improvement, with an error that is consistently 1/3 that of the randomly sampling learner.

6.4.1   Computation time

One obvious concern about the criterion described here is its computational cost. In situations where obtaining new examples may involve considerable time and expense, it is clearly wise to expend computation to ensure that those examples are as useful as possible. In other situations, however, new data may be relatively inexpensive, so the computational cost of finding optimal examples must be considered.

Table 6.1 summarises the computation times for the mixture of Gaussians. The times reported are "per reference point" and "per candidate per reference point"; overall time must be computed from the number of candidates and reference points examined.

Table 6.1 Computation times on a Sparc 10 as a function of training set size m. Mixture model had 60 Gaussians trained for 20 iterations. Reference times are per reference point; candidate times are per candidate point per reference point.

Computation times	
Training	$3.9 + 0.05m$ sec
Evaluating Reference	$15000\ \mu$sec
Evaluating Candidate	$1300\ \mu$sec

While the training time incurred by the mixture of Gaussians may make it infeasible for selecting optimal actions in real-time control, it is certainly fast enough to be used in many applications. Optimised, parallel implementations will also enhance its utility.[7]

6.5 DISCUSSION

The mixture of Gaussians is a statistical model that offers an elegant representation and efficient learning algorithm. In this chapter we have shown that it also offers the opportunity to perform active learning in an efficient and statistically correct manner. The variance-minimising criterion derived here can be computed cheaply and, for the problem tested, demonstrates good predictive power. In industrial settings, where gathering a single data point may take days and cost thousands of dollars, the techniques described here have the potential for enormous savings.

In this chapter, we have only considered function approximation problems. Problems requiring classification could be handled analogously by identifying different sets of Gaussians with different classes. The probability that an input has some class label would then be the probability that was produced by one of the Gaussians in the corresponding set. Active learning in this case then would attempt to select training examples so as to maximise discriminability between classes.

Our future work will proceed in several directions. The most important is active bias minimisation. As noted in section 6.2, the learner's error is composed of both bias and variance. The variance-minimising strategy examined here ignores the bias component, which can lead to significant errors when the learner's bias is non-negligible. Work in progress examines effective ways of measuring and optimally eliminating bias (Cohn 1995); future work will examine how to jointly minimise both bias and variance to produce a criterion that truly minimises the learner's expected error.

Another direction for future research is the derivation of variance-minimising (and bias-minimising) techniques for other statistical learning models. Of particular interest is the class of models known as "belief networks" or "Bayesian networks" (Pearl 1988, Heckerman *et al.* 1994). These models have the advantage of allowing inclusion of domain knowledge and prior constraints while still adhering to a statistically sound framework. Current research in belief networks focuses on algorithms for efficient inference and learning; it would be an important step to derive the proper criteria for learning *actively* with these models.

---

[7]It is worth mentioning that approximately half of the training time for the mixture of Gaussians is spent computing the correction factor in equation (6.16). Without the correction, the learner still computes $P(y|x)$, but does so by modelling the training set distribution rather than the reference distribution. We have found however, that for the problems examined, the performance of such "uncorrected" learners does not differ appreciably from that of the "corrected" learners.

6.6 NOTATION

	General
X	input space
Y	output space
$\mathbf{x}$	an arbitrary point in the input space
$\mathbf{y}$	true output value corresponding to input $\mathbf{x}$
$\hat{\mathbf{y}}$	predicted output value corresponding to input $\mathbf{x}$
$\mathbf{x}_i$	"input" part of example i
$\mathbf{y}_i$	"output" part of example i
m	the number of examples in the training set
$\tilde{\mathbf{x}}$	specified input of a query
$\tilde{\mathbf{y}}$	the (possibly not yet known) output of query $\tilde{\mathbf{x}}$
$\Sigma_{\hat{\mathbf{y}}}$	estimated variance of $\hat{\mathbf{y}}$
$\tilde{\Sigma}_{\hat{\mathbf{y}}}$	new variance of $\hat{\mathbf{y}}$, after example $(\tilde{\mathbf{x}}, \tilde{\mathbf{y}})$ has been added
$\left\langle \tilde{\Sigma}_{\hat{\mathbf{y}}} \right\rangle$	the expected value of $\tilde{\Sigma}_{\hat{\mathbf{y}}}$
$P(\mathbf{x})$	the (known) natural distribution over $\mathbf{x}$

	Neural Network
$\mathbf{w}$	a weight vector for a neural network
$\hat{\mathbf{w}}$	estimated "best" $\mathbf{w}$ given a training set
$f_{\hat{\mathbf{w}}}()$	function computed by neural network given $\hat{\mathbf{w}}$
S^2	average estimated noise in data, used as an estimate for $\Sigma_{\mathbf{y}}$

	Mixture of Gaussians
N	total number of Gaussians
g_i	Gaussian number i
n_i	total point weighting attributed to Gaussian i
$\mathbf{v}$	the concatenated input–output vectors
$\mu_{\mathbf{x},i}$	estimated x mean of Gaussian i
$\mu_{\mathbf{y},i}$	estimated y mean of Gaussian i
μ_i	the concatenated input–output means of Gaussian i
$\Sigma_{\mathbf{x},i}$	estimated x variance of Gaussian i
$\Sigma_{\mathbf{y},i}$	estimated y variance of Gaussian i
$\Sigma_{\mathbf{xy},i}$	estimated xy covariance of Gaussian i
Σ_i	the complete estimated covariance matrix of Gaussian i
$\Sigma_{\mathbf{y}|\mathbf{x},i}$	estimated y variance of Gaussian i, given $\mathbf{x}$
$P(\mathbf{x}, \mathbf{y}|i)$	joint distribution of input–output pair given Gaussian i
$P(\mathbf{x}|i)$	distribution of $\mathbf{x}$ being given Gaussian i
h_i	weight of a given point that is attributed to Gaussian i
$\tilde{h}_i$	weight of new point $(\tilde{\mathbf{x}}, \tilde{\mathbf{y}})$ that is attributed to Gaussian i

REFERENCES

ANGLUIN, D. (1988) 'Queries and concept learning'. *Machine Learning* **2**, 319–342.

ATKINSON, A. AND A. DONEV (1992) *Optimum Experimental Designs*. Clarendon Press, New York, NY.

BAUM, E. AND K. LANG (1991) 'Neural network algorithms that learn in polynomial time from examples and queries'. *IEEE Trans. Neural Networks*.

BREIMAN, L., J. H. FRIEDMAN, R. A. OLSHEN AND C. J. STONE (1984) *Classification and Regression Trees*. Wadsworth International Group. Belmont, CA.

CHEESEMAN, P., M. SELF, J. KELLY, W. TAYLOR, D. FREEMAN AND J. STUTZ (1988) Bayesian classification. In 'AAAI 88, The 7th National Conference on Artificial Intelligence'. AAAI Press. pp. 607–611.

COHN, D. (1995) Minimizing statistical bias with queries. AI Lab memo AIM-1552. Massachusetts Institute of Technology. Available by anonymous ftp from `publications.ai.mit.edu`.

COHN, D. (1996) 'Neural network exploration using optimal experiment design'. *Neural Networks* **9**(2), 1–13. Preliminary version available as MIT AI Lab memo 1491 by anonymous ftp to `publications.ai.mit.edu`.

COHN, D., L. ATLAS AND R. LADNER (1990) Training connectionist networks with queries and selective sampling. In D. Touretzky (Ed.). 'Advances in Neural Information Processing Systems 2'. Morgan Kaufmann, San Francisco, CA.

COHN, D., L. ATLAS AND R. LADNER (1994) 'Improving generalization with active learning'. *Machine Learning* **5**(2), 201–221.

DEMPSTER, A., N.M. LAIRD AND D.B. RUBIN (1977) 'Maximum likelihood from incomplete data via the EM algorithm'. *J. Royal Statistical Society Series B* **39**, 1–38.

FEDOROV, V. (1972) *Theory of Optimal Experiments*. Academic Press, New York, NY.

FE'LDBAUM, A. A. (1965) *Optimal control systems*. Academic Press, New York, NY.

FREUND, Y., H.S. SEUNG, E. SHAMIR AND N. TISHBY (1993) Information, prediction, and query by committee. In S. Hanson, J. Cowan and L. Giles (Eds.). 'Advances in Neural Information Processing Systems 5'. Morgan Kaufmann, San Francisco, CA.

FRIEDMAN, J. H. (1991) 'Multivariate adaptive regression splines'. *The Annals of Statistics* **19**, 1–141.

GEMAN, S., E. BIENENSTOCK AND R. DOURSAT (1992) 'Neural networks and the bias/variance dilemma'. *Neural Computation* **4**, 1–58.

GHAHRAMANI, Z. (1994) Solving inverse problems using an EM approach to density estimation. In M. Mozer, P. Smolensky, D. Touretzky, J. Elman and A. Weigend (Eds.). 'Proceedings of the 1993 Connectionist Models Summer School'. Erlbaum Associates. Hillsdale, NJ. pp. 316–323.

GHAHRAMANI, Z. AND M. JORDAN (1994) Supervised learning from incomplete data via an EM approach. In J. Cowan, G. Tesauro and J. Alspector (Eds.). 'Advances in Neural Information Processing Systems 6'. Morgan Kaufmann, San Francisco, CA.

HECKERMAN, D., D. GEIGER AND D. CHICKERING (1994) Learning Bayesian networks: the combination of knowledge and statistical data. Tech Report MSR-TR-94-09. Microsoft.

JACOBS, R., M.I. JORDAN, S.J. NOWLAN AND G.E. HINTON (1991) 'Adaptive mixture of local experts'. *Neural Computation* **3**, 79–87.

JORDAN, M. AND R.A. JACOBS (1994) 'Hierarchical mixtures of experts and the EM algorithm'. *Neural Computation* **6**, 181–214.

LEWIS, D. D. AND J. CATLETT (1994) Heterogeneous uncertainty sampling for supervised learning. In W. Cohen and H. Hirsh (Eds.). 'Proceedings of the 11th International Conference on Machine Learning'. Morgan Kaufmann. San Francisco, CA. pp. 148–156.

LINDEN, A. AND F. WEBER (1993) Implementing inner drive by competence reflection. In H. Roitblat (Ed.). 'Proceedings of the 2nd International Conference on Simulation of Adaptive Behavior'. MIT Press, Cambridge, MA.

MACKAY, D. J. (1992) 'Information-based objective functions for active data selection'. *Neural Computation* **4**(4), 590–604.

MURRAY-SMITH, R. (1994) A Local Model Network Approach to Nonlinear Modelling. PhD thesis. University of Strathclyde.

NOWLAN, S. (1991) *Soft Competitive Adaptation: Neural Network Learning Algorithms based on Fitting Statistical Mixtures*. PhD Thesis CMU-CS-91-126. School of Computer Science, Carnegie Mellon University. Pittsburgh, PA.

PAASS, G. AND J. KINDERMANN (1995) Bayesian query construction for neural network models. In G. Tesauro, D. Touretzky and T. Leen (Eds.). 'Advances in Neural Information Processing Systems 7'. MIT Press, Cambridge, MA.

PEARL, J. (1988) *Probabilistic Reasoning in Intelligent Systems*. Morgan Kaufmann, San Francisco, CA.

PLUTOWSKI, M. AND H. WHITE (1993) 'Selecting concise training sets from clean data'. *IEEE Transactions on Neural Networks* **4**, 305–318.

SCHMIDHUBER, J. AND J. STORCK (1993) Reinforcement driven information acquisition in non-deterministic environments. Technical report. Fakultät für Informatik, Technische Universität München.

SPECHT, D. F. (1991) 'A general regression neural network'. *IEEE Trans. Neural Networks* **2**(6), 568–576.

TAKAGI, T. AND M. SUGENO (1985) 'Fuzzy identification of systems and its application to modelling and control'. *IEEE Transactions on Systems, Man, and Cybernetics* **15**(1), 116–132.

THRUN, S. AND K. MÖLLER (1992) Active exploration in dynamic environments. In J. Moody, S. Hanson and R. Lippmann (Eds.). 'Advances in Neural Information Processing Systems 4'. Morgan Kaufmann, San Francisco, CA.

TITTERINGTON, D., A. SMITH AND U. MAKOV (1985) *Statistical Analysis of Finite Mixture Distributions*. John Wiley, New York.

TRESP, V., S. AHMAD AND R. NEUNEIER (1994) Training neural networks with deficient data. In J. D. Cowan, G. Tesauro and J. Alspector (Eds.). 'Advances in Neural Information Processing Systems 6'. Morgan Kaufman. San Francisco, CA.

WHITEHEAD, S. (1991) A study of cooperative mechanisms for faster reinforcement learning. Technical Report CS-365. University of Rochester, Rochester, NY.

CHAPTER SEVEN

Local Learning in Local Model Networks

RODERICK MURRAY-SMITH and TOR ARNE JOHANSEN

This chapter deals with the local/global trade-off in learning. It points out problems with global learning methods in Local Model Networks when the network is locally over-parameterised or poorly structured. The bias-variance trade-offs for local and global learning are examined, illustrating the regularising effect of local learning, which can be advantageous compared to direct maximisation of a global cost function. Smoothing analysis is used to explain the results, and graphical interpretation of the theoretical smoothing results allows more intuitive explanations of differences in learning algorithms. The methods are then related to Expectation Maximisation (EM) techniques, and the smooth-based estimates of the degrees of freedom are used to develop a new learning algorithm which minimises the generalised cross-validation (GCV) error statistic.

7.1 INTRODUCTION

Here we study parameter estimation in nonlinear models that are composed of a combination of multiple local linear models. A general and common approach to parameter estimation is to minimise some criterion that measures the difference between observed outputs and outputs predicted by the global model, i.e. the combination of the local models. In the rest of this chapter, this will be called global learning. The particular structure with multiple local models also suggests another approach, namely to estimate the parameters of each local model separately, possibly using some weighted subset of the data. This will be called local learning. Both these approaches have been applied with success. For example, the common least squares estimation approach is a global learning algorithm, while the EM algorithm used in (Jordan and Jacobs 1994) and the algorithm suggested in (Murray-Smith 1994) are local learning algorithms.

This chapter discusses the differences between the local and global learning approaches. It is demonstrated that global learning is in general more accurate when the model structure matches the system well, but it is also demonstrated that global learning may lead to spurious behaviour of the global and local models when the model structure does not match the underlying system or the complexity in the estimation data. Local learning is proven to be more robust and produce a more well-behaved model in such cases. Moreover, local learning will, in general, identify local models that are closer approximations to local linearisations of the underlying system,[1] which is not guaranteed with global learning. These aspects are of considerable importance in many applications.

The abovementioned results are illustrated using simple examples and statistical arguments such as bias/variance considerations and smoothing analysis. These mathematical and graphical tools should also be of interest to other modelling and learning approaches.

7.2 LEARNING NONLINEAR MODELS

7.2.1 The modelling problem

The basic assumption underlying the use of learning systems for modelling purposes is that the behaviour of the system can be described in terms of a training set $\mathcal{D}_N = ((\psi(1), y(1)), ..., (\psi(N), y(N)))$ consisting of its observed input vector ψ and corresponding scalar output y. We assume the system output can therefore be modelled as

$$y = f(\psi) + \varepsilon, \tag{7.1}$$

where f is a function, and ε is independent random measurement noise with zero mean and variance σ^2.

The modelling problem, as seen in this chapter is to robustly estimate the function f from observation data, having already used existing *a priori* information to pre-structure and parameterise the model structure $\hat{f}$. Typically, the model structure will not be able to exactly describe the system, so a bias $b(\psi)$ will naturally be associated with $\hat{f}$:

$$y = \hat{f}(\psi, \theta^\star) + b(\psi) + \varepsilon. \tag{7.2}$$

We define the optimal p-dimensional parameter vector $\theta^\star$ for the learning problem in equation (7.2) such that the average bias is minimised over the domain of interesting inputs

$$\theta^\star = \arg\min_\theta \int \left(f(\psi) - \hat{f}(\psi, \theta) \right)^2 w(\psi) d\psi,$$

where it is assumed that a unique minimum exists and w is a weighting function. Unfortunately, since f is not known, we cannot find these optimal parameters, but must look for an estimate based on the data $\mathcal{D}_N$.

The model function $\hat{f}$ can in general be pre-structured and parameterised in a number of ways. With a linear model, the data, learning and validation are all considered to be globally (i.e. over the entire input domain) relevant. For nonlinear models, however, it may be advantageous to partition the input domain into multiple subsets – a strategy inherent to local modelling techniques, as discussed in Chapter 1.

---

[1] See section 1.6 for further discussion of this for dynamic systems.

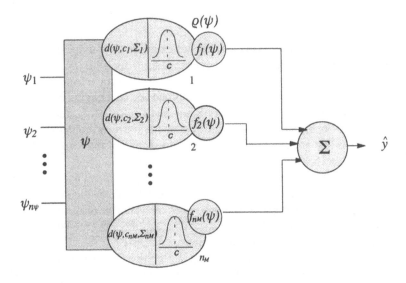

Figure 7.1 Local Model Network.

7.2.2 The Local Model Network

The *Local Model Network*, as used in Chapters 3 and 8 is:[2]

$$\hat{f}(\psi, \theta) = \sum_{i=1}^{n_{\mathcal{M}}} \hat{f}_i(\psi, \theta_i)\rho_i(\psi), \tag{7.3}$$

The total parameter vector is $\theta^T = [\theta_1^T, ..., \theta_{n_{\mathcal{M}}}^T]$. The ρ_i-functions can be seen as *gating* or *weighting* functions for the local models (which are defined on the full input space). The basis functions used are defined,

$$\rho_i(\psi) = \frac{\rho(d(\psi, \mathbf{c}_i, \Sigma_i))}{\sum_{j=1}^{n_{\mathcal{M}}} \rho(d(\psi, \mathbf{c}_j, \Sigma_j))}, \tag{7.4}$$

where $\rho(\cdot)$ is the underlying *unnormalised* basis function, e.g. a Gaussian $\rho(d) = \exp(-d/2)$, and

$$d(\psi, \mathbf{c}_i, \Sigma_i) = (\psi - \mathbf{c}_i)^T \Sigma_i^{-2} (\psi - \mathbf{c}_i) \tag{7.5}$$

is a weighted Euclidean distance metric which measures the distance of the current input ψ from the basis function's centre $\mathbf{c}_i$. The *normalised* basis functions $\rho_i(\cdot)$ now sum to unity. The local models used are linear:

$$\hat{f}_i(\psi, \theta_i) = \begin{bmatrix} 1, & \psi^T \end{bmatrix} \theta_i. \tag{7.6}$$

---

[2]For readers more familiar with neural network approaches, this can be viewed as an RBF network where the basis function coefficients have been generalised to allow not just a constant parameter to be associated with each basis function, but a more powerful function of the inputs ψ, see (Stokbro *et al.* 1990, Jones *et al.* 1989). This means that a smaller number of local models can cover larger areas of the input domain, reducing somewhat the problem of the 'curse of dimensionality' associated with RBF networks.

7.3 LOCAL AND GLOBAL LEARNING ALGORITHMS

In this chapter we mainly consider the identification of the parameters of the local models for a given model structure, i.e. we are estimating θ for an *a priori* given set of c_i, Σ_i parameters. We recognise that the problem of pre-structuring or learning the structure of the local model network is perhaps the more important and challenging one (Johansen and Foss 1995, Jordan and Jacobs 1994, Murray-Smith and Gollee 1994), and we briefly deal with this in section 7.6, but as we shall see, the parameter learning problem has some interesting characteristics that are closely related to the structuring of the local model network. This justifies a closer look at the parameter learning problem.

First, we briefly review the standard solution to this problem, using a least squares criterion that penalises mismatch between the training data and the global model output. Thereafter, we describe an alternative solution based on a number of weighted least squares criteria that penalises mismatch between the data and the local model outputs. Finally, the properties of these two different algorithms are discussed.

The general concept of local learning algorithms was suggested in the context of non-parametric models (Holcomb and Morari 1991), (Nadaraya 1964), (Watson 1964), (Benedetti 1977) and (Cleveland *et al.* 1988). The parametric model form, as considered here, is seen in (Bottou and Vapnik 1992). The main contribution of this chapter is an analysis of such an algorithm in the context of local model networks, summarising and extending some results described in (Murray-Smith 1994, Murray-Smith and Johansen 1995).

7.3.1 Global learning

The local models are assumed linear in the parameters, as in equation (7.6), so the learning problem is a straightforward application of linear regression techniques to find the parameters θ which best match the data (assuming fixed c_i, Σ_i). Stacking the data into matrices, we get the following regression model:

$$\mathbf{y} = \Phi\theta^\star + \mathbf{b} + \mathcal{E}, \tag{7.7}$$

where Φ is the design matrix, the rows of which are defined by

$$\Phi_k^T = \left[\rho_1(\psi(k))[1, \ \psi^T(k)], ..., \rho_{n_\mathcal{M}}(\psi(k))[1, \ \psi^T(k)]\right], \tag{7.8}$$

so that the design matrix Φ, vector of output measurements $\mathbf{y}$, vector of biases $\mathbf{b}$, and errors $\mathcal{E}$ are

$$\Phi = \left(\ \Phi_1, ..., \Phi_N \ \right)^T, \quad \mathbf{y} = \left(\ y(1), ..., y(N) \ \right)^T$$
$$\mathbf{b} = \left(\ b(1), ..., b(N) \ \right)^T, \quad \mathcal{E} = \left(\ \varepsilon(1), ..., \varepsilon(N) \ \right)^T.$$

The standard least squares criterion for this estimation problem is

$$J(\theta) = \frac{1}{N}(\mathbf{Y} - \Phi\theta)^T(\mathbf{Y} - \Phi\theta) \tag{7.9}$$

and the Moore–Penrose pseudoinverse of Φ, Φ^+ is used to estimate the weights:

$$\hat{\theta}_{LS} = \Phi^+\mathbf{y} = (\Phi^T\Phi)^{-1}\Phi^T\mathbf{y}, \tag{7.10}$$

assuming $\Phi^T \Phi$ is non-singular. The numerical algorithm used in the examples[3] to calculate the pseudoinverse is the *Singular Value Decomposition* (SVD). The SVD algorithm decomposes any $N \times p$ matrix Φ, such that $\Phi = USV^T$ and the pseudoinverse of Φ is:

$$\Phi^+ = VS^+U^T. \tag{7.11}$$

Once the singular values have been zeroed, the parameters solving the regression problem in equation (7.7) can be calculated.

$$\hat{\theta}_{LS} = VS^+U^T\mathbf{y}. \tag{7.12}$$

7.3.2 Local learning

The global learning approach is based on the assumption that all of the parameters θ would be estimated in a single regression operation. This may not always be computationally feasible for problems with a large amount of data, or where many local models are needed (see section 7.5.4). Perhaps more worrying, the global nature of the learning also means that the parameters of the local models cannot be interpreted independently of neighbouring local models, which means that they cannot be seen as local approximations to the underlying system (see section 7.4). An alternative to global learning which is less prone to these disadvantages is to locally estimate the parameters of each of the local models (as defined in equation (7.6)) independently.[4] The parameters of the local models are then estimated using a set of locally weighted estimation criteria for the ith local model

$$J_i(\theta_i) = \frac{1}{N}(\mathbf{y} - \Psi\theta_i)^T \mathbf{Q}_i (\mathbf{y} - \Psi\theta_i), \tag{7.13}$$

where $i = 1, ..., n_\mathcal{M}$. $\mathbf{Q}_i$ is an $N \times N$ diagonal weighting matrix, where the diagonal elements are weights $\alpha_i(\psi(1)), ..., \alpha_i(\psi(N))$, which are used to weight the importance of the different samples in the training set on the ith local model. The matrix Ψ is the local model design matrix, defined by

$$\Psi = \begin{pmatrix} 1, & \psi^T(1) \\ 1, & \psi^T(2) \\ \vdots & \vdots \\ 1, & \psi^T(N) \end{pmatrix}. \tag{7.14}$$

To achieve local learning it is necessary to define a set of local criteria. Our confidence in a given observation regarding its relevance for the ith local model is directly reflected in the ith basis function. A plausible local weighting function is therefore $\alpha_i(\psi) = \rho_i(\psi)$, which results in

$$\mathbf{Q}_i = \text{diag}(\rho_i(\psi(1)), ..., \rho_i(\psi(N))). \tag{7.15}$$

In this case the locally weighted least squares estimate of the local model parameter vector θ_i is given by the minimum of J_i. In matrix terms the operation is now

$$\hat{\theta}_{LLS} = \left(\hat{\theta}_{LLS,1}^T, ..., \hat{\theta}_{LLS,n_\mathcal{M}}^T\right)^T$$

$$\hat{\theta}_{LLS,i} = \left(\Psi^T\mathbf{Q}_i\Psi\right)^{-1}\Psi^T\mathbf{Q}_i\,\mathbf{y}, \quad i = 1, 2, ..., n_\mathcal{M}.$$

---

[3]We used the MATLAB function svd(). For more details on SVD see the general treatment in books such as (Golub and van Loan 1989), or (Press *et al.* 1988). Diagonal elements of S that are less than a preset tolerance, are zeroed, effectively reducing the degrees of freedom in the model.

[4]This assumes that the basis functions achieve a partition of unity.

A side-effect of local learning is that it also allows more flexibility in the use of learning algorithms, which will be especially useful with heterogeneous local model networks which use a variety of optimisation algorithms (possibly also not linear in the parameters), which are locally applied, each suited to the individual local model type.

7.4 EXPERIMENTS ON A 1-D NOISY FUNCTION

Local and global learning may lead to considerably differing results. To show the effect of the two learning methods we use a simple one-dimensional example. The arbitrarily chosen nonlinear function is

$$y = \cos(6\psi^2) + \varepsilon(\psi), \tag{7.16}$$

where the additive noise term $\varepsilon(\psi)$ is Gaussian with a varying standard deviation of $\sigma(\psi) = 0.4 \exp\left(-4.6 \left|\psi - \frac{1}{2}\right|\right)$. The unnormalised basis-function is the Gaussian. The function is shown in Figure 7.2. Global learning gives an 'oscillatory' model, and local models that are not accurate local approximations. Local learning avoids these undesirable phenomena by using a cost function which forces local models to approximate the target function, while global learning has no such cost function – only the sum of the local models is important.

Both examples used the SVD algorithm, zeroing the inverse of singular values smaller than 10^{-5}. For smaller training sets (Figure 7.2), the relative robustness of the local learning is immediately obvious, both in the smoother response, and in the lower error on the fit to the test function. As the amount of data increases, the global method's accuracy improves beyond the accuracy of local learning, but the 'oscillating' local models remain.

This example is intended as an illustration of how sensitive global learning can be, and the same phenomenon has been observed in a number of other examples (Chapter 3 uses local learning in a modelling application. Murray-Smith (1994) provides further examples).

7.4.1   Effect of overlap factor on condition

These experimental results lead us to ask whether the local model network structure is particularly prone to ill-conditioning due to over-parameterisation. The fact that the bases of a local model net are often effectively the same inputs weighted differently by the basis functions (see equation (7.8)), means that the higher the level of overlap between basis-functions in local model nets, the higher level of correlation of any model's inputs with its neighbours. The level of overlap thus becomes a major factor in the poor conditioning of the design matrix in local model nets. Reducing the overlap, however, leads to undesirable non-smooth transitions between models. Figure 7.3(a) shows the increase in condition number with increasing numbers of local models, when the relative overlap remains identical. To better understand the role of overlap, Figure 7.3(b) shows the increase in condition number in the example used earlier with a fixed number of local models (seven) when the overlap is increased. It should be mentioned that the high condition numbers are partially due to suboptimal scaling of the parameters, in particular for linear and quadratic local models. See also Chapter 8 for further analysis of condition in local model networks.

A major effect of local learning on the cost functional is to remove the interaction of neighbouring local models' parameters from the design matrix. See section 7.5.3 for a theoretical analysis of the effect of the overlap factor on learning.

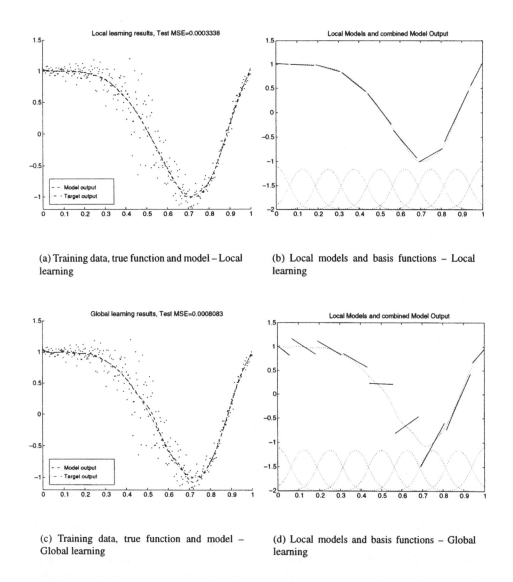

(a) Training data, true function and model – Local learning

(b) Local models and basis functions – Local learning

(c) Training data, true function and model – Global learning

(d) Local models and basis functions – Global learning

Figure 7.2 Experimental comparison of global and local learning for 300 training points. The left-hand side shows the true function, the noisy training data and the trained network's response. The validation criterion in the figure titles is the standard mean square error, evaluated on the model's deviation from the true function (not the training data). The right-hand side of each figure shows the normalised basis functions and the associated local linear models.

7.5 DIFFERENCES BETWEEN LOCAL AND GLOBAL LEARNING ALGORITHMS

7.5.1 Causes for ill-conditioning in global learning

In local model networks using global learning, even with numerically robust estimation algorithms, we sometimes observe, as in Figure 7.2, that the local models do not change

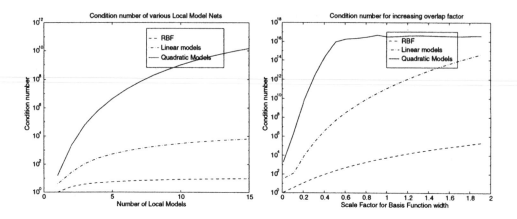

(a) Constant relative overlap, with uniform basis functions evenly spaced, $\Sigma = |c_1 - c_2|$, but increasing numbers of local models.

(b) Varying overlap. The x-axis shows the scaling factor for the basis function width. A scaling factor of 1 represents $\Sigma = |c_1 - c_2|$.

Figure 7.3 Condition number of design matrix for global learning increasing with number of local models or with overlap.

smoothly as a function of the input. Also, they cannot be viewed as local approximations of the systems. This phenomenon of "oscillating" local models, where the negative contribution of one local model is compensated for by the positive contribution of the neighbouring local models, leads to a delicate balance which may minimise the global error on the training set, but need not necessarily be robust when confronted with new samples, i.e. the model generalises poorly, and is certainly more difficult to interpret. The use of such non-smooth models can also be highly disadvantageous in many applications.[5] We have identified two potential causes of such ill-conditioned behaviour:

1. Even the optimal parameters θ^* may give such behaviour, when there is a *fundamental structural mismatch* between the model $\hat{f}(\cdot, \theta^*)$ and the true system f. One example of such a structural mismatch is a discontinuous function f together with smooth basis-functions, as in Figure 7.4. Another example of structural mismatch is a high concentration of local models in regions of the input space where the system has low complexity, rather than in regions with high complexity.[6] The local model parameters in regions corresponding to low system complexity are effectively utilised to improve the fit in regions with high system complexity and low model complexity, which causes ill-conditioning because the corresponding basis-functions are close to zero, giving very low sensitivity. Such problems are well known phenomena from approximation theory, cf. Gibbs phenomena in Fourier series, and the instability of high-order polynomial approximations which motivated the introduction of splines.

2. Alternatively, the cause may be *over-parameterisation*, which gives an *ill-conditioned*

---

[5]For example, models are often applied to formulate optimisation criteria. If, however, the model is not smooth but highly oscillatory, then optimisation based on gradient-type search methods could be subject to many local minima and it could lead to an unreliable solution.

[6]This is also an argument against unsupervised learning methods for basis function optimisation, as these simply approximate the distribution of the input data, and do not take system complexity into account.

$\Phi^T \Phi$ and a larger parameter estimate variance, which obviously may appear as large variability among neighbouring local models' parameters. For example, an increasing number of local models (increasing model structure complexity) will typically lead to higher condition number of $\Phi^T \Phi$, and hence, larger variance on the parameter estimate (see below). On the other hand, the effect of an increasing number of local models on the prediction performance depends on whether the decrease in bias is more significant than the increase in variance, as we shall see later.

7.5.2 A statistical view on ill-conditioning in global learning

In this section we investigate from a statistical perspective the observed ill-conditioning using global learning, and show that any ill-conditioning is expected to be reduced with local learning.

In order to make the following analysis simple and transparent, we assume that the inputs $\psi(k)$ in the design matrix Φ are deterministic.[7] With global learning (the least squares estimator), we have (e.g. (Ljung 1987))

$$E\hat{\theta}_{LS} = \theta^* + (\Phi^T \Phi)^{-1} \Phi^T \mathbf{b} \tag{7.17}$$

$$E(\hat{\theta}_{LS} - E\hat{\theta}_{LS})(\hat{\theta}_{LS} - E\hat{\theta}_{LS})^T = (\Phi^T \Phi)^{-1} \sigma^2, \tag{7.18}$$

where E is the expectation with respect to the probability distribution of ε. Moreover, we get the following bias/variance decomposition of the expected squared prediction error (e.g. (Hastie and Tibshirani 1990))

$$
\begin{aligned}
\text{PSE}_{LS} &= \frac{1}{N} E \left(\mathbf{y} - \Phi\hat{\theta}_{LS} \right)^T \left(\mathbf{y} - \Phi\hat{\theta}_{LS} \right) \\
&= \frac{1}{N} E (\mathbf{y} - \Phi E\hat{\theta}_{LS})^T (\mathbf{y} - \Phi E\hat{\theta}_{LS}) \\
&\quad + \frac{1}{N} \text{tr} \left(\Phi E \left(\hat{\theta}_{LS} - E\hat{\theta}_{LS} \right) \left(\hat{\theta}_{LS} - E\hat{\theta}_{LS} \right)^T \Phi^T \right) \\
&= \frac{1}{N} \mathbf{b}^T (I - \Phi(\Phi^T \Phi)^{-1} \Phi^T) \mathbf{b} + \sigma^2 \\
&\quad + \frac{1}{N} \text{tr} \left(E(\hat{\theta}_{LS} - E\hat{\theta}_{LS})(\hat{\theta}_{LS} - E\hat{\theta}_{LS})^T \Phi^T \Phi \right).
\end{aligned}
$$

Substituting (7.18) into the last term, and observing that the trace of the resulting identity matrix simply reduces to its dimension, $p = \dim(\theta)$, it follows that with the least squares estimator we get, e.g. (Hastie and Tibshirani 1990)

$$\text{PSE}_{LS} = \beta^2 + \sigma^2 + \sigma^2 p/N. \tag{7.19}$$

The bias term

$$\beta^2 = \frac{1}{N} \text{tr} \left((I - \Phi(\Phi^T \Phi)^{-1} \Phi^T) \mathbf{b}\mathbf{b}^T \right)$$

is the average bias, while $\sigma^2 p/N$ is the effect of the parameter estimator variance on the prediction. Hence, we have the well known bias/variance trade-off (Geman et al. 1992),

---

[7] Notice that qualitatively similar results also hold with a random design matrix, at least asymptotically, although the analysis is considerably more complicated.

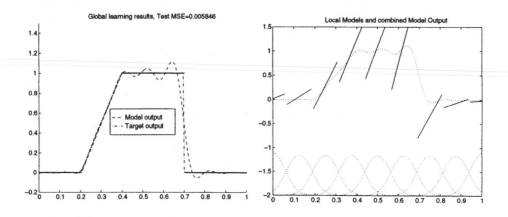

(a) Training data, true function and model – Global learning of model parameters, with fixed basis functions

(b) Local models and basis functions – Global learning

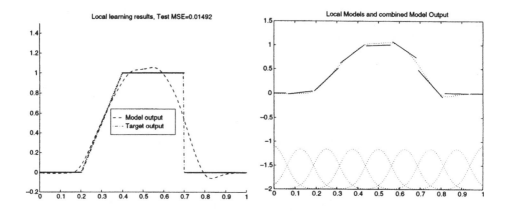

(c) Training data, true function and model – Local learning of model parameters, with fixed basis functions

(d) Local models and basis functions – Local learning

Figure 7.4 Global learning is ill-conditioned for learning problems with fundamental model mismatch, even when as in this case, there is no variance in the data and the data set is very large and uniformly distributed on the input space. As one would expect, given the different cost functions, local learning produces local models which are close to the target function locally, but global learning has a smaller mean square error (MSE) (a global cost measure), meaning that the sum of the local models is closer to the target function in the globally trained case.

which is essentially that the model variance can be reduced at the cost of increased bias by reducing the degrees of freedom in the model.

It is clear from (7.18) that an ill-conditioned matrix $\Phi^T\Phi$ will lead to a large variability in some directions in the parameter space.[8] However, whether the matrix $\Phi^T\Phi$ is ill-conditioned or not has no impact on the predicted squared error (PSE), cf. (7.19) where only the number of parameters relative to the number of training samples p/N is of importance.[9] It also follows from (7.19) that the amount of overlap does not affect the variance part of the PSE, while it does greatly affect the variance of the parameter estimate, see Figure 7.3(b) and (7.18). There is also empirical evidence that the amount of overlap is important for the bias, since it is our experience that "reasonably wide and smooth" basis functions tend to give less PSE, with global learning and a well-behaved, smooth function to be estimated.

7.5.3 Regularising effect of local learning

Following the classic regularisation method as defined in *regularisation theory* (Tikhonov and Arsenin 1977), the standard quadratic error criterion is extended to become a cost-complexity operator,

$$J(\boldsymbol{\theta}) = \frac{1}{N}\sum_{k=1}^{N}(y(k) - \hat{f}(\psi(k), \boldsymbol{\theta}))^2 + \lambda R(\boldsymbol{\theta}), \tag{7.20}$$

including a nonnegative penalty function R which includes *a priori* information, such as smoothness, which makes the learning problem better conditioned. This forces the optimisation to find a regularised solution (λ is a weighting, which defines the relative amount of regularity enforced on the estimated parameters), which is likely to reduce the variance of the solution, at the cost of introduced bias, e.g. (Bishop 1991). Many neural network learning algorithms have implicitly (often unplanned!) had a regularisation effect, in that they do not find the 'optimal' (in the least squares sense) solution to the posed optimisation problem. Methods such as weight decay, stopping learning early (Sjöberg and Ljung 1992), network pruning, learning with noise (Bishop 1994) are all examples of *ad hoc* attempts to produce a regularisation effect. Local learning is an alternative method that can be applied to eliminate, or at least reduce the ill-conditioning and oscillating local model phenomena that may be caused by the reasons mentioned above.

To understand the effect of local learning on the expected squared prediction error, we examine the change in effective degrees of freedom in the model.[10] Notice that

$$\hat{\mathbf{y}} = \sum_{i=1}^{n_\mathcal{M}}\mathbf{Q}_i\boldsymbol{\Psi}\hat{\boldsymbol{\theta}}_{LLS,i} = \mathbf{S}\mathbf{y}, \tag{7.21}$$

where the smoothing matrix $\mathbf{S}$ for locally estimated weights is defined by

$$\mathbf{S} = \sum_{i=1}^{n_\mathcal{M}}\mathbf{Q}_i\boldsymbol{\Psi}(\boldsymbol{\Psi}^T\mathbf{Q}_i\boldsymbol{\Psi})^{-1}\boldsymbol{\Psi}^T\mathbf{Q}_i. \tag{7.22}$$

---

[8]The *condition* of the $\Phi^T\Phi$ is in general important for the robustness of the learning process. The *condition number* of a square matrix is defined to be the ratio between its largest and smallest singular values. The larger the condition number, the larger the effect of slight changes in the matrix Φ on the solution of the pseudoinverse Φ^+. As the parameters are dependent on Φ^+, a slight change in data would lead to different weights, so generalisation is likely to be poor.

[9]Notice that when the inputs ψ are viewed as stochastic variables, the condition of $\Phi^T\Phi$ may have some effect on the PSE.

[10]Also used in neural network literature, e.g. (Moody 1992, Wu and Moody 1996).

It is straightforward to show that

$$\text{PSE}_{LLS} \;=\; \tilde{\beta}^2 + \sigma^2 + \sigma^2 \tilde{p}/N, \tag{7.23}$$

where the average bias is redefined by

$$\tilde{\beta}^2 \;=\; \frac{1}{N}\text{tr}\left((\mathbf{I}-\mathbf{S})(\mathbf{I}-\mathbf{S})\mathbf{bb}^T\right),$$

and the degrees of freedom are defined by

$$
\begin{aligned}
\tilde{p} \;&= \text{tr}\left(\mathbf{SS}^T\right) \\
&= \sum_{i=1}^{n_\mathcal{M}} \text{tr}\left((\boldsymbol{\Psi}^T\mathbf{Q}_i\boldsymbol{\Psi})^{-1}\boldsymbol{\Psi}^T\mathbf{Q}_i\mathbf{Q}_i\boldsymbol{\Psi}(\boldsymbol{\Psi}^T\mathbf{Q}_i\boldsymbol{\Psi})^{-1}\boldsymbol{\Psi}^T\mathbf{Q}_i\mathbf{Q}_i\boldsymbol{\Psi}\right) \\
&= \sum_{i=1}^{n_\mathcal{M}} \text{tr}\left((\boldsymbol{\Psi}^T\mathbf{Q}_i\boldsymbol{\Psi})^{-1}\boldsymbol{\Psi}^T\mathbf{Q}_i\boldsymbol{\Psi}(\boldsymbol{\Psi}^T\mathbf{Q}_i\boldsymbol{\Psi})^{-1}\boldsymbol{\Psi}^T\mathbf{Q}_i\boldsymbol{\Psi}\right. \\
&\qquad -(\boldsymbol{\Psi}^T\mathbf{Q}_i\boldsymbol{\Psi})^{-1}\boldsymbol{\Psi}^T\mathbf{D}_i\boldsymbol{\Psi}(\boldsymbol{\Psi}^T\mathbf{Q}_i\boldsymbol{\Psi})^{-1}\boldsymbol{\Psi}^T\mathbf{Q}_i\mathbf{Q}_i\boldsymbol{\Psi} \\
&\qquad \left. - (\boldsymbol{\Psi}^T\mathbf{Q}_i\boldsymbol{\Psi})^{-1}\boldsymbol{\Psi}^T\mathbf{Q}_i\boldsymbol{\Psi}(\boldsymbol{\Psi}^T\mathbf{Q}_i\boldsymbol{\Psi})^{-1}\boldsymbol{\Psi}^T\mathbf{D}_i\boldsymbol{\Psi}\right) \\
&= p - \sum_{i=1}^{n_\mathcal{M}} \text{tr}\left((\boldsymbol{\Psi}^T\mathbf{Q}_i\boldsymbol{\Psi})^{-1}\boldsymbol{\Psi}^T\mathbf{D}_i\boldsymbol{\Psi}(\boldsymbol{\Psi}^T\mathbf{Q}_i\boldsymbol{\Psi})^{-1}\boldsymbol{\Psi}^T\mathbf{Q}_i\mathbf{Q}_i\boldsymbol{\Psi}\right. \\
&\qquad \left. + (\boldsymbol{\Psi}^T\mathbf{Q}_i\boldsymbol{\Psi})^{-1}\boldsymbol{\Psi}^T\mathbf{D}_i\boldsymbol{\Psi}\right),
\end{aligned}
$$

where $\mathbf{D}_i \;=\; \mathbf{Q}_i - \mathbf{Q}_i\mathbf{Q}_i$, which will always be positive semi-definite, and $\tilde{p} \leq p$. The degrees of freedom in the model are therefore less with local learning than with global learning. This gives a reduced variance $\sigma^2\tilde{p}/N$, at the possible cost of an increased bias $\tilde{\beta}^2$, compared to global learning. This is the well known bias/variance dilemma discussed in (Geman *et al.* 1992) and (Hastie and Tibshirani 1990).

Overlap and regularisation

In the case when the overlap becomes very small, it is clear that the $\rho_k(\cdot)$ functions approach step-functions, and $\mathbf{Q}_i\mathbf{Q}_i \to \mathbf{Q}_i$. Hence, it follows from the expression for $\tilde{p}$ that $\tilde{p} \to p$ as the overlap parameter σ_i goes to zero. This can also be seen intuitively, since there will be no interaction between the parameters of the different local models. In the opposite case, when the overlap parameter σ_i approaches infinity, it is easily seen that all the local models are given the same weight uniformly over the input domain, i.e. $\mathbf{Q}_i \to \mathbf{I}/N$, where $\mathbf{I}$ is the $N \times N$ identity matrix. We see from the equation for $\tilde{p}$ that $\tilde{p} \to p/N = 1 + n_\psi$. Again, this can be seen intuitively from the fact that the uniform weighting of the $n_\mathcal{M}$ local models effectively corresponds to only one local model which covers the whole input domain.

The amount of regularisation and overlap is controlled by the width parameters[11] $\boldsymbol{\Sigma}_i$, which can be interpreted as regularisation parameters or local capacity control parameters (Bottou and Vapnik 1992). Standard methods for choosing such parameters can therefore be used, (Hastie and Tibshirani 1990), and this is shown in section 7.6.1.

---

[11]The centres also help determine the level of overlap between given units, but we shall concentrate on the width parameters in this analysis.

7.5.4 Computational effort

One reason for using local learning methods is that the computational effort may be reduced. The effort needed to find the pseudoinverse using SVD for a $(N \times p)$ matrix[12] is roughly (Noble and Daniel 1988),

$$O_{gl} = O\left(N^2 p + N p^2 + \min(N, p)^3\right), \tag{7.24}$$

where $p = n_{\mathcal{M}}(n_\psi + 1) = \dim(\theta)$ is the number of parameters in the model, which is clearly the crucial factor with regard to computational effort. The local effort O_{lo} is repeated $n_{\mathcal{M}}$ times,

$$O_{lo} = O\left(n_{\mathcal{M}}\left(N_i^2 n_\psi + N_i n_\psi^2 + \min(N_i, n_\psi)^3\right)\right), \tag{7.25}$$

where N_i is usually significantly smaller than N.

7.6 INTERACTION BETWEEN MODEL STRUCTURE AND LEARNING METHOD

In summary, the simple analysis in the previous section has suggested that local learning has a regularising effect, characterised by

1   A reduction of the degrees of freedom in the model structure.
2   Reduced variance at the possible cost of increased bias in the squared prediction error PSE, compare (7.19) with (7.23).
3   Reduced variance at the cost of increased bias in the parameter estimate.

With some model structures there is a trade-off between smoothly changing local model behaviour, and small expected squared prediction error (PSE). These model structures are typically characterised by over-parameterisation or fundamental structural mismatch. If the cause of the ill-conditioning in an estimated local model net is due to *over-parameterisation*, local learning has the beneficial effect of reducing the variance and therefore minimising the PSE (generalisation error), since the degrees of freedom in the model structure have effectively been reduced by the implicit regularisation in local learning. If, on the other hand, there is a *fundamental structural mismatch* between the model and the underlying system, the use of local learning will lead to extra bias being introduced and if this is more significant than the reduction in variance, then local learning will lead to an increase in PSE. However, local learning *will* reduce the variability of the parameter estimate, avoiding the oscillations and leading to more transparent local models.

We still, despite our local representation, need to approximate the model's global fit to the data. The improved understanding of the interdependence of estimation method and model structure (relative to the underlying process and available data) should help design model structures as well as the learning algorithms which adapt local models and their basis functions. This section will illustrate the methods on two learning algorithms for basis function optimisation.

7.6.1 Adapting model structures

In the examples used above, the model structure had uniform complexity (or capacity) over the input space, due to the constant overlap factors. The example used in this section shows

---

[12] $p = n_{\mathcal{M}}(n_\psi + 1)$ is equivalent to the cost of a homogeneous local model net with linear local models, where n_ψ represents the dimension of the model's input space.

the effect of adapting the overlap between the basis functions to reduce the problems with ill-conditioning, because the local capacity (or complexity) of the model is adapted to match the local complexity of the true system. Both learning algorithms used, a GCV minimising algorithm and an Expectation Minimisation (EM – see section 7.8.2 for details) approach, lead to models with non-uniform local complexity. To illustrate the differing effects of the learning algorithms, global learning, local learning and EM learning were used on the same local model network. The learning task is the one-dimensional function shown in Figure 7.4, which illustrates the difference between local learning and global learning.

Overlap parameter optimisation

First we describe a method which optimises the basis function overlap, using the effective degree of freedom estimates developed earlier in the chapter. Let Σ be the matrix containing the basis function overlap parameters Σ_i, which for example in the case of Gaussian basis functions would be parameters from the basis functions' covariance matrices. If the dependence on Σ is written explicitly, then the smoothing matrix defined by (7.22) is written

$$S(\Sigma) \;=\; \sum_{i=1}^{n_{\mathcal{M}}} Q_i(\Sigma)\Psi(\Psi^T Q_i(\Sigma)\Psi)^{-1}\Psi^T Q_i(\Sigma)$$

and the degrees of freedom are

$$\tilde{p}(\Sigma) \;=\; \mathrm{tr}(S(\Sigma)S(\Sigma)^T).$$

Now, the expected squared prediction error can be estimated by for example using the Generalised Cross Validation statistic (Craven and Wahba 1979)

$$\mathrm{GCV}(\Sigma) \;=\; \frac{1}{(1 - \tilde{p}(\Sigma)/N)^2}\frac{1}{N}(y - \Phi\hat{\theta}_{LLS}(\Sigma))^T(y - \Phi\hat{\theta}_{LLS}(\Sigma))$$

using a standard nonlinear programming algorithm. Similar to the standard procedure in non-parametric regression and smoothing, we suggest that minimising GCV with respect to Σ is one possible alternative for choosing overlap parameters that are close to optimal in terms of global prediction performance, while still using local learning to fit the local model parameters. Hence, this approach lies somewhat between local and global learning, since the local model parameters are fitted to minimise local model mismatch criteria, while the overlap parameters are fitted to minimise a global model mismatch criterion.

The GCV optimisation[13] approach to selection of overlap parameters typically leads to basis functions with little overlap in areas with high system complexity, and large levels of overlap in areas with low system complexity when the model is pre-structured with uniform complexity. From the discussion in section 7.5.3 it is clear that large overlap in an area means locally low model capacity in that area, and vice versa. Hence, the local capacity of the model is adapted to match the local complexity of the true system, and the fundamental structural model mismatch is reduced.

EM optimisation of basis functions

The EM algorithm for local model networks is described in section 7.8.2, and the local model parameter identification stage has many similarities to local learning, but uses a different weighting function – the posterior instead of the prior. Although the local model

---

[13]The optimisation was done using the standard MATLAB function `constr()`.

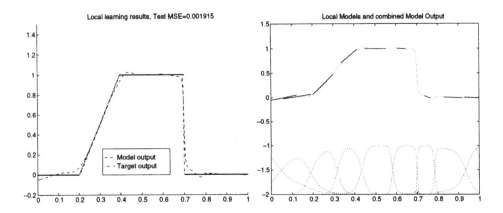

(a) Training data, true function and model – Local and GCV learning of model parameters and basis functions

(b) Local models and basis functions – GCV and Local learning

Figure 7.5 Trained Local Model Network using Local and GCV learning. Results on this experiment are better than any of the other approaches in this chapter, but this is not necessarily generally so.

parameter identification phase of EM is initially identical to local learning, the iterative nature of EM makes the approach more global, as after the first iteration the posterior will differ from the prior, and the global information about model fit will affect the optimal local model parameters, producing dependencies between neighbouring local models.

The models resulting from EM learning are plotted in Figure 7.6. From the MSE results in the figures, it is clear that, as expected, basis function adaptation reduces the error, and it is interesting to note that the GCV approach compares favourably with EM, even though it was not free to adapt the basis function centres.

7.6.2 Visualisation of the *effective* smoothing kernels

The basis functions in a local model net are usually visualised as the individual ρ_k, but using the smoothing formalism described in equation (7.21), we can plot the effective smoothing function at each point in the training set as in (Lowe 1995). (He also points out that global basis functions can still lead to local effective kernels), and as with the *equivalent kernels* analysis used in (Hastie and Tibshirani 1990). These basis functions are therefore data-dependent $\Psi_k(\mathcal{D})$, and are combined linearly with the training set outputs $\mathbf{y}$.

In the global learning case,

$$\hat{\mathbf{y}} = \Phi\hat{\theta}_{LS} = \Phi\Phi^+\mathbf{y} \tag{7.26}$$

$$= \mathbf{Sy}, \tag{7.27}$$

where, for global least squares,

$$\mathbf{S} = \Phi(\Phi^T\Phi)^{-1}\Phi^T, \tag{7.28}$$

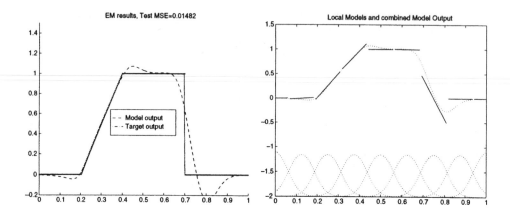

(a) Training data, true function and model – EM learning of model parameters, with fixed basis functions

(b) Local models and basis functions – EM learning of local model parameters, with fixed basis functions

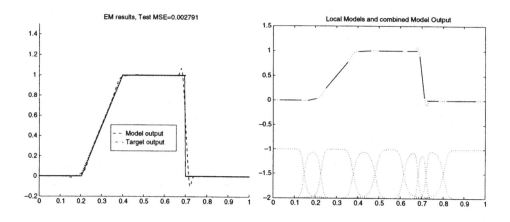

(c) Training data, true function and model – EM learning of all parameters

(d) Local models and basis functions – EM learning

Figure 7.6   Trained Local Model Network using EM learning. Compare with Figures 7.4 and 7.5. EM learning of local model parameters produces a slight improvement over local learning, and further improvement is gained by optimising basis function parameters.

there are N effective kernels (one for each data point). The kernel associated with the kth input vector in the training set is defined as

$$\Psi_k = \Phi_k(\Phi^T\Phi)^{-1}\Phi^T, \tag{7.29}$$

where Φ_k is the kth row of the design matrix, and the model output is:

$$\hat{y}_k = \Psi_k\mathbf{y}. \tag{7.30}$$

In the local learning case,

$$\hat{\mathbf{y}} = \sum_{i=1}^{n_M} \mathbf{Q}_i \Psi \hat{\theta}_{LLS,i} = \mathbf{Sy} \tag{7.31}$$

$$\mathbf{S} = \sum_{i=1}^{n_M} \mathbf{Q}_i \Psi (\Psi^T \mathbf{Q}_i \Psi)^{-1} \Psi^T \mathbf{Q}_i \tag{7.32}$$

$$\Psi_k = \sum_{i=1}^{n_M} \rho_i(\psi(k))[1, \psi^T(k)]^T (\Psi^T \mathbf{Q}_i \Psi)^{-1} \Psi^T \mathbf{Q}_i. \tag{7.33}$$

For EM learning, the equations are slightly different, as the weighting term is no longer the prior $\mathbf{Q}_i$, but the posterior $\text{diag}(h_i^t)$ at learning iteration t.

$$\mathbf{S}^t = \sum_{i=1}^{n_M} \mathbf{Q}_i^t \Psi (\Psi^T \text{diag}(h_i^t) \Psi)^{-1} \Psi^T \text{diag}(h_i^t) \tag{7.34}$$

$$\Psi_k = \sum_{i=1}^{n_M} \rho_i(\psi(k))[1, \psi^T(k)]^T (\Psi^T \text{diag}(h_i^t) \Psi)^{-1} \Psi^T \text{diag}(h_i^t). \tag{7.35}$$

The effective kernels for the trained networks in Figures 7.4 and 7.6 are plotted in Figure 7.7 for data points spread out over the input space. This is basically looking at slices of the smoothing matrix (assuming the inputs in the training set were ordered and evenly distributed).

To better view the smoothing effect we can plot the smoothing matrices for the above example, where the input data is ordered, and evenly spread from 0 to 1. The three-dimensional plots in Figures 7.8(a) and 7.8(b) are basically giving the same information as the kernels, so again we see that the smoothing matrix associated with global learning has oscillations far from the given input point, pointing out the non-locality of the resulting model in a very clear manner. The local smoothing matrix, in contrast, has far fewer oscillations and is more localised around a given kernel.

7.6.3 Local degree of freedom estimates

The earlier sections discussed the effective degrees of freedom measure as an interesting theoretical tool which gave us some insight into the behaviour of different learning algorithms. Given that our architecture is local, and that adaptation of the basis functions leads to different densities of basis functions and levels of overlap, it would be interesting to estimate local measures of available degrees of freedom. Consider the following decomposition of the DOF,

$$\tilde{p} = \text{trace}(\mathbf{SS}^T)$$

$$= \text{trace}\left(\sum_{i=1}^{n_M} \sum_{j=1}^{n_M} \mathbf{Q}_i \Psi (\Psi^T \mathbf{Q}_i \Psi)^{-1} \Psi^T \mathbf{Q}_i \mathbf{Q}_j \Psi (\Psi^T \mathbf{Q}_j \Psi)^{-1} \Psi^T \mathbf{Q}_i \right)$$

$$= \sum_{i=1}^{n_M} \sum_{j=1}^{n_M} \text{trace}\left(\mathbf{Q}_i \Psi (\Psi^T \mathbf{Q}_i \Psi)^{-1} \Psi^T \mathbf{Q}_i \mathbf{Q}_j \Psi (\Psi^T \mathbf{Q}_j \Psi)^{-1} \Psi^T \mathbf{Q}_i \right)$$

$$= \sum_{i=1}^{n_M} \tilde{p}_i,$$

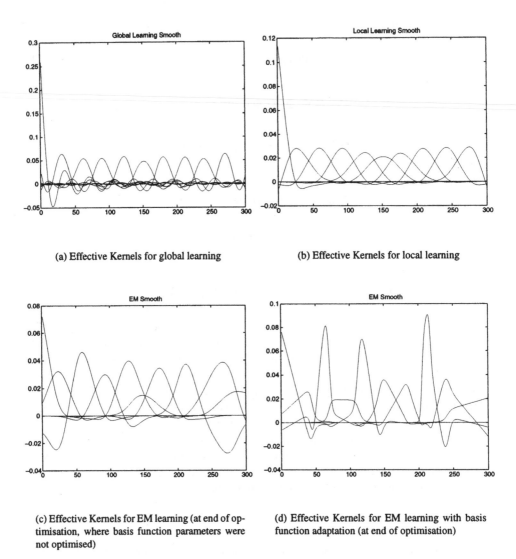

(a) Effective Kernels for global learning

(b) Effective Kernels for local learning

(c) Effective Kernels for EM learning (at end of optimisation, where basis function parameters were not optimised)

(d) Effective Kernels for EM learning with basis function adaptation (at end of optimisation)

Figure 7.7 Effective kernels for different learning algorithms. The effective smoothing kernels reveal the effect of the parameter estimation method used, as well as the shape of the original basis functions. Note also, that these kernels are no longer positive definite. The kernels associated with global learning in Figure 7.7(a) give further indication of the problems associated with the method, as they oscillate through a wide range of the input space. This shows the coupling between widely spread data points in the training set on the model's performance in a given locale. Figure 7.7(b) shows the smoother smooth produced by local learning. The kernels for EM learning in Figure 7.7(c) show the effect of the use of the posterior as a weighting function in the parameter estimation process (even though the priors are identical to those shown in 7.7(b).

where $\bar{p}_i$, the 'local degrees of freedom' associated with local model i, are

$$\bar{p}_i = \sum_{j=1}^{n_{\mathcal{M}}} \text{trace}\left(\mathbf{Q}_i\mathbf{\Psi}(\mathbf{\Psi}^T\mathbf{Q}_i\mathbf{\Psi})^{-1}\mathbf{\Psi}^T\mathbf{Q}_i\mathbf{Q}_j\mathbf{\Psi}(\mathbf{\Psi}^T\mathbf{Q}_j\mathbf{\Psi})^{-1}\mathbf{\Psi}^T\mathbf{Q}_i\right). \qquad (7.36)$$

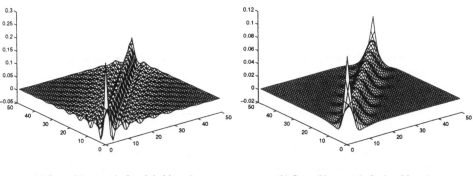

(a) Smoothing matrix for global learning

(b) Smoothing matrix for local learning

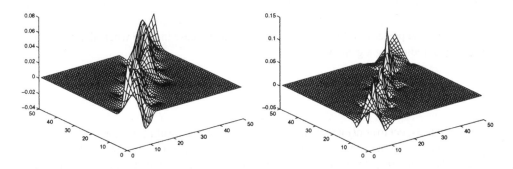

(c) Smoothing matrix for EM learning with fixed basis functions (at end of optimisation)

(d) Smoothing matrix for EM learning with structure adaptation (at end of optimisation)

Figure 7.8 Smoothing matrices for different learning algorithms. Note the extended oscillations in the global case.

These can then be used to interpret the model. The 'average' degrees of freedom of the local models used at any given point in the input space can be defined as

$$\tilde{p}_{av}(\psi) \;=\; \sum_{i=1}^{n_{\mathcal{M}}} \tilde{p}_i \rho_i(\psi), \tag{7.37}$$

and from this we notice in Figure 7.9(a) that the higher overlap often found in more densely populated areas of the input space leads to local models having on average fewer degrees of freedom, even though the full model obviously has a higher effective DOF in such areas. A more useful measure would be a local 'DOF density'. The 'density' of estimated DOF in a volume V of input space can be defined as,

$$\tilde{p}_{dens} = \frac{d\tilde{p}}{dV}, \tag{7.38}$$

so for a local model representation, with the total relevant input space defined as V_T,

$$\tilde{p}_{dens}(\psi) \;=\; \sum_{i=1}^{n_{\mathcal{M}}} \tilde{p}_i \frac{\rho_i(\psi)}{\int_{V_T} \rho_i(\mathbf{x})\,d\mathbf{x}}, \tag{7.39}$$

and the available DOF in any given volume of the input space V can be found by integrating over the density $\tilde{p}_{dens}$:

$$\tilde{p}_V \;=\; \int_V \sum_{i=1}^{n_{\mathcal{M}}} \tilde{p}_i \frac{\rho_i(\psi)}{\int_{V_T} \rho_i(\mathbf{x})\,d\mathbf{x}}\,d\psi. \tag{7.40}$$

This leads to the plots such as Figure 7.9(b), which match intuitive expectations. The size of V will act as a resolution parameter on the resulting visualisation. Plotting the DOF density function during learning can also give us insight into the learning algorithm, and an example of this is shown in Figure 7.9(c).

The progression of the total estimated degrees of freedom in the model during learning, can also be instructive. Figure 7.10(a) shows the increase in degrees of freedom when basis functions are not adapted. The change in DOF illustrates the difference in weighting between priors and posteriors. Figure 7.10(b) shows the increase in DOF when all parameters are adapted, indicating a general decrease in overlap.

7.7 CONCLUSIONS

An important aspect to consider when developing learning algorithms for local model structures is the trade-off between achieving a good *global* fit to the training data while still utilising the *local* nature of the representation. Global learning methods for local model structures were found to be computationally expensive, and non-robust when applied to over-parameterised or poorly structured local model networks. This leads to local models which cannot be interpreted locally and can also lead to poor prediction performance. On the other hand, it should be stressed that *when the model structure is well chosen, global learning is more accurate*. Local learning of the local model parameters is, however, very relevant for the practical application of the Local Model network, as the induction of a model structure from training data will often lead to non-optimal, locally over-parameterised structures. It is obviously important that parameter estimation methods should be robust with regards to poor model structure. The trade-off is therefore between the use of the less variable local learning with a necessarily extended model structure, and a more powerful, expensive and variable global learning method, with a reduced model structure. The analysis of *individual* local models as local approximations to the underlying process is in general only valid with local learning.

Examination of the computational complexity shows that local learning is faster than global learning for large problems, where the effort increases cubically with the number of models. Even ignoring the speed-up gained by the reduced number of points (as $N_i \leq N$), the local variant will be faster. The effort for local learning also increases linearly in $n_{\mathcal{M}}$ as opposed to the cube of $(n_{\mathcal{M}}(n_\psi + 1))$, which makes local learning more suitable for larger problems.

Analysis of local learning showed that it can be seen as a simple form of regularisation, such that local methods often produce models with higher accuracy, and greater robustness than global learning methods. The level of overlap between local models was found to

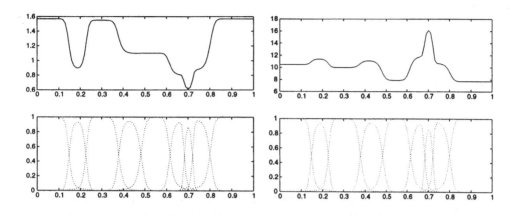

(a) Local average Degrees of Freedom estimate after EM learning with structure adaptation. Note how the average DOF (upper curve) actually *decreases* in the area of greatest complexity, due to the increase in overlap there. Notice also the edge effects, due to lower levels of overlap with edge basis functions.

(b) Effective Degrees of Freedom density estimate after EM learning with structure adaptation. The upper curve shows the density for the basis functions shown below.

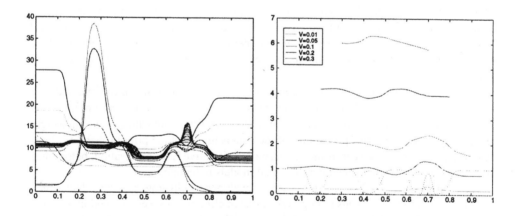

(c) Changes in Local Degrees of Freedom density estimate during EM learning with basis function adaptation. Note the transition from very smooth densities via 'glitch-like' densities to the final estimate which is shown in 7.9(b).

(d) Local estimates of DOF available, by integrating DOF density over a number of window sizes (V) over the input space. Larger values of V obviously correspond to the higher curves.

Figure 7.9 Local estimates of DOF for the local model network with the given basis functions, trained using EM on the test function. The changes in DOF locally and during learning can be used to give us insight into the structure of the learned model, and the learning algorithm.

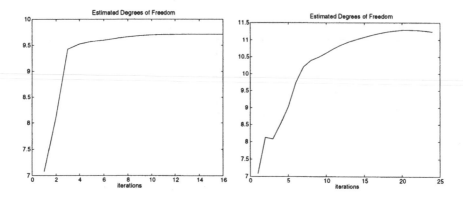

(a) Degrees of Freedom estimate during EM
learning without adaptation of the basis functions.

(b) Degrees of Freedom estimate during EM
learning. Note the increase in DOF as the
overlap between basis functions decreases.

Figure 7.10 Degrees of Freedom estimates during EM learning. Note that even when the
basis functions are not adapted, the effective DOF increases, due to the improving model fit
'sharpening' the posterior distributions. This effect is naturally more prominent in the case
where all parameters are optimised.

play a major role in the ill-conditioning of the global learning problem, and the overlap
also determines the amount of regularisation for local learning (see also the related work
in Chapter 8). Smoothing analysis was used to better understand the effects, and graphical
representations of the smoothing matrices were found to be illustrative in explaining the
processes involved.

The use of adaptation of the basis functions combined with local learning methods was
examined, using EM and GCV based algorithms. The stage of EM learning for optimisation
of the local models' parameters is analogous to local learning, but with a changing weighting function – the posterior probability of the data is used, not the prior probability. Local
learning can therefore be seen as the first step of the iterative EM optimisation, before there
is any interaction between neighbouring local models in learning.

'Local degree-of-freedom' estimates were developed to illustrate how the adaptation of
localised basis functions leads to large variations in effective degrees-of-freedom throughout the input space. The extension of the use of this tool to more complex models with
hierarchical structure and heterogeneous local models may prove to be interesting.

7.8 APPENDIX

7.8.1 Probabilistic Interpretation of Local Model Nets

Methods used in probabilistic frameworks such as the *Mixtures of Experts* architecture[14] are
also interesting, e.g. (Jacobs *et al.* 1991, Jordan and Jacobs 1994). The use of Expectation-

---

[14]The architecture is functionally identical to Local Model Nets – the theoretical interpretation differs.

Maximisation (EM) algorithms (Dempster *et al.* 1977) to maximise the total likelihood of the mixtures has become increasingly popular.

The probabilistic interpretation of the basic local model structure is based on the assumption that the data were generated by a decision process which selects a regressive process (the local model) mapping ψ to y. The basis function $\rho_i(\psi)$ therefore represents the prior probability of local model i being chosen. The local model is represented as $P(y|\psi, \theta_i)$, the conditional probability, given that model i is chosen. Assuming the probabilistic component of the local models to be Gaussian, i.e.

$$P(y|\psi, \theta_i) = \frac{1}{(2\pi)^{n/2}\sigma_i^n} \exp\left(-\frac{1}{2\sigma_i^2}(y - \hat{f}_i(\psi, \theta_i))^2\right), \qquad (7.41)$$

where σ_i^2 represents the covariance of the noise on the output y.

The total probability of generating y from ψ is the mixture of the probabilities of generating y from each of the local density functions, and the mixing functions are the basis functions:

$$P(y|\psi, \theta) = \sum_{i=1}^{n_M} \rho_i(\psi)P(y|\psi, \theta_i). \qquad (7.42)$$

The log likelihood of the parameters θ for a given data set $\mathcal{D}$ is

$$l(\theta; \mathcal{D}) = \sum_{k=1}^{N} \ln \sum_{i=1}^{n_M} \rho_i(\psi(k))P(y(k)|\psi(k), \theta_i) \qquad (7.43)$$

and the optimisation task is to maximise $l(\theta; \mathcal{D})$. This is discussed in section 7.8.2.

7.8.2 EM Learning

The Expectation-Maximisation (EM) method is an iterative approach to maximum likelihood estimation, where each iteration of the algorithm is composed of an Estimation (E) step followed by a Maximisation (M) step (it is also used in Chapters 5 and 6). The M step involves the maximisation of a likelihood function redefined in each iteration in the E step. This is aided by assuming the existence of 'missing' variables which would simplify the optimisation problem. So the observed data $\mathcal{D}$ are termed 'incomplete' compared to the 'complete' data set $\mathcal{C}$ which includes the missing variables $\mathcal{Z}$, which are labels pointing out which is the 'correct' local model for the given input. The fictive missing variables are linked to the data by a probability model $P(y, z|\psi, \theta)$.

To apply EM to Local Model nets we have to define the appropriate 'missing data', so as in (Jordan and Jacobs 1994), we define indicator variables z_i such that one, and only one of the z_i is 1 and all others are 0. The correct functions relating the z_i to the inputs ψ are unknown, and must be estimated using the data $\mathcal{D}$.

$$P(y(k), z_i(k)|\psi(k), \theta) = \rho_i(\psi(k))P(y(k)|\psi(k), \theta_i).$$

The logarithm of this probability model provides us with the complete-data likelihood:

$$l_c(\theta; \mathcal{C}) = \sum_{k}^{N} \sum_{i}^{n_M} z_i(k) (\ln \rho_i(\psi(k)) + \ln P(y(k)|\psi(k), \theta_i)) \qquad (7.44)$$

which has simplified the maximisation process by decoupling the problem.

The E step finds the expected value of the complete-data likelihood $l_c(\theta; C)$, given the observed data $\mathcal{D}$ and the current model θ^t:

$$L(\theta, \theta^t) = E\left[l_c(\theta; C)|\mathcal{D}, \theta^t\right].$$ (7.45)

The M step then maximises this function with respect to θ to find the new parameter estimates θ^{t+1},

$$\theta^{t+1} = \arg\max_{\theta} L(\theta, \theta^t).$$ (7.46)

The E step is repeated to produce an improved estimate of the complete likelihood, and the process iterates.

Although the EM algorithm increases expected complete likelihood, this procedure is guaranteed to increase the incomplete likelihood $l(\theta; \mathcal{D})$ to a local maximum (Dempster *et al.* 1977, Jordan and Jacobs 1994).

E step in Local Model nets

The E step is therefore the expectation of the complete likelihood,

$$L(\theta, \theta^t) = \sum_k \sum_i h_i(k, \theta^t) \left(\ln \rho_i(\psi(k)) + \ln P(y(k)|\psi(k), \theta_i)\right),$$ (7.47)

where the posterior estimate $h_i(k, \theta^t)$ has replaced the indicator variable, as

$$
\begin{aligned}
E[z_i(k)|\mathcal{D}, \theta^t] &= P(z_i(k) = 1|y(k), \psi(k), \theta^t) & (7.48)\\
&= \frac{P(y(k)|z_i(k) = 1, \psi(k), \theta^t)P(z_i(k) = 1|\psi(k), \theta^t)}{P(y(k)|\psi(k), \theta^t)} & (7.49)\\
&= \frac{P(y(k)|\psi(k), \theta_i^t)\rho_i(\psi(k))}{\sum_i \rho_i(\psi(k))P(y(k)|\psi(k), \theta_i^t)} & (7.50)\\
&= h_i(k, \theta^t). & (7.51)
\end{aligned}
$$

M step in Local Model nets

Examining equation (7.47) we see that the maximisation process can be decoupled, as the local model parameters influence L only through the term $\ln P(y(k)|\psi(k), \theta_i)$ and the basis functions are only influenced by $\ln \rho_i(\psi(k))$.

To maximise the log likelihood, the local models' parameters are optimised by

$$\theta_i^{t+1} = \arg\max_{\theta_i} \sum_k h_i(k, \theta^t) \ln P(y(k)|\psi(k), \theta_i),$$ (7.52)

where the h_is are the posterior probabilities (defined once the input and the output are known) of the output y, given inputs ψ and parameters θ. The posterior $h_i(k, \theta^t)$ is the probability that local model i can be considered to have generated the output $y(k)$ given input $\psi(k)$ and parameters θ^t. The optimisation of the local expert i's parameters θ_i is therefore a weighted maximum likelihood for a GLIM, and if the weighting functions (see the $\alpha_i(\psi)$ weightings in section 7.3.2) $\rho_i(\psi(k))$ in equation (7.53) are local, this becomes very similar to the local learning described earlier, but as the process is iterative, information about the global fit comes in via the posterior probability weightings h_i, which are re-estimated after each iteration.

If all the model parameters (including c_i and Σ_i) are free to adapt,

$$c_i^{t+1} = \arg\max_{c_i} \sum_k h_i(k, \theta^t) \ln \rho_i(\psi(k)) = \sum_k h_i(k, \theta^t)\psi(k), \qquad (7.53)$$

$$\Sigma_i^{t+1} = \arg\max_{\Sigma_i} \sum_k h_i(k, \theta^t) \ln \rho_i(\psi(k)) = \sum_k h_i(k, \theta^t)\|\psi(k) - c_i^t\|, \qquad (7.54)$$

the tendency will be for the use of posteriors to lead to competition between neighbouring local models. The experiments in section 7.4 are focused on the stage of local model parameter optimisation, described in equation (7.52), whereas those in section 7.6.1 also include equations (7.53) and (7.54).

REFERENCES

BENEDETTI, J. K. (1977) 'On the nonparametric estimation of regression functions'. *J. Royal Stat. Soc., Ser B* **39**, 248–253.

BISHOP, C. (1991) 'Improving the generalization properties of Radial Basis Function neural networks'. *Neural Computation* **3**(4), 579–588.

BISHOP, C. M. (1994) Training with noise is equivalent to Tikhonov Regularization. Submitted for publication.

BOTTOU, L. AND V. VAPNIK (1992) 'Local learning algorithms'. *Neural Computation* **4**(6), 888–900.

CLEVELAND, W. S., S. J. DEVLIN AND E. GROSSE (1988) 'Regression by local fitting'. *Journal of Econometrics* **37**, 87–114.

CRAVEN, P. AND G. WAHBA (1979) 'Smoothing noisy data with spline functions. Estimating the correct degree of smoothing by the method of generalized cross-validation'. *Numerical Math.* **31**, 317–403.

DEMPSTER, A. P., N. M. LAIRD AND D. B. RUBIN (1977) 'Maximum likelihood from incomplete data via the EM algorithm'. *J. Royal Statistical Society Series B* **39**, 1–38.

GEMAN, S., E. BIENENSTOCK AND R. DOURSAT (1992) 'Neural networks and the bias/variance dilemma'. *Neural Computation* **4**(1), 1–58.

GOLUB, G. H. AND C. F. VAN LOAN (1989) *Matrix Computations.* Johns Hopkins University Press.

HASTIE, T. J. AND R. J. TIBSHIRANI (1990) *Generalized Additive Models.* Monographs on Statistics and Applied Probability 43. Chapman and Hall. London.

HOLCOMB, T. AND MANFRED MORARI (1991) 'Local training for Radial Basis Function networks: Towards solving the hidden unit problem'. *Neural Networks* **4**, 361–369.

JACOBS, R. A., M. I. JORDAN, S. J. NOWLAN AND G. E. HINTON (1991) 'Adaptive mixtures of local experts'. *Neural Computation* **3**(1), 79–87.

JOHANSEN, T. A. AND B. A. FOSS (1995) 'Identification of non-linear system structure and parameters using regime decomposition'. *Automatica* **31**, 321–326.

JONES, R. D., Y. C. LEE, C. W. BARNES, G. W. FLAKE, K. LEE, P. S. LEWIS AND S. QIAN (1989) Function approximation and time series prediction with neural networks. Technical Report 90-21. Los Alamos National Lab., New Mexico.

JORDAN, M. AND R. A. JACOBS (1994) 'Hierarchical mixtures of experts and the EM algorithm'. *Neural Computation* **6**, 181–214.

LJUNG, L. (1987) *System Identification — Theory for the User*. Prentice Hall. Englewood Cliffs, New Jersey, USA.

LOWE, D. (1995) On the use of nonlocal and non positive definite basis functions in radial basis function networks. In 'Proc. IEE 3rd Int. Conf. on Artificial Neural Networks'. IEE. Cambridge. pp. 206–211.

MOODY, J. (1992) The effective number of parameters: An analysis of generalization and regularization in nonlinear learning systems. In 'Advances in Neural Information Processings Systems 4'. Morgan Kaufmann Publishers, San Mateo, CA.

MURRAY-SMITH, R. (1994) A Local Model Network Approach to Nonlinear Modelling. PhD Thesis. Department of Computer Science, University of Strathclyde. Glasgow, Scotland.

MURRAY-SMITH, R. AND H. GOLLEE (1994) A constructive learning algorithm for local model networks. In 'Proc. IEEE Workshop on Computer-intensive methods in control and signal processing, Prague, Czech Republic'. pp. 21–29.

MURRAY-SMITH, R. AND TOR ARNE JOHANSEN (1995) Local learning in local model networks. In '4th IEE Intern. Conf. on Artificial Neural Networks'. pp. p40–46.

NADARAYA, E. A. (1964) 'On estimating regression'. *Theory of Probability and Its Applications* **9**, 141–132.

NOBLE, B. AND J. W. DANIEL (1988) *Applied linear algebra*. 3rd edn. Prentice–Hall Int.

PRESS, W. H., B. P. FLANNERY, S. A. TEUKOLSKY AND W. T. VETTERLING (1988) *Numerical Recipes (C): The Art of Scientific Computing*. Cambridge Press. UK.

SJÖBERG, J. AND L. LJUNG (1992) Overtraining, regularization, and searching for minimum in neural networks. In 'Proc. IFAC Symposium on Adaptive Systems in Control and Signal Processing, Grenoble, France.'. pp. 669–674.

STOKBRO, K., D. K. UMBERGER AND J. A. HERTZ (1990) 'Exploiting neurons with localized receptive fields to learn chaos'. *Complex Systems* **4**(3), 603–622.

TIKHONOV, A. N. AND V. Y. ARSENIN (1977) *Solutions of Ill-posed problems*. Winston, Washington DC.

WATSON, G. S. (1964) 'Smooth regression analysis'. *Sankhya, Ser. A* **26**, 359–372.

WU, L. AND J. MOODY (1996) 'A smoothing regularizer for feedforward and recurrent neural networks'. *Neural Computation* **8**(3), 461–489.

CHAPTER EIGHT

Side-Effects of Normalising Basis Functions in Local Model Networks

ROBERT SHORTEN and RODERICK MURRAY-SMITH

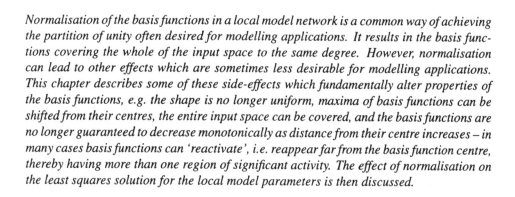

Normalisation of the basis functions in a local model network is a common way of achieving the partition of unity often desired for modelling applications. It results in the basis functions covering the whole of the input space to the same degree. However, normalisation can lead to other effects which are sometimes less desirable for modelling applications. This chapter describes some of these side-effects which fundamentally alter properties of the basis functions, e.g. the shape is no longer uniform, maxima of basis functions can be shifted from their centres, the entire input space can be covered, and the basis functions are no longer guaranteed to decrease monotonically as distance from their centre increases – in many cases basis functions can 'reactivate', i.e. reappear far from the basis function centre, thereby having more than one region of significant activity. The effect of normalisation on the least squares solution for the local model parameters is then discussed.

8.1 INTRODUCTORY REMARKS

This paper discusses the effect of normalisation, as used in (Moody and Darken 1989, Barnes *et al.* 1991, Jones *et al.* 1989, Johansen 1994), on the behaviour of radial basis functions in local model networks, and extends the results described in (Shorten and Murray-Smith 1996). Within a local model structure (equation (8.1)), where the f_i's are local approximations to the system (Johansen 1994, Murray-Smith 1994), a partition of unity ensures that it is possible for the network to represent the target system, given an exact structural match between the system and the local models (the $f_i(\psi)$'s) associated with each basis function.[1]

In other contexts normalisation is sometimes desired because it results in every point in the input space being covered by the basis functions to the same degree, i.e. the basis functions sum to unity at every point. When this is the case a *partition of unity* across the

---

[1]This assumes a network where the $f_i(\psi)$'s have the same model structure.

input space is said to have been achieved. A partition of unity is an important property for basis function networks for a number of reasons. It often results in a structure which is less sensitive to poor basis function parameter selection (centres and widths). Normalisation of basis functions arises naturally when they are viewed in a probabilistic framework, as in Chapter 6, and is also used in fuzzy logic systems (see also equation (2.7) for an example of the use of normalisation in the defuzzification of a Takagi–Sugeno fuzzy model (Takagi and Sugeno 1985)).

Normalisation is most relevant for radial basis functions, as opposed to basis functions which partition the input space in an axis-orthogonal manner (e.g. B-Splines). Such axis-orthogonal basis functions can be designed to achieve a partition of unity without normalisation. Fuzzy systems can also be designed to achieve the partition of unity, depending on the choice of membership function. While the approximation capabilities of normalised networks have been analysed (Benaim 1994), the side-effects of normalisation were first considered in detail in (Shorten and Murray-Smith 1994). This chapter extends our earlier work on RBF networks to the more general local model network case.

This chapter also emphasises the need for easy-to-use tools for model analysis and visualisation in multiple model systems. Such tools highlight often counter intuitive behaviour as a result of normalisation. This chapter should therefore not be seen as a pro- or contra normalisation discussion, but rather as a discussion of the effects on the model representation due to it, and a discussion of the methods for analysing and preventing some of the undesirable side-effects.

8.2 MODELLING WITH LOCAL MODEL NETWORKS

The output of a local model network (see Figure 7.1) is described by,

$$\hat{f}(\psi, \theta) = \sum_{i=1}^{n_{\mathcal{M}}} \hat{f}_i(\psi, \theta_i)\rho_i(\psi), \tag{8.1}$$

where y is the network output,[2] ψ is the $n_{\psi} \times 1$ dimensional vector of input variables, c_i is the centre of the ith basis function, Σ_i is the width of the ith basis function, $n_{\mathcal{M}}$ is the number of local models, θ_i is the vector of local model parameters associated with local model i, $d(\cdot)$ denotes some distance metric, and ρ is the nonlinear activation function before normalisation.

We denote the normalised form of the basis function by ρ_k, for the basis function associated with local model k, where

$$\rho_k(\psi) = \frac{\rho(d(\psi, c_i, \Sigma_i))}{\sum_{j=1}^{n_{\mathcal{M}}} \rho(d(\psi, c_j, \Sigma_j))}. \tag{8.2}$$

In principle, the basis function can be any nonlinear function. However, many authors use local basis functions for a number of practical reasons discussed in Chapter 1 ('local' implying that the basis function is 'significantly active' for some limited range of the input). Local basis functions are advantageous because of the increased interpretability of the network, the ability to produce locally accurate confidence limits (Leonard et al. 1992), and because locality can also be utilised to improve computational efficiency.

---

[2]This chapter considers single output systems only.

A common choice for ρ_i takes the form of a Gaussian, although other activation functions have been proposed (Hlaváčková and Neruda 1993). The Gaussian activation function used in this chapter[3] takes the form,

$$d\left(\psi, \mathbf{c}_i, \, \mathbf{\Sigma}_i\right) = \left(\psi - \mathbf{c}_i\right)^T \mathbf{\Sigma}_i^{-2}(\psi - \mathbf{c}_i), \tag{8.3}$$

$$\rho_i(d(\psi; \mathbf{c}_i, \, \mathbf{\Sigma}_i)) = \exp(-d(\psi; \mathbf{c}_i, \, \mathbf{\Sigma}_i)). \tag{8.4}$$

Normalisation of the basis functions is often motivated by the desire to achieve a partition of unity across the input space. By partition of unity it is meant that, at any point in the input space the sum of all normalised basis functions equals unity, i.e.

$$\sum_{i=1}^{n_{\mathcal{M}}} \rho_i(\psi) = 1. \tag{8.5}$$

This has the effect of covering every point of the input space to the same degree, unlike the un-normalised case where the total weighting over the input space is nonuniform. The approximation capabilities of such networks have been considered in detail in (Benaim 1994) and it has been shown that normalised RBFs are capable of *universal approximation* (Kaplan 1991) in a satisfactory sense. Werntges discusses the advantages of a partition of unity produced by normalisation of RBF nets in (Werntges 1993), but without considering the side-effects discussed in this chapter.

8.3 EFFECT OF NORMALISATION ON THE MODEL PARTITION

Basis functions represent the partition of the input space in a local model framework. In order to achieve a partition of unity for many types of basis function it is necessary to normalise. However, as described above, normalisation also leads to a number of important side-effects which can have important consequences for the resulting network. In this section we describe these side-effects.

8.3.1 Loss of independence and change of shape of basis function

The un-normalised basis functions are often homogeneous basis functions, sometimes with differing parameters. Once normalised, however, this is not the case – the shape of the basis functions is usually quite different from the un-normalised basis function, and the shape is influenced not only by the basis function's width, but also by the proximity of the other basis functions in the network. Note the decrease in basis function maxima in the normalised case shown in Figure 8.1(b). As the width of the basis function decreases the normalised network becomes less smooth, and tends towards a crisp nearest-neighbour classifier. The effect in two dimensions is shown by the contour plot in Figure 8.5 and the surface plot in Figure 8.6.

It can also be seen from equation (8.2) that each normalised basis function is a function of *all* the original basis functions; normalisation introduces a coupling between the original basis functions. Therefore, changing the parameters of one basis function affects all other normalised basis functions in the network. This can have important consequences for

---

[3]This ignores the scaling term.

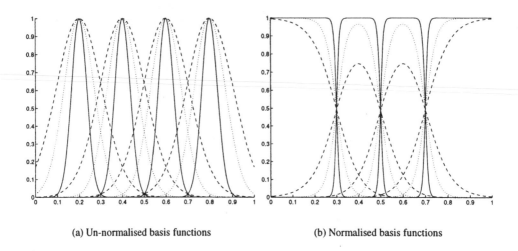

(a) Un-normalised basis functions   (b) Normalised basis functions

Figure 8.1   Change in shape due to normalisation. As the original functions become wider the evenly spaced normalised basis functions become less square and their maxima are reduced. Edge basis functions tend to unity as they move to the limits of their support.

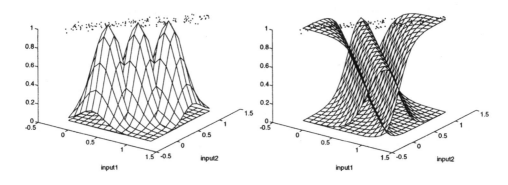

Figure 8.2   Covering of input space due to normalisation. The basis functions can now be significantly active over their entire support.

on-line applications where the model parameters are updated with each new data point, i.e., a (local) change in the parameters of a single basis function affects all other basis functions in the network (i.e. has a global effect).

8.3.2   Covering of the input space

In the case where the basis function used is non-compact in nature, for example when Gaussians are used, then normalisation results in the *whole* of the input space being covered and not just the region of the input space populated by the training data. It can be seen from Figures 8.1(b) and 8.2 that in the normalised case the activation tends toward unity at the edges of the space.

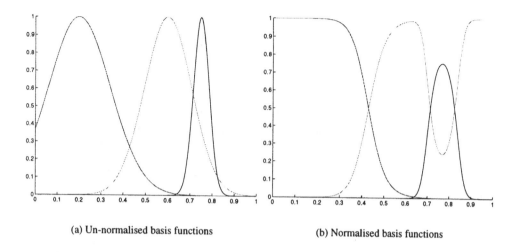

(a) Un-normalised basis functions (b) Normalised basis functions

Figure 8.3 Shift in maxima and reactivation. Note the reactivation of the centre basis function, the reduced maximum of the right-hand basis function, and the shift in maximum for all three functions. The vertical lines show the positions of the centres of basis functions to emphasise the centre-shift effect.

8.3.3 Irregularly spaced basis functions: reactivation and shift in maxima

A further difficulty with normalised basis functions involves two phenomena which occur due to normalisation. If centres are not uniformly spaced, or if basis functions of differing widths are used, the maximum of the basis function may no longer be at its centre. A further effect of varying basis widths is that the basis function can become multi-modal, meaning that it can now also increase as the distance function increases, instead of continuously decreasing – the basis function 'reactivates'. These effects are shown in Figure 8.3. This reactivation far from a model's centre has obvious repercussions for the interpretation techniques discussed in section 1.6. The supposedly 'local' basis functions can now have a far wider support (the area where the basis function is non-zero) than before normalisation, reactivation leading to local models becoming significantly active in regions in which they were never intended to operate.

Reactivation

Reactivation occurs when neighbouring basis functions have differing widths. A one-dimensional example shown in Figure 8.4 illustrates how the phenomenon occurs. The reactivation point x, assuming monotonically decreasing basis functions, is the point at which the distance metric d_1 (equation (8.3)) for the first basis function is no longer smaller than d_2 for the second basis function. It can therefore be determined from the functions' centres and widths.

For a Euclidean distance metric this implies,

$$\left(\frac{x - c_1}{\sigma_1}\right)^2 < \left(\frac{x - c_2}{\sigma_2}\right)^2,$$

$$(8.6)$$

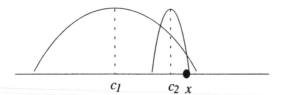

c_1 c_2 x

Figure 8.4 One-dimensional reactivation example for two basis functions. The point in the input space x where the basis function reactivates can be determined from the basis functions' centres (c_1 and c_2, where c_1 is furthest from the input x) and their widths (σ_1 and σ_2).

which results in a condition for reactivation given by,

$$\frac{\sigma_2}{\sigma_1} < \frac{|x - c_2|}{|x - c_1|}. \tag{8.7}$$

Equation (8.7) shows that reactivation only occurs when the ratio between σ_1 and σ_2 is less than the ratio of the unweighted distances from the centres. In models with uniformly wide basis functions reactivation cannot occur. The shift in the position of the activation function's maximum occurs when neighbouring basis functions are either unevenly spaced or have differing widths.

Equation (8.7) could therefore be used as a design criterion in any basis function/fuzzy logic representation whose basis functions were produced in a tensor product fashion. To ensure that local models are interpretable, any reactivation should occur well outside the area covered by the training data (i.e. the region of the input space the model is expected to be accurate in). This could be especially useful in adaptive fuzzy systems where the parameters of the membership functions are learned from training data.

8.3.4 Effects of normalisation on multi-dimensional problems

The effects of normalisation tend to become more pronounced as the input dimension increases (two-dimensional contour plots are shown below in Figure 8.5). A given level of smoothness in representation leads to a need for wider operating regimes in higher dimensional spaces. Wider operating regimes lead to an increased number of neighbouring local models overlapping with any given individual local model. This means that the cumulative activation in a given region tends to increase with dimension, leading to normalised basis functions often having dramatically reduced maxima. This reduction in basis function magnitude is quite important for multiple model representations, as it implies that a larger number of 'local' representations are being combined at any given point in the input space than for low-dimensional spaces.

8.4 LOCAL MODEL PARAMETERS AND CONDITION NUMBER

The concepts of model robustness (with regards to variance of the optimisation process and generalisation ability for new data) are closely related to the magnitude of the local model parameters θ and the condition number κ of the design matrix (Φ, defined below). The condition number is a useful indicator of how robust the solution for the local model

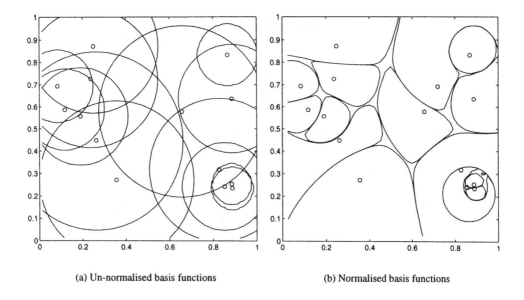

(a) Un-normalised basis functions   (b) Normalised basis functions

Figure 8.5 Normalised basis functions. Contours (drawn at a level of 0.5) of the basis functions are shown for a two-dimensional problem. The matrix of basis function centres and distance metrics is the same in each case.

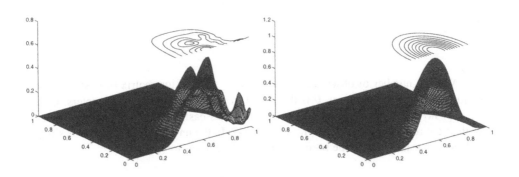

Figure 8.6 Effect of normalisation on basis function shape. The normalised basis function has a far more complex surface than before, with many local maxima and minima.

parameters actually is, i.e. how the solution for θ is affected by slightly perturbing the training data. Similarly, large model parameters can lead to non-robust models. It is therefore interesting to examine the effect of normalisation on the *least squares* weight solution for the local model parameters, and the effect of normalisation on the condition number of the local model network's design matrix. This chapter deals primarily with global least squares, as opposed to local least squares (see Chapter 7 for details on local learning).

The condition number of the design matrix is strongly dependent on the degree of 'similarity'[4] of overlapping basis functions. As the overlap increases (as basis functions become wider, or closer together) the condition number increases. Figure 8.7 shows the effect of basis function width on the condition number of regular basis functions before and after normalisation.

8.4.1 Normalisation and the least squares solution

This section discusses the effect of normalisation on the least squares solution for the local model parameters and on the condition number of the design matrix. We consider approximating the function $y = f(\psi)$ by a basis function expansion of the form of equation (8.1), where the basis functions are normalised as in equation (8.2).

If all of the observations are grouped into vector form with $\mathbf{y}^T$ defined as,

$$\mathbf{y}^T = [y_1, ..., y_N], \tag{8.8}$$

and

$$\gamma_i{}^T = \left[\rho(d(\psi_i; \mathbf{c}_1, \Sigma_1))[1, \psi] ... \rho(d(\psi_i; \mathbf{c}_{n_\mathcal{M}}, \Sigma_{n_\mathcal{M}}))[1, \psi] \right], \tag{8.9}$$

with

$$\Phi = \begin{pmatrix} \gamma_1^T \\ . \\ . \\ \gamma_N^T \end{pmatrix}, \tag{8.10}$$

then,

$$\hat{\mathbf{y}} = \mathbf{C}\Phi\theta, \tag{8.11}$$

where N is the number of observations, Φ is the $N \times L$ design matrix of basis function activations from the training set, θ is the $L \times 1$ vector of local model parameters, $\mathbf{C}$ is a $N \times N$ positive definite diagonal matrix (this assumes basis functions which are positive over their entire support) and $L = (n_\psi + 1 \times n_\mathcal{M})$. In the un-normalised case $\mathbf{C}$ is simply the identity matrix, while in the normalised case $\mathbf{C}$'s entries are given by

$$c_{kk} = \left(\sum_{i=1}^{n_\mathcal{M}} \rho(d(\psi_k; \mathbf{c}_i, \Sigma_i)) \right)^{-1}, \quad \forall \ 1 \le k \le N, \ C_{ik} = 0, \ \forall i \ne k. \tag{8.12}$$

The solution to the least squares minimisation problem is then given by finding a θ which minimises

$$(\hat{\mathbf{y}} - \mathbf{y})^T (\hat{\mathbf{y}} - \mathbf{y}). \tag{8.13}$$

Minimisation of (8.13) yields an expression for θ given by,

$$\theta = (\mathbf{C}\Phi)^+ \mathbf{y}, \tag{8.14}$$

---

[4]The degree of similarity is quantified by the linear correlation of basis function responses in their region of overlap.

where $(\cdot)^+$ denotes the Moore–Penrose pseudoinverse and is equal to (Noble and Daniel 1988, Golub and Van Loan 1989),

$$(\mathbf{C\Phi})^+ = (\mathbf{\Phi}^T \mathbf{C}^T \mathbf{C\Phi})^{-1} \mathbf{\Phi}^T \mathbf{C}, \tag{8.15}$$

when the rank of Φ is L. In the following analysis we wish to examine the effect of normalisation on the local model parameters and on the condition number of the design matrix $\mathbf{C\Phi}$. Before proceeding some definitions and basic results are necessary:

Definition 8.4.1 *(Noble and Daniel 1988) The 2 norm of a real matrix* $\mathbf{J}$ *is defined,*

$$
\begin{aligned}
\| \mathbf{J} \|_2 &= \sigma_{max}(\mathbf{J}), \\
&= \sqrt{\lambda_{max}(\mathbf{J}^T \mathbf{J})},
\end{aligned}
\tag{8.16}
$$

where $\sigma(\mathbf{J})$ *denotes a singular value of* $\mathbf{J}$ *and* $\lambda(\mathbf{J})$ *denotes an eigenvalue of* $\mathbf{J}$.

Definition 8.4.2 *(Golub and Van Loan 1989) The condition number* $\kappa(\mathbf{J})$ *of a rectangular matrix* $\mathbf{J}$ *is given by,*

$$\kappa(\mathbf{J}) = \frac{\sigma_{max}(\mathbf{J})}{\sigma_{min}(\mathbf{J})}, \tag{8.17}$$

where $\sigma_{max}(\mathbf{J})$ *denotes the maximum singular value of* $\mathbf{J}$ *and where* $\sigma_{min}(\mathbf{J})$ *denotes the minimum singular value of* $\mathbf{J}$.

Lemma 8.4.1 *(Söderström and Stoica 1989)*

$$\| \mathbf{J}^+ \|_2 = \frac{1}{\sigma_{min}(\mathbf{J})}. \tag{8.18}$$

Proof
The above result follows directly from the definition of the pseudoinverse.

Theorem 8.4.1 *Let* $\mathbf{Z} = \mathbf{C\Phi}$, *where* $\mathbf{C}$ *is an* $N \times N$ *diagonal matrix whose diagonal entries are real and positive, and* Φ *is a* $N \times M$ *real matrix. Let* $\sigma(\mathbf{J})$ *denote the singular values of some matrix* $\mathbf{J}$. *Then,*

$$\sigma_{min}^2(\mathbf{Z}) \geq c_{min}^2 \sigma_{min}^2(\Phi), \tag{8.19}$$

$$\sigma_{max}^2(\mathbf{Z}) \leq c_{max}^2 \sigma_{max}^2(\Phi), \tag{8.20}$$

where c_{min} *is the minimum entry of* $\mathbf{C}$.

Proof
The singular values of a real matrix $\mathbf{J}$ *satisfy,*

$$\sigma^2(\mathbf{J}) = \lambda(\mathbf{J}^T \mathbf{J}).$$

From the Raleigh quotient (Noble and Daniel 1988),

$$\sigma_{min}^2(\mathbf{Z}) = \frac{\mathbf{x}_{min}^T \mathbf{\Phi}^T C^2 \mathbf{\Phi} \mathbf{x}_{min}}{\mathbf{x}_{min}^T \mathbf{x}_{min}},$$

when $\mathbf{x}_{min} \neq 0$ *is the eigenvector of* $\mathbf{Z}^T\mathbf{Z}$ *corresponding to the minimum eigenvalue of* $\mathbf{Z}^T\mathbf{Z}$. *Then,*

$$\sigma^2_{min}(\mathbf{Z}) \geq c^2_{min} \frac{\mathbf{x}^T_{min}\mathbf{\Phi}^T\mathbf{\Phi}\mathbf{x}_{min}}{\mathbf{x}^T_{min}\mathbf{x}_{min}}.$$

But from the Raleigh inequality,

$$\sigma^2_{min}(\mathbf{\Phi}) \leq \frac{\mathbf{x}^T\mathbf{\Phi}^T\mathbf{\Phi}\mathbf{x}}{\mathbf{x}^T\mathbf{x}} \ \forall \ \mathbf{x}.$$

Therefore,

$$\sigma^2_{min}(\mathbf{Z}) \geq c^2_{min}\sigma^2_{min}(\mathbf{\Phi}).$$

The inequality

$$\sigma^2_{max}(\mathbf{Z}) \leq c^2_{max}\sigma^2_{max}(\mathbf{\Phi}),$$

follows using the same arguments as above for the maximum singular values.

Theorem 8.4.2 *Let* $\mathbf{w} = \mathbf{H}\mathbf{x}$, *where* $\mathbf{w}$ *is a* $M \times 1$ *vector,* $\mathbf{H}$ *is a* $M \times N$ *matrix of rank* $M \leq N$, *and* $\mathbf{x}$ *is a* $N \times 1$ *vector. Then,*

$$\| \mathbf{w} \|_2 \geq \frac{\| \mathbf{x} \|_2}{\| \mathbf{H}^+ \|_2}. \tag{8.21}$$

Proof

A bound on the local model parameters can be written,

$$\begin{aligned} \mathbf{w}^T\mathbf{w} &= \mathbf{y}^T\mathbf{H}^T\mathbf{H}\mathbf{y} \\ &\geq \lambda_{min}(\mathbf{H}^T\mathbf{H})\mathbf{y}^T\mathbf{y} \\ &= \frac{\mathbf{y}^T\mathbf{y}}{\sigma^2_{max}(\mathbf{H}^+)}. \end{aligned}$$

8.4.2   Bounds on local model parameters

Now consider equation (8.14). An upper bound for the local model parameters is given by,

$$\begin{aligned} \| \theta \|_2 &\leq \| (\mathbf{C}\mathbf{\Phi})^+ \|_2 \| \mathbf{y} \|_2 \\ &= \frac{1}{\sigma_{min}(\mathbf{C}\mathbf{\Phi})} \| \mathbf{y} \|_2, \text{ from lemma 8.4.1,} \\ &\leq \frac{1}{c_{min}\sigma_{min}(\mathbf{\Phi})}, \text{ from theorem 8.4.1,} \\ &= \frac{1}{c_{min}} \| \mathbf{\Phi}^+ \|_2 . \end{aligned} \tag{8.22}$$

Similarly, using theorem 8.4.2 and equation (8.20) from theorem 8.4.1 the lower bound for the local model parameters can be written,

$$
\begin{aligned}
\| \, \theta \, \|_2 \quad &\geq \quad \frac{\| \, \mathbf{y} \, \|_2}{\| \, (\mathbf{C}\Phi) \, \|_2}, \\
&= \quad \frac{\| \, \mathbf{y} \, \|_2}{\sigma_{max}(\mathbf{C}\Phi)} \text{ from definition 8.4.1,} \\
&\geq \quad \frac{\| \, \mathbf{y} \, \|_2}{c_{max} \, \| \, (\Phi) \, \|_2} \text{ from theorem 8.4.1.} \quad (8.23)
\end{aligned}
$$

Equations[5] (8.22) and (8.23) demonstrate that the bounds on the local model parameters are dependent on the entries of $\mathbf{C}$ (i.e. the inverse sum of the basis functions). Note that c_{min} and c_{max} can be very small when there is a large amount of overlap between the local model basis functions. In this case the bound on the local model parameters will increase significantly as a result of normalisation. An increase in the parameters typically occurs when the widths of the un-normalised basis functions are large, whereas a decrease in weight magnitude tends to be associated with small widths. In multidimensional cases, the effect of large basis functions becomes even more dramatic, for the reasons described in section 8.3.4.

8.4.3 Normalisation and design matrix condition number

Finally, note that equations (8.20) and (8.19) can be used to find an upper bound for the condition number (as discussed in Chapter 7) of the design matrix $\mathbf{C}\Phi$,

$$
\begin{aligned}
\kappa(\mathbf{C}\Phi) \quad &= \quad \frac{\sigma_{max}(\mathbf{C}\Phi)}{\sigma_{min}(\mathbf{C}\Phi)}, \\
&\leq \quad \frac{c_{max}\sigma_{max}(\Phi)}{c_{min}\sigma_{min}(\Phi)} \text{ from theorem 8.4.1 .} \quad (8.24)
\end{aligned}
$$

Equation (8.24) suggests that the ratio $\frac{c_{max}}{c_{min}}$ is a useful indication of changes in the condition number of $\mathbf{C}\Phi$. Note that for large amounts of overlap between the un-normalised basis functions, both c_{max} and c_{min} can be very small, whereas the ratio $\frac{c_{max}}{c_{min}}$ need not be significantly greater than unity ($\frac{c_{max}}{c_{min}} = 1$ implies that the un-normalised basis functions form a uniform partition in the area covered by the input data). This implies that although the local model parameters could increase significantly, this increase does not imply a significant increase in model condition number with normalisation. The converse is true for models with basis functions which do not overlap significantly; a significant increase in condition number can correspond to a decrease in local model parameters.

In Figures 8.7 and 8.8 we illustrate the effect of overlap on the condition of the design matrix for six basis functions spread uniformly on a single input dimension from 0 to 1. The condition numbers of the design matrices are calculated for basis function widths ranging from $\Sigma^2 = 0.0001...0.04$, and 1000 data points spread uniformly from 0 to 1. In both Figures, once the basis functions start to overlap significantly, the condition number increases rapidly with Σ in both cases. The change in condition due to normalisation in these examples is unlikely to have a major effect on learning, but the figures illustrate the processes involved. The effect of normalisation on matrix condition becomes more important in higher dimensional spaces because of the increased level of overlap, and in irregularly spaced, or overlapping basis functions, where reactivation often increases the coupling between basis functions.

---

[5]These equations are trivial to derive when both Φ and $\mathbf{C}$ are square matrices.

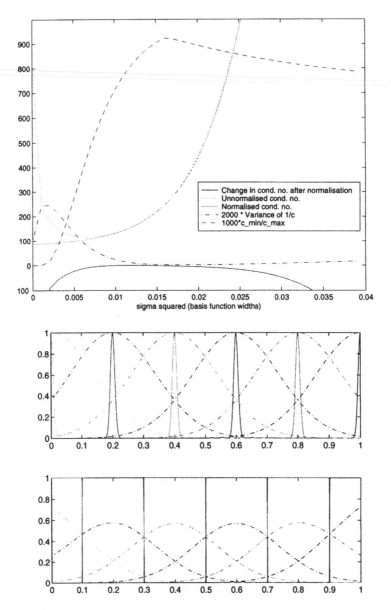

Figure 8.7 The difference in un-normalised and normalised condition number as a function of overlap between neighbouring basis functions. The ratio $\frac{c_{max}}{c_{min}}$ and the variance of the cumulative basis function activation (indicating how close the basis functions are to a uniform partition of the input space) are also depicted. The lower plots show the two extremes of basis function width in the plot, before and after normalisation. We note that, as expected, the condition number decreases with normalisation for low levels of overlap. However, unlike the RBF case (Shorten and Murray-Smith 1996), the condition number always decreases with normalisation. For large levels of overlap normalisation has little effect on the condition number.

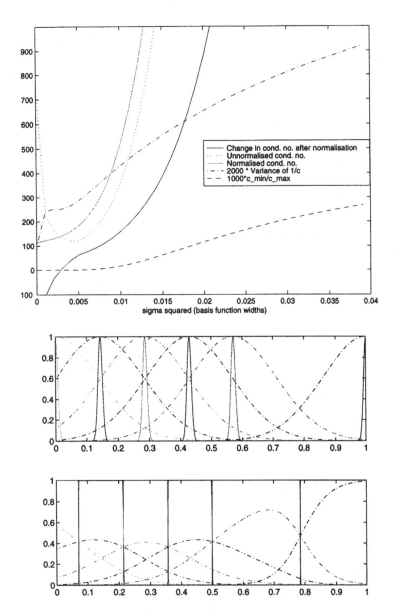

Figure 8.8   As in Figure 8.7, the difference in un-normalised and normalised condition number as a function of overlap between neighbouring basis functions. In this figure, however, the basis functions are spread unevenly, with one basis function far from the others. The differences in overlap lead to the steady increase in variance of $\frac{1}{c_{kk}}$ (the variance of the sum of the basis functions) as the width is increased. The partition of unity is also not approximated in the un-normalised case – basis functions should be optimised to better approximate the partition of unity. Again we note that, as expected, the condition number decreases with normalisation for low levels of overlap, and for large levels of overlap normalisation has little effect on the condition number. However, at intermediate levels of overlap the effect of normalisation on the condition is more complicated than for regular networks.

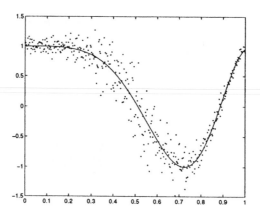

Figure 8.9   Target function and training data.

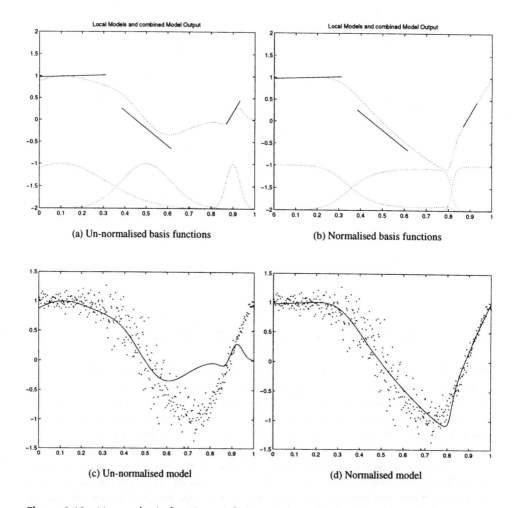

(a) Un-normalised basis functions   (b) Normalised basis functions

(c) Un-normalised model   (d) Normalised model

Figure 8.10   Narrow basis functions (relative to Figure 8.11). Note the problems of not attaining a partition of unity in the un-normalised case.

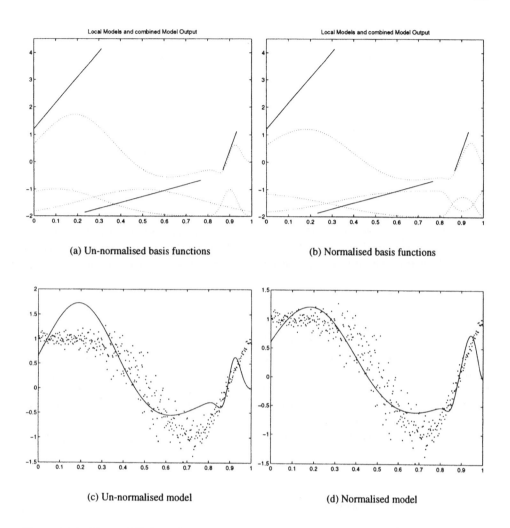

(a) Un-normalised basis functions

(b) Normalised basis functions

(c) Un-normalised model

(d) Normalised model

Figure 8.11 The middle basis function is enlarged. The shapes of the basis functions change due to normalisation. Note the reactivation effects on the right-hand side of the basis function plot for the normalised case. This causes the 'dip' in the curve close to 1, whereas in the un-normalised case the dip is caused by the partition of unity not being achieved.

8.5 ILLUSTRATIVE EXAMPLE

In order to illustrate the effects noted in the previous sections a simple one-dimensional example of modelling with a local model network is presented. The function to be modelled is depicted in Figure 8.9. Uniform white noise of amplitude 0.2 was superimposed on the target as depicted in Figure 8.9.

A local model network consisting of three basis functions (centred at [0.3, 0.5, 0.9]) was used for the modelling task. The widths of the various basis functions were varied to illustrate the effects noted above. The performance of the model with narrow basis functions is depicted in Figure 8.10. The basis functions are deliberately chosen to have a low level of

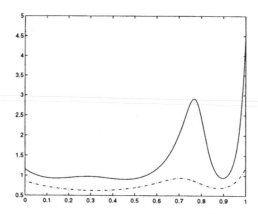

Figure 8.12 Diagonal elements of **C** matrix for wide (dotted line) and narrow (solid line) basis functions from the examples shown in Figure 8.9, with ordered inputs. Note the large variation of c_{kk} for the narrower basis functions, meaning that the ratio $\frac{c_{max}}{c_{min}}$ is greater for the case of the narrow basis functions.

overlap in order to exaggerate the effects of normalisation. The reactivation and covering of the input space effects can be clearly seen in Figure 8.10. In the un-normalised case, the approximation is very poor because the basis functions are too narrow. The effect of wide basis functions is presented in Figure 8.11. It can be clearly seen that the approximation is better than with narrow basis functions, due to the increased level of overlap. Reactivation still occurs in the normalised case.

The diagonal elements of the **C** matrix for the above example are depicted in Figure 8.12.

8.6   CONCLUDING REMARKS

This chapter describes the side-effects of achieving a partition of unity by normalising the basis functions in local model networks. These effects can be summarised as follows:

1   Normalisation leads to a change in shape of basis function.
2   If non-compact basis functions are used, normalisation leads to the whole of the input space being covered. This can result in stability problems at the edges of the space populated by the training data, when networks are used to represent dynamic systems.[6]
3   For irregular basis functions the maxima can shift away from the centres, and the local models can reactivate in other parts of the input space. Reactivation, and the resulting non-localised behaviour of individual basis functions means that the very assumptions implicit in operating regime-based models such as local model nets, i.e. *localised behaviour*, is no longer guaranteed. The condition for reactivation in one-dimensional cases is described, providing a diagnostic tool for reactivation in basis function models where the basis function is the product of a number of one-dimensional basis functions (e.g. fuzzy membership functions).

---

[6]Ideally, a model should also return a confidence estimate which rejects data points outside the area covered by the training data, but in practice this is often difficult to achieve.

4   The analysis in section 8.4.1 shows how normalisation can affect both the condition number of the design matrix and the magnitude of the optimal local model parameters θ for a given network. The condition of the design matrix can subsequently affect the robustness of the learning procedure leading to high variance in the resulting parameters of the network. Large parameter magnitudes can also lead to non-robust models.

5   Effects 1–4 become more pronounced as the input dimension increases, due to the larger number of neighbouring basis functions and the consequent increase in overlap required by the 'curse of dimensionality'.

A partition of unity is an inherent assumption of all of the local modelling and control approaches described in this book. Often a partition of unity can only be achieved by normalising the local models' basis functions. This chapter should not be seen as a criticism of the technique, but rather as providing some insight into its side-effects and tools for their analysis. Researchers and users of local model approaches, probabilistic methods, and fuzzy systems should consider these effects when designing, interpreting and validating models and controllers as well as their associated learning algorithms. It is often easier to work with the parameters of the un-normalised basis functions, and as an example of this we provide a condition for the detection of reactivation in one-dimensional basis functions which would be of interest to users of tensor product basis functions, e.g. fuzzy models. The heuristic of trying to minimise the effect of normalisation by placing the un-normalised basis functions such that the $\frac{c_{max}}{c_{min}}$ measure is close to one, is also one which could be applied to learning algorithms.

ACKNOWLEDGEMENTS

This work was in part sponsored by the European Union Human Capital and Mobility programme (HCM), contract number ERBCHBICT930711. The authors also wish to acknowledge the many helpful comments made by D. Neumerkel, J. Franz, R. Bjørgan and J. Kalkkuhl.

REFERENCES

BARNES, C., S. BROWN, G. FLAKE, R. JONES, M. O'ROURKE AND Y. C. LEE (1991) Applications of neural networks to process control and modelling. In 'Artificial Neural Networks, Proceedings of 1991 Internat. Conf. Artif. Neur. Nets'. Vol. 1. pp. 321–326.

BENAIM, M. (1994) 'On functional approximation with normalised Gaussian units'. *Neural Computation* **6**(1), 319–333.

GOLUB, G. AND C. VAN LOAN (1989) *Matrix Computations*. John Hopkins University Press.

HLAVÁČKOVÁ, K. AND R. NERUDA (1993) 'Radial Basis Function networks'. *Neural Network World* **1**, 93–101.

JOHANSEN, T. A. (1994) Operating Regime Based Process Modelling and Identification. 94-109-W. Dr. Ing. thesis. Norges Tekniske Høgskole, Department of Engineering Cybernetics. Trondheim, Norway. http://www.itk.unit.no/ansatte/Johansen_Tor.Arne/dring_taj.ps.gz.

JONES, R. D., Y. C. LEE, C. W. BARNES, G. W. FLAKE, K. LEE, P. S. LEWIS AND S. QIAN (1989) Function approximation and time series prediction with neural networks. Technical Report 90-21. Los Alamos National Lab., New Mexico.

KAPLAN, W. (1991) *Advanced Calculus*. Addison-Wesley.

LEONARD, J. A., M. A. KRAMER AND L. H. UNGAR (1992) A neural network architecture that computes its own reliability. MIT Industrial Liaison Report 3-7-92. MIT, Dept. of Chemical Engineering.

MOODY, J. AND C. DARKEN (1989) 'Fast-learning in networks of locally-tuned processing units'. *Neural Computation* **1**, 281–294.

MURRAY-SMITH, R. (1994) A Local Model Network Approach to Nonlinear Modelling. PhD Thesis. Department of Computer Science, University of Strathclyde. Glasgow, Scotland.

NOBLE, B. AND J. W. DANIEL (1988) *Applied linear algebra*. 3rd edn. Prentice–Hall Int.

SHORTEN, R. AND R. MURRAY-SMITH (1994) On normalised basis function networks. Proceedings of the 4th Irish Conference on Neural Networks, pp. 212-219.

SHORTEN, R. AND R. MURRAY-SMITH (1996) 'Side-effects of Basis Function Normalisation in Radial Basis Function Networks'. *International Journal of Neural Systems* **7**(2), 167–179.

SÖDERSTRÖM, T. AND P. STOICA (1989) *System Identification*. Prentice Hall, Englewood Cliffs, NJ.

TAKAGI, T. AND M. SUGENO (1985) 'Fuzzy identification of systems and its applications for modeling and control'. *IEEE Trans. on Systems, Man and Cybernetics* **15**(1), 116–132.

WERNTGES, H. W. (1993) Partitions of unity improve neural function approximation. In 'Proc. IEEE Int. Conf. Neural Networks'. San Francisco, CA. pp. 914–918. Vol. 2.

PART THREE

Control

CHAPTER NINE

The Composition and Validation of Heterogeneous Control Laws

BENJAMIN KUIPERS and KARL ÅSTRÖM

We present[1] a method for creating and validating a nonlinear controller by the composition of heterogeneous local control laws appropriate to different operating regions. Like fuzzy logic control, these methods apply even in the presence of incomplete knowledge of the structure of the system, the boundaries of the operating regions, or even the control action to take. Unlike fuzzy logic control, these methods can be analysed by a combination of classical and qualitative methods. Each operating region of the system has a classical control law, which provides high-resolution control and can be analysed by classical methods. Operating regions are defined by fuzzy set membership functions. The global control law is the weighted average of the local control laws, where the weights are provided by the operating region membership functions. A heterogeneous control law can be analysed, even in the presence of incomplete knowledge, by representing it as a qualitative differential equation and using qualitative simulation to predict the set of possible behaviours of the system. By expressing the desired guarantee as a statement in a modal temporal logic, the validity of the guarantee can be automatically checked against the set of possible behaviours. We demonstrate heterogeneous controllers and our qualitative methods for proving their properties, first for a simple level controller for a water tank, and second for a highly nonlinear chemical reactor.

9.1 INTRODUCTION

Much control theory is based on linear models. This works very well for steady-state regulation at a fixed operating point. To make a control system that can operate over wide regions it is, however, necessary to introduce nonlinearities. There are several ways to do this. Linear feedback control can be combined with logic for switching between several linear feedback laws. Selectors that choose between different control laws depending on signal levels can be introduced. Systems of these types are common in industry, where their design is based on engineering experience combined with extensive simulation. Classical control theory (e.g., (Franklin *et al.* 1994)) provides a rich set of methods for local analysis

---

[1] This chapter reprints the following article, with a few updated references added. B. J. Kuipers and K. Åström (1994). The composition and validation of heterogeneous control laws. *Automatica* **30**(2): 233–249.

of the individual control laws and for describing their behaviour, but theoretical analysis of combined laws has proved to be much more difficult.

Fuzzy logic control (Zadeh 1973, Mamdani 1974) is another approach to obtain nonlinear control systems, especially in the presence of incomplete knowledge of the plant or even of the precise control action appropriate to a given situation. In this approach the measured variables are represented as fuzzy variables. A representation of the control signal as a fuzzy variable is computed from the measurements using fuzzy logic. The fuzzy variable is converted to a real variable using some type of "defuzzification". Again, design and validation of these control laws is based primarily on experience and extensive simulation.

In this chapter, we take a new look at the problem of specifying, analysing, and verifying the behaviour of nonlinear control laws, especially in the presence of incomplete knowledge. We focus our attention on *heterogeneous* control laws, which are composed of classical (typically linear) control laws defined over different operating regions (possibly with fuzzy boundaries). The control signal from a heterogeneous controller is the average of the signals from the local control laws, each weighted by the value of its operating region membership function. This approach to fuzzy control was pioneered by Takagi and Sugeno (Takagi and Sugeno 1985) and Sugeno and Kang (Sugeno and Kang 1986).

The analysis and verification of a heterogeneous control law is complicated by the fact that such a controller is normally designed to cope with incomplete knowledge of the structure of the plant, the boundaries of the operating regions, and even the desired control action. Qualitative simulation (Kuipers 1986, Kuipers 1989, Kuipers 1994) addresses this issue by making it possible to predict the behaviours consistent with qualitative knowledge about a dynamical system and its initial state. Specifically, given a qualitative differential equation (QDE) and a qualitative description of an initial state ($QState(t_0)$), the QSIM algorithm predicts a set of possible behaviours,

$$QDE \wedge QState(t_0) \rightarrow or(Beh_1, \ldots Beh_n),$$

that is guaranteed to include a description of the solution to any ordinary differential equation and initial state matching the qualitative description. Finally, using a model-checking algorithm for statements in modal temporal logic (Emerson 1990), we can automatically determine whether a given specification is guaranteed to hold within the predicted set of qualitative behaviours (Kuipers and Shults 1994).

The basic concepts of heterogeneous control will be introduced with a simple level controller for a water tank. We then discuss qualitative simulation and modal temporal logic, leading up to semi-automated proofs of properties of the heterogeneous controller. Finally, we present a heterogeneous controller for a highly nonlinear chemical reactor (Economou *et al.* 1986).

9.2 FUZZY CONTROL AND OTHER HETEROGENEOUS METHODS

Fuzzy sets were originally developed by Zadeh (Zadeh 1965, Yager *et al.* 1987) to formalise qualitative concepts without precise boundaries. For example, when describing values of a continuous scalar quantity such as the amount of water in a tank, there are no meaningful landmark values representing the boundaries between *low* and *normal*, or between *normal* and *high*.

(Zadeh 1965) formalises linguistic terms such as these as referring to *fuzzy sets* of numbers. A fuzzy set, S, within a domain, D, is represented by a *membership function*,

$s : D \rightarrow [0, 1]$. We interpret the value of $s(x)$, for $x \in D$, as a measure of the *appropriateness* of describing x with the descriptor S. Figure 9.2 includes three membership functions defining the appropriateness of applying the qualitative descriptors {*low, medium, high*} to quantitatively defined levels.[2]

Fuzzy control is a family of methods for expressing and applying control laws, using fuzzy sets to provide several benefits. First, they provide the ability to express and use incomplete knowledge of the system being controlled and of the control law itself. Second, they allow one to specify a complex control law as the composition of simple components. Third, fuzzy set membership functions provide smooth transitions from region to region. There are at least two distinct approaches to fuzzy control:

- *Fuzzy logic control* determines the control action by a combination of fuzzy logic rules.
- *Heterogeneous control* determines the control action as the weighted average of classical control laws.

A *fuzzy logic controller* (Zadeh 1973, Mamdani 1974, Michie and Chambers 1968) consists of a collection of simple control laws whose inputs and outputs are both fuzzy values. For example,

> If water level is *high*, then set drain opening to *wide*;

where *high* and *wide* are qualitative terms described by fuzzy sets over their quantitative domains.

All controller rules are fired in parallel, and the recommended actions are combined according to fuzzy value combination rules, weighted (or bounded) by the degree of satisfaction of the antecedent. Some process of "defuzzification" is required to convert the resulting fuzzy set description of an action into a scalar value for a control variable.

A *heterogeneous controller* decomposes the state space into multiple, possibly overlapping, operating regions. The domain of each operating region is characterised by a fuzzy set membership function. This makes it possible to express smooth transitions between adjacent regions. Each operating region is associated with a qualitative description of the system state, e.g. the low, normal, or high level of water in a tank. The fuzzy set membership functions may be regarded as a measure of the *appropriateness* of applying a given qualitative description to the system state. It may be assumed that, for any given system state, the appropriateness measures sum to 1.0.

Each region is associated with a control law. The control signal applied to the plant is a weighted average of the control signals for each region, where the weights are provided by the membership functions of each region.

The idea of combining simple linear feedback units with operations such as average, min, max, etc, is widely used industrially. The intent of this chapter is to provide a mathematical basis for the local and global analysis of these systems. The heterogeneous control approach decomposes the design of a controller into two relatively independent decisions: (1) the specification of natural, qualitatively distinct operating regions, and (2) the specification of a control law for each region. The weighted sum combination method provides smooth transitions from one region to another, and facilitates local and global analysis.

Heterogeneous control is also related to gain scheduling. There are however some differences. In gain scheduling a specific control law is selected for a given operating region

---

[2]*Appropriateness measure* is technically synonymous with the terms *membership function* and *possibility measure* as used in the fuzzy research community. However, for our purposes, the English connotation of *appropriateness measure* seems better to capture the relationship between a linguistic term and a scalar quantity.

and the parameters of the controller are changed with the region. In heterogeneous control
the values of the control signal for different regions are computed and averaged.

Sliding control (Utkin 1977, Utkin 1991, Slotine and Li 1991) is another method for constructing nonlinear controllers by combining the effects of simpler controllers. In its pure
form, the sliding control law changes discontinuously at a boundary, possibly leading to
rapid "chattering" between the two control surfaces but yielding extremely good ("perfect")
tracking of disturbances. To avoid chattering, the discontinuity can be smoothed by taking a
weighted average of the two control laws in a narrow band near the boundary, at some cost
in tracking error. Essentially, this replaces "crisp" by "fuzzy" operating regions for two
classical control laws, and thus is in the spirit of heterogeneous fuzzy control. However,
the design philosophy behind fuzzy control typically treats overlap regions as transitory,
existing primarily to provide smooth transitions between pure control regions. In contrast,
sliding control works exactly by exploiting the joint action of two control laws to confine
the system to a very small overlap region. Cross-fertilisation between these two approaches
should be of value.

9.3 A HETEROGENEOUS CONTROLLER FOR THE WATER TANK

9.3.1 The water tank

Consider control of the amount x of water in a tank, where the inflow rate q may vary, and
the area u of the drain opening is the control variable. The function $p(x)$ is a monotonically
increasing function of x whose exact form is not known to the designer. In this case, $p(x)$ is
approximately proportional to the square root of the pressure, whose relation to x depends
on the geometry of the tank.

$$
\begin{aligned}
x &= \text{amount in tank} \\
q &= \text{inflow into tank} \\
u &= \text{drain area} \\
p(x) &= \text{influence of pressure at drain}
\end{aligned}
$$

The dynamic behaviour of the system is described by:

$$\dot{x} = f(x, u) = q - u \cdot p(x).$$

9.3.2 Overlapping operating regions

The system has separate control laws in three operating regions, Low, $Normal$, and $High$,
with overlapping appropriateness measures, as shown in Figure 9.2.

For an implemented controller, the operating region appropriateness measures must be
specified as real-valued functions of a real variable. However, for purposes of analysis, the
appropriateness measures $l(x)$, $n(x)$, and $h(x)$, for the three operating regions need not be
completely specified. All that is known is that they rise or fall smoothly and monotonically
between their plateaus, where the boundaries of the plateaus are characterised by the landmark values, a, b, c, and d. They are normalised, so that $l(x) + n(x) + h(x) = 1$. Note

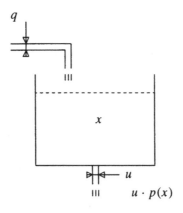

$$\dot{x} = f(x, u) = q - u \cdot p(x).$$

Figure 9.1   The water tank.

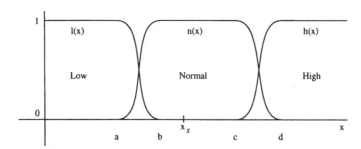

Figure 9.2   Three operating regions and their appropriateness measures.

that there is a "pure" region over each interval $[0, a]$, $[b, c]$, and $[d, \infty)$, and overlapping regions on (a, b) and (c, d). We assume that the setpoint x_s is in (b, c).

Because we specify the appropriateness measures qualitatively, and depend only on properties of the qualitative class, the conclusions we derive apply to every member of the class. The remaining degrees of freedom are available to the designer to meet other implementation requirements.

9.3.3   Heterogeneous control laws

The control laws for the three regions are:

$$
\begin{aligned}
x \in Low &\implies u_l(x) = 0 \\
x \in Normal &\implies u_n(x) = k(x - x_s) + u_s \\
x \in High &\implies u_h(x) = u_{max}
\end{aligned}
$$

where $0 \leq u \leq u_{max}$, and the bias term u_s is adjusted to give the desired set point x_s for a nominal inflow q_s. We are assuming for this example that the state variable x is directly observable, rather than separating out measurements, $y = g(x, u)$.

The water tank:

$$\dot{x} = q - u \cdot p(x), \quad p \in M_0{}^+$$

q : the flow into the tank (exogenous)
x : the level in the tank (sensed)
u : the drain opening (controlled)

The operating regions and their appropriateness measures:

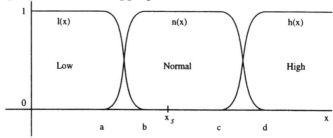

The local control laws:

$$x \in Low \quad \Rightarrow \quad u_l(x) = 0$$
$$x \in Normal \quad \Rightarrow \quad u_n(x) = k(x - x_s) + u_s$$
$$x \in High \quad \Rightarrow \quad u_h(x) = u_{max}$$

The global control law:

$$u(x) = l(x)u_l(x) + n(x)u_n(x) + h(x)u_h(x).$$

The discrete abstraction:

$$\boxed{\text{Low}} \longrightarrow \boxed{\boxed{\text{Normal}}} \longleftarrow \boxed{\text{High}}.$$

Figure 9.3   A heterogeneous controller for the water tank.

The global heterogeneous control law is the average of the individual control laws, weighted by the appropriateness measures of their regions:

$$u(x) = l(x) \cdot 0 + n(x) \cdot [k(x - x_s) + u_s] + h(x) \cdot u_{max}$$

The design goal for the controller is a simple discrete abstraction: the system will move from the High or Low region into the Normal region, and reach equilibrium there. Our qualitative analysis, described below, helps us identify the additional constraints necessary to provide this guarantee. Figure 9.3 summarises the heterogeneous controller for the water tank.

9.3.4   Simulation results

By numerical simulation, we can illustrate the performance of this heterogeneous controller on a water tank, in comparison with a proportional controller.

The capacity of the tank is 1000 litres of water. The nominal inflow rate is 100 litres/minute. The setpoint, x_s, is 700 litres. The offset u_s in the *Normal* control law u_n is set so that the steady state is at the setpoint when inflow is nominal. The gain k is set so that $u_n(0) = 0$. The proportional controller simply uses u_n as the global control law. The comparison is for illustration only, since the proportional controller has an unrealistically low gain. With a higher gain, however, the physical limits on the valve make the proportional controller behave like a heterogeneous controller, but without smooth transitions or explicit design and validation.

The operating regions for the heterogeneous (HC) controller are specified as in Figure 9.3, with $a = 600$, $b = 650$, $c = 750$, $d = 800$, and $u_{max} = 50$.

Figure 9.4(a) compares the two control laws $u(x)$ and $u_n(x)$. Figure 9.4(b) contrasts the behaviour of the two controllers at constant nominal inflow, starting from the initial states $x(t_0) = 0$ and $x(t_0) = 1000$. Figure 9.5 shows the responses of the two controllers to random variation in inflow between zero, nominal, and twice nominal.

9.4 METHODS FOR PROVIDING GUARANTEES

We want to prove that the heterogeneous controller brings the system back to the *Normal* operating region under some range of disturbances, and that an equilibrium in the region is obtained for constant disturbances. More importantly, we want to determine any quantitative constraints on the design of the controller (e.g. the value for u_{max}), and the range of possible disturbances on q that the controller can handle.

There are two methods for doing this (Figure 9.6), which are elaborated on below.

1 (a) Determine the qualitative behaviour of the system within each operating region.
   (b) Combine the qualitative descriptions.
2 (a) Combine the local laws into a global law using the weighted average combination rule.
   (b) Determine the qualitative behaviour of the global system.

First, however, we must introduce the ordinal, landmark-based methods of qualitative representation and the QSIM algorithm for qualitative simulation (Kuipers 1986, Kuipers 1989, Kuipers 1994).

Qualitative categories may be described by fuzzy set membership functions where they lack meaningful boundaries, or by ordinal relations with landmark values where precise boundaries are meaningful. By using the landmark-based representation to describe a restricted class of fuzzy set membership functions, we can combine the performance benefits of smooth fuzzy-set transitions with the analytic power of landmark-based qualitative simulation.

9.4.1 Landmark-based representation

Qualitative categories may be defined by *landmark values*: precise boundary points separating qualitatively distinct regions of a continuum. For example, angles in a triangle can be described in the following qualitative terms:

$$\textit{Zero} \quad \cdots\cdots \quad \textit{Right} \quad \cdots\cdots \quad \textit{Straight}$$
$$(\textit{acute}) \qquad\qquad (\textit{obtuse})$$

A value can be described qualitatively either as equal to a landmark value or in an open interval bounded by two landmark values, even when numerical information is unavailable.

U vs Un comparison

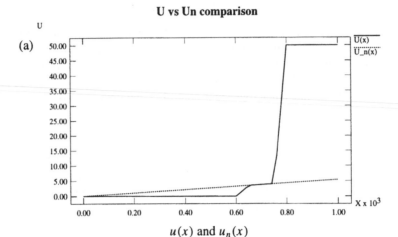

$u(x)$ and $u_n(x)$

HC vs P comparison

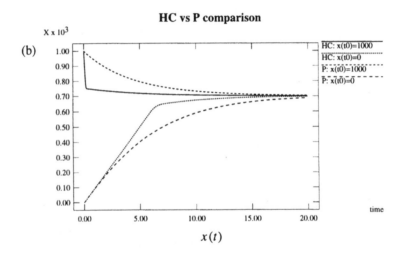

$x(t)$

Figure 9.4   Comparison between P and HC controllers.

(a) The heterogeneous control law $u(x)$, and the proportional controller $u_n(x)$ are identical in the Normal region.

(b) The behaviours, $x(t)$, of the P- and HC-controllers, starting with the tank empty or full, with constant q at the nominal rate, so that steady state is at the setpoint.

It is often easier to obtain and justify the qualitative description of a quantity than its numerical value, particularly when knowledge is incomplete. Human perception, memory, and similarity judgements often reflect underlying landmark-based qualitative representations (Goldmeier 1972). Fortunately, landmark-based descriptions support *qualitative simulation*, to derive qualitative descriptions of the possible behaviours of a system from a qualitative description of its structure (Kuipers 1986, Kuipers 1989, Kuipers 1994).

An ordinary differential equation describes a system in terms of a set of variables which vary continuously over time, along with constraints among those variables such as addition, multiplication, and differentiation. A *qualitative differential equation* (QDE) describes a

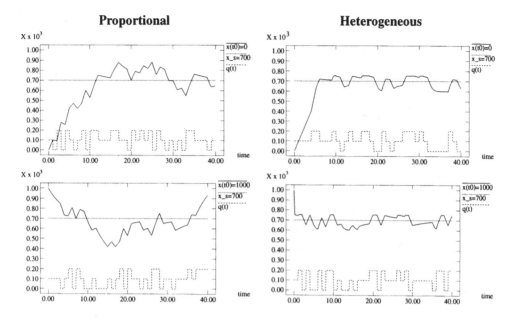

Figure 9.5 The effect of random inflow variation on P and HC controllers.
Inflow, q, plotted at the bottom of each graph, varies randomly between zero, nominal (100 litres/minute), and twice nominal. This figure shows the proportional controller (left column) and the heterogeneous controller (right column), with the tank initially empty (top row) or full (bottom row).

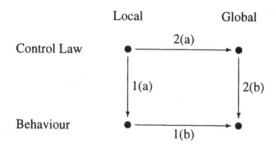

Figure 9.6 Two approaches to analysing a heterogeneous controller.

system in much the same terms, except that (1) the values of variables are described qualitatively, and (2) certain functional relationships between variables may be incompletely known. For example, air resistance on a moving body increases monotonically with velocity, and flow of water through an orifice increases monotonically with pressure. Both of these relations are nonlinear, but useful qualitative conclusions can be drawn purely from monotonicity. It is useful to define the class M^+ of monotonic functions, and the class S^+ of monotonic functions with saturation.

- A *reasonable function* is a continuously differentiable function defined on a closed interval (including the extended real number line $\Re^* = [-\infty, +\infty]$), with only isolated critical points, and derivatives continuous at the endpoints of the domain. See (Kuipers 1986).

■ M^+ is the set of reasonable functions f such that $f' > 0$ on the interior of its domain. In a QDE, we may write $M^+(pressure, outflow)$ or $outflow = M^+(pressure)$ to mean that there is some $f \in M^+$ such that $outflow = f(pressure)$. M_0^+ is the subset of M^+ such that $f(0) = 0$, and M^- is the set of f such that $-f \in M^+$.

■ S^+ is the set of reasonable functions f such that, for specified pairs of landmark values (x_1, y_1) and (x_2, y_2),

   – $f(x) = y_1$ for all $x \leq x_1$,
   – $f(x) = y_2$ for all $x \geq x_2$,
   – $f'(x) > 0$ for all $x_1 < x < x_2$.

The turning points (x_1, y_1) and (x_2, y_2) must be specified as landmark values whenever the S^+ constraint is used. The subset of S^+ with turning points at (a, c) and (b, d) is called $S_{(a,c),(b,d)}^+$. S^- is the set of f such that $-f \in S^+$.

The qualitative structure of the appropriateness measures in Figure 9.2 can be expressed in terms of the S^+ constraint by introducing the two functions,

$$s_1(x) \in S_{(a,0),(b,1)}^+ \qquad\qquad s_2(x) \in S_{(c,0),(d,1)}^+$$

such that

$$
\begin{aligned}
l(x) &= 1 - s_1(x) \\
n(x) &= s_1(x)(1 - s_2(x)) \\
h(x) &= s_2(x).
\end{aligned}
$$

This qualitative description expresses a state of incomplete knowledge about the operating regions and their boundaries. We represent our knowledge of the system, the operating regions, and the local control laws, as a qualitative differential equation.

QSIM (Kuipers 1986, Kuipers 1989, Kuipers 1994) allows us to specify a qualitative differential equation (QDE) and a description of an initial state ($QState(t_0)$), and predicts a set of possible behaviours, such that

$$QDE \wedge QState(t_0) \rightarrow or(Beh_1, \ldots Beh_n).$$

The inference done by QSIM is sound, so the set of behaviours $\{Beh_1, \ldots Beh_n\}$ includes all possible behaviours of systems described by QDE and $QState(t_0)$.

Each behaviour Beh_i is a sequence of qualitative states representing alternating time-points and qualitatively uniform time-intervals:

$$QState(t_0), QState(t_0, t_1), QState(t_1), \ldots QState(t_{k-1}, t_k), QState(t_k).$$

The set of behaviours is represented by a tree of qualitative states linked by successor relations, so each time-point state $QState(t_i)$ is followed by all possible immediately succeeding time-interval states $QState(t_i, t_{i+1})$, and each time-interval state $QState(t_i, t_{i+1})$ is followed by all possible immediately succeeding time-point states $QState(t_{i+1})$.[3] This tree of states can be viewed as a branching-time description of a set of possible futures.

The tree of possible behaviours of a qualitatively described system can be a powerful analytical tool. In particular, if a qualitative property (e.g. stability or zero-offset) holds on every branch of the tree, it must hold for every behaviour of every fully specified instance of the system. The importance of the *qualitative* level of description is that the tree of behaviours for a given QDE may be finite, whereas the corresponding set of ordinary differential equations and their solutions is infinite.

---

[3] Provisions exist in QSIM for identifying and representing cycles and other repeated states, so the behaviour tree may actually be a transition graph.

9.4.2 QSIM and temporal logic

In order to state and prove properties of a continuous dynamic system, we use QSIM to create a symbolic description of the set of all possible behaviours in the form of a tree of qualitative states, then state the desired properties as propositions in a modal temporal logic, and finally check that the propositions are true of the tree of states. We summarise here work presented in more detail by Kuipers and Shults (Kuipers and Shults 1994).

Temporal and modal logics have been developed for expressing and inferring properties of time-varying systems (Emerson 1990). They have primarily been applied to discrete-time rather than continuous-time models of the world, such as might arise in verification of computing systems. However, as we have seen, QSIM provides a discrete tree of qualitative states that describes the behaviours of a continuous system. Therefore, we define a modal temporal logic (an instance of Computational Tree Logic (CTL) (Emerson 1990)) customised for application to QSIM behaviour trees.

Following Emerson (Emerson 1990), we define:

- A *temporal structure* is $M = \langle S, R, L \rangle$, where

    - S is a set of states;
    - R is a binary successor relation defined on $S \times S$;
    - L is a labelling, associating with each state $s \in S$ a set of atomic propositions.

- The behaviour tree resulting from QSIM simulation of QDE and $QState(t_0)$ can be viewed as a temporal structure $M = \langle S, R, L \rangle$, where S is the set of states in the tree, R is the union of the QSIM successor and transition relations, and L labels states in S with atomic propositions as follows.

    - $status(s, tag)$, where s is a state, and tag is one of { *quiescent, stable, unstable, transition* }. The proposition is true of a state s in M if tag is an element of s.status (an element of the qualitative state description).
    - $qval(s, v, qmag, qdir)$, where s is a state, v is a variable appearing in QDE, $qmag$ is a landmark or interval defined by a pair of landmarks in the quantity space for v, and $qdir$ is one of $\{inc, std, dec\}$ representing the sign of the derivative of v. The proposition is true of a state s in M if the value of v in s satisfies the description $\langle qmag, qdir \rangle$.

- *State formulae* and *path formulae* are built up from atomic propositions. A *path* is a sequence of states in M linked by the successor relation R.
- State formulae have the following syntax and semantics.

    - (S1) Each atomic proposition is a state formula. Its truth conditions have already been defined.
    - (S2) If p, q are state formulae, then so are $p \wedge q$, $p \vee q$, $\neg p$, and $p \rightarrow q$, with the usual rules defining the truth values of composed expressions.
    - (S3) If p is a path formula, then (necessarily p) is a state formula, which is true of a state s in M if p is true of *every* path starting at s. Similarly, (possibly p) is a state formula, which is true of a state s in M if p is true of *any* path starting at s.

- Path formulae have the following syntax and semantics.

    - (P0) If p, q are state formulae, then (until p q) is a path formula, which is true of a path if q is true for some state in the path, and p is true of every previous state in that path. Similarly, (next p) is a path formula, which is true of p if it is true of the path starting with the second element of p.

- We can define the useful forms (eventually p) as (until true p), and (always p) as (not (eventually (not p))).

Basically, CTL allows a path quantifier (possibly or necessarily) to be followed by a single temporal operator (until, next, always, or eventually).

■ An extended language, CTL*, allows arbitrary boolean combinations or nestings of the temporal operators. It can be defined by replacing (P0) with the following syntax and semantics for path formulae.

  - (P1) Each state formula is also a path formula, and is true of the path if it is true of the first state in the path.
  - (P2) If p, q are path formulae, then so are $p \wedge q$, $p \vee q$, $\neg p$, and $p \rightarrow q$, with the usual rules defining the truth values of composed expressions.
  - (P3) If p, q are path formulae, then so is (until p q), which is true of a path x if q is true of some suffix path x^i of x, and p is true of every longer suffix path x^j containing x^i. Similarly, (next p) is a path formula, which is true of p if it is true of the path starting with the second element of p.

■ Based on the above definitions, we have written algorithms to check whether a state s in the temporal structure M, derived from a QSIM behavioural prediction, satisfies a given temporal logic proposition P in CTL or CTL*.

■ Emerson (Emerson 1990) provides the following complexity results for the slightly more difficult *model checking* problem, which determines the satisfiability of P for every s in M.

  - The model-checking problem for CTL is in deterministic polynomial time.
  - The model-checking problem for CTL* is PSPACE-complete.

Therefore, while CTL* is a more expressive language than CTL, model-checking for CTL* is potentially vastly more computationally expensive. CTL appears to be sufficiently expressive for our purposes.

■ Interesting statements are naturally expressible in modal temporal logic, and hence can be automatically checked against a qualitative behaviour tree. For example,

  - (necessarily (always P)) means that P is true throughout the dynamic behaviour of the system.
  - (necessarily (eventually
    (and (status quiescent) (status stable)))) and
    (necessarily (always (implies (status stable) P)))
    together imply that the dynamical system has a quasi-equilibrium view, and that P is true in that view.

Figure 9.7 illustrates the checking of statements in CTL against a partial behaviour tree produced by QSIM simulation of a simple nonlinear PI controller.

QSIM guarantees coverage of every possible behaviour by the predictions in the behaviour tree. However, its inference capabilities may not be powerful enough to refute every impossible behaviour. Therefore, caution is required before concluding that a proposition is true of the behaviour of a dynamical system, given the fact that it is true of a QSIM behaviour tree.

■ If (necessarily P) is true of the behaviour tree, then it is true of every dynamical system matching the QDE and initial state description.

■ If (possibly P) is true of the behaviour tree, it may *not* be true of corresponding dynamical systems, since P might have been true only of spurious behaviours in the tree.

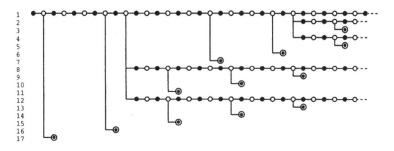

(a)

modal	temporal	logical	QSIM
necessarily	*until*	*and*	*qval*
possibly	*next*	*or*	*status*
	eventually	*implies*	
	always	*not*	

(b)

```
Simulating KJA PI controller.
Behaviour tree rooted at S-0, with 1 initial states and 17 behaviours.

Some behaviours don't terminate...
Checking:  (EVENTUALLY (STATUS QUIESCENT)).
Validity = (NIL NIL T NIL T T T NIL T T T NIL T T T T T).

...but all that terminate have zero error.
Checking:  (NECESSARILY (ALWAYS (IMPLIES (STATUS QUIESCENT) (QVAL E (0 STD)))))).
Validity = T.

Every fixed point is stable.
Checking:  (NECESSARILY (ALWAYS (IMPLIES (STATUS QUIESCENT) (STATUS STABLE))))).
Validity = T.
```

(c)

Figure 9.7   Qualitative behaviour trees and temporal logic.

(a)   The possible behaviours of a QDE model of a simplified nonlinear PI controller $-\dot{e} + f(e) + \int e = 0$, where $f \in M^+$ – are described by a tree of qualitative state descriptions.

The tree grows – and time passes – from left to right, with each state linked to its possible direct successors. Filled and open circles represent time points and open intervals, respectively. A circled dot represents a fixed point, not necessarily stable. An ellipsis represents an incomplete branch of the tree.

(b)   The operators of modal temporal logic provide an appropriate language for querying and describing the behaviour tree.

(c)   For this model of a simple PI controller, we can conclude that every fixed-point is stable and represents zero error. (There are methods in QSIM for eliminating the non-terminating behaviour descriptions, but they are not yet integrated with the CTL validity-checker.)

Fortunately, in all of our examples, we have needed only statements of the form (necessarily P), so their validity applies both to the QSIM behaviour tree and to the dynamical systems being reasoned about.

9.5 GUARANTEES FOR THE WATER TANK CONTROLLER

9.5.1 Qualitative combination of local properties

Figure 9.8 summarises a qualitative analysis of the water tank controller taking the first approach described in Figure 9.6. The analysis begins by determining the direction of motion of the system as specified by each control law individually. The properties of the appropriateness measures are not required. Then, in the regions of overlap, if the directions of change agree, the global law for the heterogeneous controller must give motion in the same direction. If the different control laws give motion in opposite directions in the overlap regions, a deeper analysis combining qualitative simulation with order-of-magnitude (Mavrovouniotis and Stephanopoulos 1988) or semi-quantitative constraints (Kuipers and Berleant 1988, Berleant and Kuipers 1992, Kay and Kuipers 1993) may be able to reduce ambiguity about the system's behaviour.

In order to guarantee that the directions of change agree on the overlap regions (a, b) and (c, d), and therefore that the system always ends up within the "pure" operating region (b, c) of the *Normal* controller, we need to impose constraints on (1) the range of inflow perturbations to be handled, and (2) the magnitude of the *High* response.

1 From the *Normal* model:

$$q_b < q < q_c, \tag{9.1}$$

where q_b (respectively, q_c) is the value of q that results in steady state at $x = b$ (respectively, $x = c$).

2 From the *High* model:

$$q < u_{max} \cdot p(c). \tag{9.2}$$

The individual steps of the analysis depicted in Figure 9.8(b) are accomplished by simulating qualitative models of the water tank with each individual local controller, over the region where its appropriateness measure is nonzero. With the constraints (9.1) and (9.2) incorporated into the models, the predictions have the desired properties. As shown in Figure 9.9(top), QSIM simulation of the water tank with control law u_l gives three qualitatively distinct behaviours; u_n gives 34; and u_h gives 21.

Once all possible behaviours have been determined, specifications in CTL of the desired properties of each local controller are checked for validity against the behaviour tree. In Figure 9.9(bot), we see that the CTL model-checker determines that each of the given specifications is a valid description of the corresponding tree of possible behaviours.

Finally, since the direction of motion of the global system is the average of the directions of motion of the local systems, with non-negative weights, the global system must show the desired qualitative behaviour: motion from any point in the state space to a stable fixed-point in the interval (b, c). Note that this conclusion does not depend on other constraints, in particular on the shapes of the appropriateness measures.

9.5.2 Abstracting behaviour to a transition graph

These qualitative properties of the behaviour of the system and its heterogeneous controller can be expressed as a finite transition graph in which the nodes correspond to the operating regions, and the directed edges correspond to transitions between regions.

$$\boxed{\text{Low}} \longrightarrow \boxed{\boxed{\text{Normal}}} \longleftarrow \boxed{\text{High}}$$

where the double box signifies that the *Normal* region includes a steady state, and so can persist indefinitely, while the other regions can persist only for a finite time.

The abstraction relation is defined as follows:

- The state of the system corresponds to a node of the transition graph if it is in the *interior* of the corresponding "pure" operating region, where its appropriateness measure is equal to 1.
- There is a directed link between two nodes in the transition graph if the system state moves monotonically from one node to the other (i.e. there is no quiescent state in the overlap region between the two pure regions). During behaviour corresponding to a directed link, the appropriateness measure of one region decreases monotonically, while the other increases monotonically.
- There may be no other behaviours that intersect the overlap regions.

9.5.3 Human and automated inference

It is important to be clear about which steps in the analysis are and are not automated. Qualitative simulation of the controller models and CTL checking of the specifications are both automated. The specifications themselves are provided by the human analyst. The need for additional constraints is determined automatically, since QSIM simulation produces a behaviour tree for which the specifications are not valid. The constraints themselves are determined by the human analyst. Assembly of the local specifications into a guarantee for the heterogeneous controller is done by the human analyst.

In principle, it should be possible to automate these steps as well, but this will require advances to the state of the art in qualitative model-building, symbolic and algebraic manipulation, and temporal logic theorem-proving.

9.5.4 Qualitative analysis of the global control law

Following the second path in Figure 9.6, a global analysis of the heterogeneous system is possible when we can establish suitable relations among the individual control laws.

Suppose we can establish that the global control law $u(x)$ is a monotonic function of x. Then the closed-loop system can be described as

$$\dot{x} = q - u(x)\,p(x) = q - f(x), \text{ for some } f \in M^+.$$

Since this is a first-order system, the analysis is straightforward. An equilibrium exists if q is in the range of f. The solution is unique since f is monotone. The solution is stable because $f' > 0$, since $f \in M^+$.

It is necessary to introduce some compatibility conditions in order to avoid pathological behaviour of the system. To see this, consider the case where only two controllers are combined (e.g., the *Normal* and *High* controllers over the range (b, ∞) in the water-tank example). The control signal is then

$$u(x) = n(x)\,u_n(x) + h(x)\,u_h(x).$$

■ (a) Overlapping operating regions for the local laws.

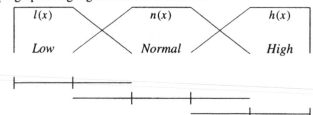

■ (b) Predict qualitative behaviours; require agreement where local laws overlap.

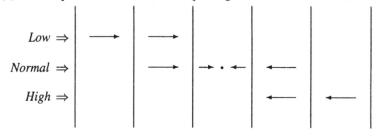

■ (c) Determine constraints to guarantee monotonic behaviour in overlap regions.

$$Low \quad \Rightarrow \quad q > 0$$
$$Normal \quad \Rightarrow \quad q_b < q < q_c$$
$$High \quad \Rightarrow \quad q < u_{max} \cdot p(c)$$

■ (d) Abstract the control law to a finite transition diagram.

$$\frac{dl(x)}{dt} < 0 \qquad\qquad \frac{dh(x)}{dt} < 0$$

$$\frac{dn(x)}{dt} > 0 \qquad\qquad \frac{dn(x)}{dt} > 0$$

| Low | ⟶ | Normal | ⟵ | High |

Figure 9.8   Qualitative combination of properties of local laws.

It is natural to have controllers such that

$$\frac{du_n}{dx} \geq 0 \text{ and } \frac{du_h}{dx} \geq 0.$$

Unfortunately, these conditions do not guarantee that u is monotone. To obtain this, some auxiliary conditions are required.

Consider

$$u' = n\,u'_n + n'\,u_n + h\,u'_h + h'\,u_h$$

$$n + h = 1$$
$$n' + h' = 0.$$

The problem is that n' is negative. However, we can conclude:

$$u' = n\,u'_n + h\,u'_h + h'(u_h - u_n).$$

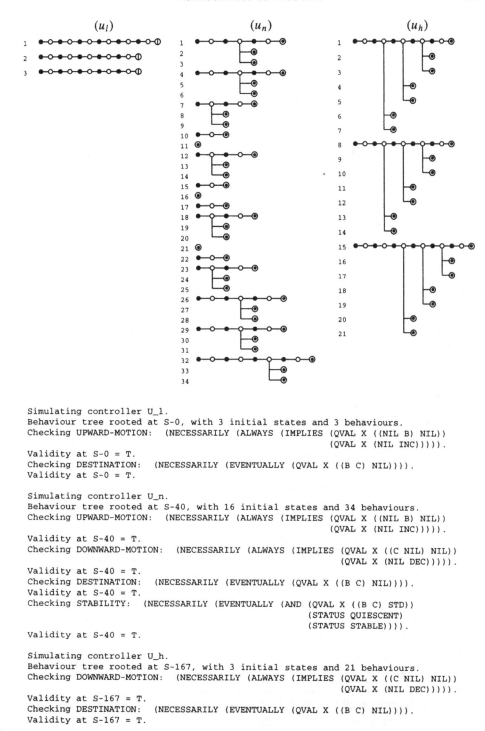

```
Simulating controller U_1.
Behaviour tree rooted at S-0, with 3 initial states and 3 behaviours.
Checking UPWARD-MOTION:  (NECESSARILY (ALWAYS (IMPLIES (QVAL X ((NIL B) NIL))
                                                       (QVAL X (NIL INC)))))).
Validity at S-0 = T.
Checking DESTINATION:  (NECESSARILY (EVENTUALLY (QVAL X ((B C) NIL)))).
Validity at S-0 = T.


Simulating controller U_n.
Behaviour tree rooted at S-40, with 16 initial states and 34 behaviours.
Checking UPWARD-MOTION:  (NECESSARILY (ALWAYS (IMPLIES (QVAL X ((NIL B) NIL))
                                                       (QVAL X (NIL INC)))))).
Validity at S-40 = T.
Checking DOWNWARD-MOTION:  (NECESSARILY (ALWAYS (IMPLIES (QVAL X ((C NIL) NIL))
                                                         (QVAL X (NIL DEC)))))).
Validity at S-40 = T.
Checking DESTINATION:  (NECESSARILY (EVENTUALLY (QVAL X ((B C) NIL)))).
Validity at S-40 = T.
Checking STABILITY:  (NECESSARILY (EVENTUALLY (AND (QVAL X ((B C) STD))
                                                   (STATUS QUIESCENT)
                                                   (STATUS STABLE)))).
Validity at S-40 = T.


Simulating controller U_h.
Behaviour tree rooted at S-167, with 3 initial states and 21 behaviours.
Checking DOWNWARD-MOTION:  (NECESSARILY (ALWAYS (IMPLIES (QVAL X ((C NIL) NIL))
                                                         (QVAL X (NIL DEC)))))).
Validity at S-167 = T.
Checking DESTINATION:  (NECESSARILY (EVENTUALLY (QVAL X ((B C) NIL)))).
Validity at S-167 = T.
```

Figure 9.9 Local analysis of heterogeneous level-controller.

QSIM behaviour trees representing the possible behaviours of the water tank controlled by each local law, along with CTL statements implying qualitative agreement among the local laws and justifying the discrete abstraction in Figure 9.8.

This assures us that $u' > 0$, and hence that $f(x) = u(x)\, p(x)$ is in M^+, if we impose the natural condition

$$u_n(x) \leq u_h(x).$$

This condition needs to hold only for x where the two regions overlap. The argument obviously extends to more complex heterogeneous controllers, such as the water tank, where no more than two regions overlap at any point.

9.5.5   Integral action

The bias term in the proportional controller was introduced to make it possible for the controller to keep the level at the set point. Integral action may be viewed as an automatic adjustment of the bias term [see Figure 2.2 in (Åström and Hägglund 1988)]. For a simple PI controller the bias is adjusted according to

$$T\frac{du_s}{dt} + u_s = ke + u_s$$

or

$$T\frac{du_s}{dt} = u_p = ke, \tag{9.3}$$

where u_p is the output of the PI controller, e is the error $x - x_s$, k is the proportional gain, and T is the integration time. For a composite controller like the one used in heterogeneous control, u_p should be replaced by the output of the heterogeneous controller.

Analysis of a controller with integral action is more complicated because the closed-loop system is described by a second-order differential equation and a simple monotonicity argument like the one used previously does not apply directly.

There were two alternative approaches to the qualitative analysis of the heterogeneous "proportional" controller. Similarly, there appear to be three basic approaches to analysing the "integral" component of a heterogeneous controller.

1   The bias term u_s is adjusted, at a slower time-scale, by a heterogeneous P-controller as a function of the steady-state error, $x_s - x(\infty)$ as discussed below.
2   Local control laws, even with integral action, can be analysed qualitatively, and associated with overlapping operating regions in the phase plane. If the directions of flow in the overlap regions are compatible, the qualitative descriptions can be combined into a discrete transition-graph representing behaviour in the phase plane (Sacks 1990).
3   The local laws may be combined into a global control law using the weighted average combination rule, which may then be analysed qualitatively.

One possibility is to exploit the fact that integral action is a slow process. The idea of time scale separation introduced in (Kuipers 1987) can then be applied. The full details will be given elsewhere. Let us just outline the ideas of the reasoning. Provided that the integration time T is sufficiently small the closed-loop system can be decomposed into a fast system, where the bias term is considered constant, and a slow system, where the fast system is considered as a static system. The previous analysis then applies to the fast system. It follows from this analysis that the level goes to an equilibrium which may be different from the set point. At equilibrium the fast system can be described by

$$u_p = -f(u_s), \tag{9.4}$$

where the function f belongs to M^+. The slow system is described by (9.4) and (9.3), i.e.

$$T\frac{du_s}{dt} = u_p = -f(u_s).$$

Since f is monotone this equation has a unique stable equilibrium

$$u_p = ke = 0,$$

which implies that the error e must be zero when the slow system reaches equilibrium.

9.6 EXAMPLE: A NONLINEAR CHEMICAL REACTOR

9.6.1 The system

In order to demonstrate heterogeneous control in a somewhat more realistic context, we consider a simple, but highly nonlinear chemical reactor: a reversible exothermic reaction

$$A \underset{k_{-1}}{\overset{k}{\rightleftharpoons}} R$$

in which the extent of the reaction increases with reactor temperature at lower temperatures and decreases at higher temperatures. This causes the gain of the plant (the dependence of the concentration of R, which is the controlled variable, on the inlet temperature, T_i) to change sign as a function of reactor temperature, making the reaction difficult to control (Economou *et al.* 1986).

The process can be modelled by the following system of ordinary differential equations, where A_i and R_i represent the feed species concentrations of the reactant and the product, and the state variables A, R and T represent the outlet species concentrations of the (unreacted) reactant and the product, and the reactor's temperature, respectively. The system is manipulated by varying the inlet feed temperature T_i.

■ Reactant mass balance:

$$\frac{dA}{dt} = \frac{1}{\tau}(A_i - A) - k_1 A + k_{-1} R \tag{9.5}$$

■ Product mass balance:

$$\frac{dR}{dt} = \frac{1}{\tau}(R_i - R) + k_1 A - k_{-1} R \tag{9.6}$$

■ Energy balance:

$$\frac{dT}{dt} = \left(\frac{-\Delta H_R}{\rho C_p}\right)(k_1 A - k_{-1} R) + \frac{1}{\tau}(T_i - T), \tag{9.7}$$

where $k_1 = C_1 \exp(-Q_1/RT)$ and $k_{-1} = C_{-1} \exp(-Q_{-1}/RT)$.

The equilibrium conversion from A to R is a function of temperature with a well-defined maximum, as shown in Figure 9.10. We want to operate the system at the setpoint $R = R_s$ when that is physically possible given the input; otherwise, as close as possible to the maximum conversion. The problem, of course, is that the steady-state gain at the maximum is zero, so manipulation of inlet feed temperature T_i has no effect on the system at that point.

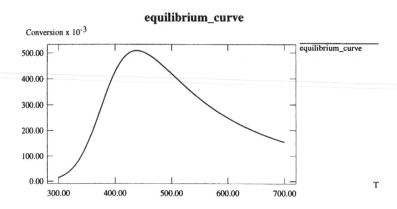

Figure 9.10 The equilibrium conversion from A to R is a non-monotonic U^- function of temperature, with a well-defined maximum.

The non-monotonic reactor:

R	:	the outlet product concentration (controlled)
R_s	:	the setpoint for R
T_i	:	the inlet temperature (manipulated)
T	:	reactor temperature (determines appropriateness)

The operating regions and their appropriateness measures:

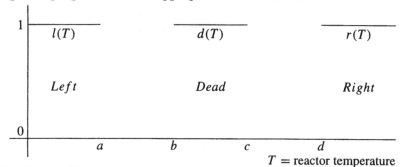

The local control laws:

$$T \in Left \implies u_l(R) = k_l(R - R_s) + u_l, \quad k_l < 0$$
$$T \in Dead \implies u_d(R) = 0$$
$$T \in Right \implies u_r(R) = k_r(R - R_s) + u_h, \quad k_r > 0$$

The global control law:

$$u(T, R) = l(T)u_l(R) + d(T)u_d(R) + r(T)u_r(R).$$

Figure 9.11 A heterogeneous controller for the non-monotonic reaction.

increase A_i decrease A_i

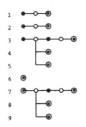

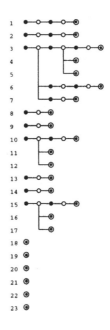

Behaviour tree rooted at S-2,
  with 3 initial states and 5 behaviours.
Validity at S-2 = T.

Behaviour tree rooted at S-26,
  with 1 initial states and 1 behaviours.
Validity at S-26 = T.

Behaviour tree rooted at S-29,
  with 1 initial states and 3 behaviours.
Validity at S-29 = T.

Behaviour tree rooted at S-2,
  with 9 initial states and 17 behaviours.
Validity at S-2 = T.

Behaviour tree rooted at S-312,
  with 3 initial states and 3 behaviours.
Validity at S-312 = T.

Behaviour tree rooted at S-322,
  with 3 initial states and 3 behaviours.
Validity at S-322 = T.

Figure 9.12 Under both perturbations, the desired stability property holds on all paths. The final states on all paths are quiescent and stable. Either the system reaches its setpoint ($E = R - R_s = 0$) or the dead zone ($T_o \in [b, c]$).

9.6.2 The controller

We will cope with the nonlinearity of the system by defining three operating regions: Left and Right, having linear control laws with gains of opposite signs; and Dead, a region around the peak of the curve where control actions have no effect, so none is taken.

The heterogeneous controller is described in Figure 9.11.

9.6.3 The proof

A QSIM model was constructed corresponding to equations (9.5), (9.6), and (9.7), and the heterogeneous controller described in Figure 9.11. The model predicted the response of the system to a change in the inlet reactant concentration, A_i.

A certain amount of effort was required to find a level of abstraction at which the QSIM

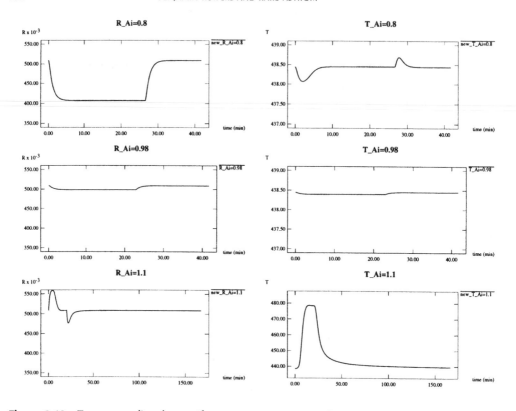

Figure 9.13 Temporary disturbance (from $t = 0$ to 20 min) of -20%, -2%, and $+10\%$ in A_i, the inlet reactant concentration. ($k_l = -1.5$, $k_h = 3.0$, $a = 420.0$, $b = 438.4$, $c = 438.5$, $d = 445.0$.)

■ -20%. Temperature at the beginning drops because A_i dropped. Then the controller brings the temperature up, towards the peak, to increase conversion, i.e. keep R high. Of course it fails to keep R to the set point (there is simply not enough input to get that much output), but it keeps it as high as possible keeping conversion near the max. When the disturbance is over ($t = 20$), A_i returns to 1.0, so T starts increasing for a while, till the controller lowers it to get back to the peak of the curve. Note that the total T variation is about 0.5 degree. (Similar to Figure 12 in (Economou *et al.* 1986), but for a much larger disturbance.)

■ -2%. Equivalent to Figure 12 in (Economou *et al.* 1986).

■ $+10\%$. Temperature is increased in order to lower R to R_s, in which it succeeds before $t = 20$, and after $t = 20$ the controller drops T to bring the system back to the set point. The time-axis on this graph is extended to demonstrate that the temperature returns to nominal.

behaviour prediction was tractable. This included a quasi-equilibrium assumption: the controller assumes that the reaction is always near equilibrium. The set of possible behaviours is shown in Figure 9.12.

In response to a change in inlet reactant concentration A_i, the desired behaviour is that the output R returns to the setpoint R_s. In case of decreasing A_i this may be impossible, in which case the system should move into the dead zone, $T \in [b, c]$. Within the dead zone, no control action is possible, so we cannot ensure that the system attains the absolute maximum conversion, but since the system's gain is very low in that region, it cannot be very far from the maximum.

We express the desired behaviour as the following CTL proposition, stating that the system necessarily moves to a stable quiescent state, either at zero error or in the dead zone.

```
STABILITY: (:NECESSARILY (:EVENTUALLY (:AND (:STATUS QUIESCENT)
                                            (:STATUS STABLE)
                                    (:OR (:QVAL E (0 STD))
                                         (:QVAL TO (B STD))
                                         (:QVAL TO (C STD))
                                         (:QVAL TO ((B C) STD))))))
```

Figure 9.12 shows the result of using QSIM and CTL to check the validity of this proposition on two scenarios, in which the system starts in equilibrium at optimal conversion and there is an upward or downward perturbation to the inlet reactant concentration A_i.

To illustrate these conclusions by numerical simulation, Figure 9.13 shows the effect on R and T of temporary but substantial (-20%, -2%, and $+10\%$) perturbations to A_i. The reaction is modelled with the same numerical parameters used in (Economou *et al.* 1986), and the heterogeneous controller uses the values $k_l = -1.5$, $k_h = 3.0$, $a = 420.0$, $b = 438.4$, $c = 438.5$, $d = 445.0$.

9.7 CONCLUSION

We have demonstrated a method for composing heterogeneous control laws from simple classical elements. We have also demonstrated a method for validating the composed laws, even in the presence of incomplete knowledge, using qualitative simulation to predict the tree of all possible behaviours of the system and modal temporal logic to check that a desired guarantee holds for that tree.

9.7.1 Relation to fuzzy logic control

Heterogeneous control is a kind of fuzzy control. It shares many goals with, and draws much inspiration from, fuzzy logic control (Zadeh 1973, Mamdani 1974, Kosko 1992). However, there are important differences between heterogeneous and fuzzy logic control. Within the framework of fuzzy logic control, it is difficult to exploit, or even relate to, the methods or results of traditional control theory. Our approach uses landmark-based qualitative reasoning to combine the benefits of fuzzy control with the analysis methods of traditional control theory.

Granularity. A fuzzy logic controller is typically specified as a relatively fine-grained set of (fuzzy) regions, with a constant (fuzzy) action associated with each region. In heterogeneous control, the design for a controller specifies a smaller set of possibly overlapping operating regions, but with a classical control law associated with each region. The net result of these two differences is that a heterogeneous controller requires a simpler specification, while providing the higher-precision control characteristic of classical control laws.

Validation. The concepts underlying fuzzy logic control are relatively difficult to map into the classical framework, making it difficult to exploit existing methods for providing guarantees for the properties of fuzzy logic controllers.

In heterogeneous control, qualitative simulation can be used to analyse the local laws, and to combine their properties to provide guarantees on the global laws.

Ontology. Heterogeneous control does not treat "linguistic values" or "linguistic variables" as objects in either the domain or range of functions. Rather, the fundamental objects in heterogeneous control are real-valued, continuously differentiable functions, and sets of such functions defined by qualitative descriptions.

Linguistic terms are treated simply as names for the operating regions of the mechanism. The specifications for the operating regions are evaluated by qualitative analysis methods.

"Defuzzification." The output of a fuzzy logic controller is an action with a fuzzy magnitude. The fuzzy magnitude must then be mapped to a real value for output. Since heterogeneous control laws are algebraic combinations of classical control laws, they provide real, not fuzzy, outputs, so "defuzzification" is not necessary.

Generality. In the *implementation* of either a heterogeneous or a fuzzy logic controller, fuzzy set membership functions must be represented as specific real-valued functions. However, the *analysis* of a heterogeneous controller relies only on its qualitative description (e.g. S^+, S^-, or $S^+ \cdot S^-$). This makes explicit the fact that a single guarantee applies to a whole class of appropriateness measures, clarifying the degrees of freedom available for implementation decisions. The goal of qualitative analysis is to define the least restrictive description of the controller that provides a given performance guarantee.

ACKNOWLEDGEMENTS

We would like to thank Evangelina Gazi, Lyle Ungar, and the anonymous referees for their help.

This work has taken place in the Qualitative Reasoning Group at the Artificial Intelligence Laboratory, The University of Texas at Austin. Research of the Qualitative Reasoning Group is supported in part by NSF grants IRI-8905494, IRI-8904454, IRI-9017047, and IRI-9216584, and by NASA contract NCC 2-760.

REFERENCES

ÅSTRÖM, K. J. AND T. HÄGGLUND (1988) *Automatic Tuning of PID Controllers.* Instrument Society of America. Research Triangle Park, NC.

BERLEANT, D. AND B. KUIPERS (1992) Combined qualitative and numerical simulation with Q3. In B. Faltings and P. Struss (Eds.). 'Recent Advances in Qualitative Physics'. MIT Press. Cambridge, MA.

ECONOMOU, C., M. MORARI AND B. PALSSON (1986) 'Internal model control. 5. extension to nonlinear systems'. *Ind. Eng. Chem. Proc. Des. Dev.* 25, 403–411.

EMERSON, E. A. (1990) Temporal and modal logic. In J. van Leeuwen (Ed.). 'Handbook of Theoretical Computer Science'. Elsevier Science Pub. B. V./MIT Press. pp. 995–1072.

FRANKLIN, G. F., J. D. POWELL AND A. EMAMI-NAEINI (1994) *Feedback Control of Dynamic Systems.* 3rd edn. Addison-Wesley. Reading, MA.

GOLDMEIER, E. (1972) 'Similarity in visually perceived forms'. *Psychological Issues.* monograph 29. International Universities Press. New York.

KAY, H. AND B. KUIPERS (1993) Numerical behavior envelopes for qualitative models. In 'Proc. 11th National Conf. on Artificial Intelligence'. AAAI/MIT Press. Cambridge, MA. pp. 606–613.

KOSKO, B. (1992) *Neural Networks and Fuzzy Systems*. Prentice-Hall. Englewood Cliffs, NJ.

KUIPERS, B. (1986) 'Qualitative simulation'. *Artificial Intelligence* **29**, 289–338.

KUIPERS, B. (1987) Abstraction by time scale in qualitative simulation. In 'Proc. 6th National Conf. on Artificial Intelligence (AAAI-87)'. Morgan Kaufmann. San Mateo, CA. pp. 621–625.

KUIPERS, B. (1989) 'Qualitative reasoning: Modeling and simulation with incomplete knowledge'. *Automatica* **25**(4), 571–585.

KUIPERS, B. (1994) *Qualitative Reasoning: Modeling and Simulation with Incomplete Knowledge*. MIT Press. Cambridge, MA.

KUIPERS, B. AND B. SHULTS (1997) Proving properties of continuous systems: qualitative simulation and temporal logic, AIJ, in publication.

KUIPERS, B. AND D. BERLEANT (1988) 'Using incomplete quantitative knowledge in qualitative reasoning'. *Proc. 7th National Conf. on Artificial Intelligence (AAAI-88)*.

MAMDANI, E. H. (1974) 'Applications of fuzzy algorithms for control of a simple dynamic plant'. *Proc. IEE* **121**, 1585–1588.

MAVROVOUNIOTIS, M. L. AND G. STEPHANOPOULOS (1988) 'Formal order-of-magnitude reasoning in process engineering'. *Computers & Chemical Engineering* **12**, 867–880.

MICHIE, D. AND R.A. CHAMBERS (1968) BOXES: An experiment in adaptive control. In E. Dale and D. Michie (Eds.). 'Machine Intelligence 2'. Oliver and Boyd, Edinburgh. pp. 137–152.

SACKS, E. (1990) 'Automatic qualitative analysis of dynamic systems using piecewise linear approximations'. *Artificial Intelligence* **41**, 313–364.

SLOTINE, J.-J. AND W. LI (1991) *Applied Nonlinear Control*. Prentice Hall. Englewood Cliffs, NJ.

SUGENO, M. AND G. T. KANG (1986) 'Fuzzy modelling and control of multilayer incinerator'. *Fuzzy Sets and Systems* **18**, 329–346.

TAKAGI, T. AND M. SUGENO (1985) 'Fuzzy identification of systems and its applications to modeling and control'. *IEEE Trans. on Systems, Man and Cybernetics* **15**, 116–132.

UTKIN, V. I. (1977) 'Variable structure systems with sliding modes'. *IEEE Transactions on Automatic Control* **22**, 212–222.

UTKIN, V. I. (1991) *Sliding Modes in Control and Optimization*. Springer-Verlag. New York.

YAGER, R. R., S. OVCHINNIKOV, R. M. TONG AND H. T. NGUYEN (1987) *Fuzzy Sets and Applications: Selected Papers by L. A. Zadeh*. John Wiley, New York, NY.

ZADEH, L. (1965) 'Fuzzy sets'. *Information and Control* **8**, 338–353.

ZADEH, L. (1973) 'Outline of a new approach to the analysis of complex systems and decision processes'. *IEEE Trans. on Systems, Man and Cybernetics* **3**, 28–44.

CHAPTER TEN

Local Laguerre Models

DANIEL SBARBARO

This chapter presents the use of Laguerre local models to solve some nonlinear identification and control problems. The proposed architecture makes use of Laguerre filters as a state representation of the system input to model stable nonlinear systems. This architecture does not require direct estimation of the local model orders, providing an alternative representation to the traditional ARMAX local model. Its characteristics make it suitable for modelling nonlinear systems having time delays or dynamic orders which depend on operational conditions. A design of a nonlinear predictive controller based on Laguerre local representations is presented. This approach does not require sophisticated algorithms for performing the calculation of the control signals. In addition, the use of Laguerre networks provides more robust operation, because it provides a smooth transition among the different operating regimes characterised by local models with different orders. The identification and control of a simulated continuous stirred tank reactor illustrates the application of the proposed models.

10.1 INTRODUCTION

Nonlinear system modelling has advanced considerably in recent years. The use of the nonlinear autoregressive moving average with exogenous input model has provided a very useful description of the systems in terms of a nonlinear function of delayed input, output and prediction error (Chen and Billings 1983). Many researchers have used Artificial Neural Networks (ANN) as a class of functional representation, and developed identification procedures for discrete-time nonlinear systems (Narendra and Parthasarathy 1990, Chen *et al.* 1990). The parameterisation considers two types of parameters: the order of the system, i.e. the number of delayed inputs, outputs, and error values, and the parameters of the functional approximation.

The performance of the NARMAX model identification depends on the nonlinear function approximation method and on the determination of the approximate model orders. In real world applications model orders are rarely provided and they must be identified prior to the implementation of any approximation method. Even though some recent advances in developing techniques to find the order of the system have been really successful (He and Asada 1993), they assume that these parameters are constant over the training set.

In order to approximate complex dynamic systems it is necessary to use large networks, even if the system is a single-input single-output (SISO) system. For example a SISO system of higher order will require a large number of delayed signals, and as a direct consequence there is an increase in the time required for the identification algorithm to converge to a good solution. Under these circumstances, the nonlinear ANN approximation methods suffer of slow convergence rates and/or the dimensionality problem. The principle of divide and conquer can be applied to solve these problems. The decomposition of the identification problem into a set of smaller subproblems corresponding to a fixed context can simplify the search for a good solution (Yeung and Bekey 1993, Jacobs and Jordan 1993, Johansen and Foss 1993).

In general, it is necessary to avoid inflexible models, i.e. models which are closely tied to a particular parameterisation of the plant and that require significant modification if the plant is altered, or if the class of the input signals are changed. In modelling dynamical systems, the context can be defined as an operating regime, and then local ARMAX models can be used to characterise the behaviour of the system in each operating region (Johansen and Foss 1993). The main assumption concerning the identification algorithm is that the orders of the ARMAX models are available. To avoid the problem of estimating the order of the system, Laguerre networks have been proposed for the linear case (Zervos and Dumont 1988), and for the nonlinear case (Kalkkuhl and Katebi 1993). This chapter, however, proposes the use of Laguerre models as local models. The use of Laguerre models should offer a more flexible and robust parameterisation of the plant in each operating regime. The chapter is organised as follows. Section 2 describes the concept of local Laguerre models. Section 3 describes the identification algorithm. Section 4 presents the simulation of a continuous stirred tank reactor (CSTR), and some identification results obtained with this approach. Section 5 describes a predictive controller based on the Laguerre local models, and it also presents some simulation results. Finally, in section 6 some conclusions are drawn.

10.2 LOCAL LAGUERRE MODELS OF NONLINEAR DYNAMICAL SYSTEMS

Context-sensitive dynamical systems arise naturally from real world applications. In many real situations the nonlinear systems are approximated by linear models which are only valid in the neighbourhood of an operating point. As the system moves away from the region defined by the operating point, the parameters of the linear representation change, for example in a heat exchanger the parameters of a linear model depend on the water and steam flow rate.

The method presented in this chapter is an extension of the work done in (Haber and Keviczky 1985). The process input and output signals are quantised according to a time-dependent context signal $\alpha = \alpha(t)$. Let

$$
\begin{aligned}
u_j^C(t) &= u(t)\phi_j(\alpha) \\
y_j^C(t) &= y(t)\phi_j(\alpha),
\end{aligned}
\tag{10.1}
$$

where $\phi_j(\alpha)$ are the m normalised local basis functions, having the following property

$\sum_{j=1}^{m} \phi_j = 1$. Thus, the input and output signal can be expressed as

$$u(t) = \sum_{j=1}^{m} u_j^C(t)$$

$$y(t) = \sum_{j=1}^{m} y_j^C(t).$$

Then we assume that for each α the corresponding pair (y_j^C, u_j^C) defines a dynamical relationship, which can be represented by a set of orthogonal functions

$$y_j^C(t) = \sum_{i=1}^{\infty} c_i g_i(t) * u_j^C(t) + c_0 \phi_j(\alpha), \qquad (10.2)$$

where $*$ is the convolution operator, and g_i are the impulse responses of the Laguerre basis functions defined by

$$g_i(t) = \sqrt{2p} \frac{e^{pt}}{(i-1)!} \frac{d^{i-1}}{dt^{i-1}} (t^{i-1} e^{-2pt}) \qquad (10.3)$$

the subindex i is the order of the function and p is the time scale. The Laguerre functions are a set of orthonormal functions that span the function space $L_2(0, \infty)$, i.e. the space of square integrable functions on the time interval $(0, \infty)$. The inner product, to which these functions are orthonormal, is the standard time domain L_2 inner product, i.e.

$$< f, g >_t = \int_0^{\infty} f(t) g(t) dt. \qquad (10.4)$$

The Laplace transfer function of the Laguerre function, $g_i(t)$, is a rational transfer function

$$G_i(s) = \sqrt{2p} \frac{(s-p)^{i-1}}{(s+p)^i} \qquad (10.5)$$

Parseval's theorem relates the time domain inner product to the standard frequency domain inner product

$$< f, g >_t = < F, G >_s = \frac{1}{2\pi} \int_{-\infty}^{\infty} F(j\omega) G(-j\omega) d\omega \qquad (10.6)$$

where F and G are the Laplace transform of f and g respectively (Wahlberg 1991). Hence, the Laguerre functions are orthonormal in both the time and frequency domains. The set $\{g_i(t)\}$ is said to be complete if every $y(t) \in L_2(0, \infty)$ can be written as

$$y(t) = \sum_{i=1}^{\infty} < y, g_i > g_i \qquad (10.7)$$

Stable linear time-invariant systems are described by impulse responses in the function space $L_1(0, \infty)$. Thus, the above theorem means that the impulse response can be approximated with arbitrary accuracy by a series of orthonormal functions. However, in practice just a finite number of terms is needed to approximate a function $y(t)$ with a finite accuracy

(Zervos and Dumont 1988), i.e. given an n number of Laguerre functions there exists a real number $\epsilon > 0$ such that

$$\int_a^b |y(t) - \sum_{i=1}^n c_i g_i(t)| dt < \epsilon. \tag{10.8}$$

Using the Laplace transform, the equation (10.2) can be represented in a convenient form as shown in Figure 10.1

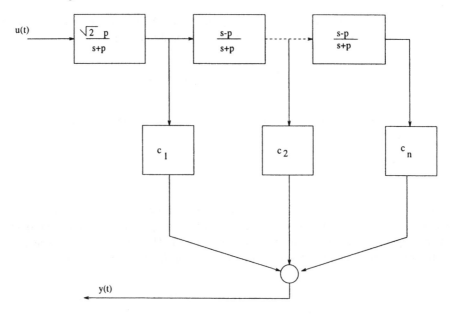

Figure 10.1   Laguerre network.

A more convenient representation is given in state-space form (Zervos and Dumont 1988). The outputs $l_i(t)$, $i = 1, \ldots, n$ from each block are taken to be the states of the Laguerre network. Defining the state vector as

$$\mathbf{l}^T = [g_1, \ldots, g_n],$$

the continuous time representation is defined by

$$\dot{\mathbf{l}}(t) = A_c \mathbf{l}(t) + B_c u(t) \tag{10.9}$$

$$A_c = -p \begin{bmatrix} 1 & 0 & \ldots & 0 \\ 2 & 1 & \ldots & 0 \\ \vdots & \vdots & \vdots & \vdots \\ 2 & \ldots & \ldots & 1 \end{bmatrix},$$

$$B_c = \sqrt{2p} \begin{bmatrix} 1 \\ 1 \\ \vdots \\ 1 \end{bmatrix}.$$

Discretising equation (10.9), we have

$$l(t + 1) = Al(t) + Bu(t),$$

where A and B are defined as

$$A = e^{-A_c T}$$
$$B = (A - I)A_c^{-1} B_c.$$

The output of the process to be modelled is then approximated by the sum of the outputs of the Laguerre filters

$$y = \mathbf{c}^T \mathbf{l},$$

this sum can be taken as projection of the plant transfer function onto the linear space whose basis lies in the orthogonal set of Laguerre functions.

The global dynamic model of the process can be found by summing the models described by equation (10.2)

$$\sum_{j=1}^{m} y_j^C = \sum_{j=1}^{m} \sum_{i=1}^{n} c_{ij} g_i * u_j^C + c_{0j} \phi_j(\alpha)$$

by using equation (10.1)

$$y = \sum_{j=1}^{m} \sum_{i=1}^{n} c_{ij} g_i(t) * u(t) \phi_j(\alpha) + c_{0j} \phi_j(\alpha),$$

and assuming that $\phi(\alpha)$ varies more slowly than $u(t)$ and $g_i(t)$, then

$$y = \sum_{j=1}^{m} \sum_{i=1}^{n} c_{ij} \phi_j(\alpha) g_i(t) * u(t) + c_{0j} \phi_j(\alpha). \qquad (10.10)$$

If we let $\mathbf{c}_i(\alpha) = \sum_{j=1}^{m} c_{ij} \phi_j(\alpha)$, then we have

$$y = \sum_{i=1}^{n} \mathbf{c}_i(\alpha) g_i(t) * u(t) + \mathbf{c}_0(\alpha). \qquad (10.11)$$

By using the state variables l_i, equation (10.11) can be written as

$$y = \sum_{i=1}^{n} \mathbf{c}_i(\alpha) l_i(u(t)) + \mathbf{c}_0(\alpha). \qquad (10.12)$$

The general block diagram for this structure is shown in Figure 10.2.

The parameterisation presented in this work has a number of nice properties, which can provide some advantages over traditional techniques when a stable nonlinear system is modelled. A property, which is shared by all the local model approaches, is the modulation of the activity of the local models by another function. This property allows the original function to be decomposed into a parameterised family of functions which are simpler than the original one. In this way, the identification algorithm can be more efficient. The use of Laguerre networks instead of ARX models enlarges this family of functions, Laguerre models can approximate better systems with varying order or time delay (Sbarbaro 1995). In addition, the Laguerre models do not require estimates of the order and time delay of the system, and they are very robust to the choice of sampling interval, which is a very important issue in nonlinear modelling (Wahlberg 1991).

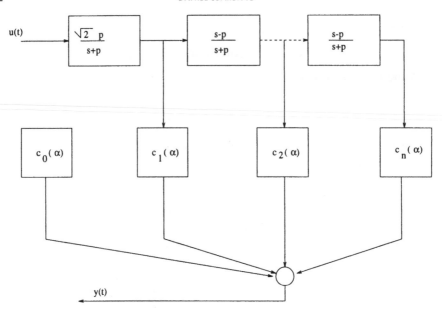

Figure 10.2   Local Laguerre network.

10.3   IDENTIFICATION ALGORITHM

The task of the identification algorithm is to estimate the parameters of the approximating model to minimise the total square error given by

$$E(t_p) = \frac{1}{2} \sum_{k=1}^{p} (y(t_k) - \sum_{i=0}^{n} \mathbf{c}_i(\alpha) \mathbf{l}_i(t_k))^2 \omega(t_k), \tag{10.13}$$

where $l_0 = 1$, and by choice of $\omega(t_k) = (1 - \lambda^{t_k})$ with $0 < \lambda < 1$, the errors are initially weighted less due to the unknown values of the initial conditions of the state variables $\mathbf{l}$, and the subsequent error signals are increasingly weighted up to 1. The variable $y(t_k)$ is the output of the real system when the input sequence $u(t_k)$ has been applied.

Let $\theta_{j+mi} = c_{ij}$ and $\psi_{j+mi} = l_i(u(t))\phi_j(\alpha)$, then equation (10.12) can be written as

$$y(t) = \sum_{j=1}^{m(n+1)} \theta_j \psi_j(\alpha, u(t)). \tag{10.14}$$

As the cost function, equation (10.13), is quadratic in the unknown parameters θ_j, the recursive solution of the minimisation problem is given by the following set of equations (Isermann 1981)

$$\hat{\theta}(t_{k+1}) = \hat{\theta}(t_k) + \frac{e(t_k) P(t_k) \psi^T(t_k)}{\lambda + \psi^T(t_k) P(t_k) \psi(t_k)},$$

$$e(t_k) = y(t_k) - \hat{\theta}_j^T \psi(t_k),$$

$$P(t_{k+1}) = \frac{1}{\lambda} (P(t_k) - \frac{P(t_k) \psi(t_k) \psi^T(t_k) P(t_k)}{\lambda + \psi^T(t_k) P(t_k) \psi(t_k)}),$$

where θ and ψ are vectors associated with θ_j and ψ_j. The forgetting factor λ is chosen as a time-varying parameter defined by the following relation

$$\lambda(t_{k+1}) = a\lambda(t_k) + (1 - a),$$

where $\lambda(t_0) < 1$, and the parameter a defines the speed of variation of the forgetting factor. The input signal should provide enough information to the local models, in order to allow a successful parameter estimation.

10.4 MODELLING A CSTR

Consider a continuous stirred tank reactor (CSTR) in which the first-order reaction $A \rightarrow B$ occurs.

$$\dot{C}_A = \frac{q}{V}(C_{Af} - C_A) - k_0 \exp(-\frac{E}{RT})C_A$$

$$\dot{T} = \frac{q}{V}(T_f - T) - k_0 \frac{-\Delta H}{C_p \rho} \exp(-\frac{E}{RT})C_A + \frac{-UA}{VC_p \rho}(T_c(t - \theta) - T),$$

where C_A is the reactor concentration, T is the reactor temperature, T_c is the coolant temperature, q is the feed flow rate, and C_{Af} is the feed concentration. The control objective is to maintain the reactor concentration C_A at its set-point by manipulating the coolant temperature T_c, which is constrained to be between 275 and 375 K (Morningred et $al.$ 1992). The nominal parameters of the model are summarised in Table 10.4.

Table 10.1 Nominal CSTR parameters.

Product concentration	C_A	0.1 mol/l
Reactor temperature	T	438.54 K
Coolant flow rate	q_c	103.41 l min^{-1}
Process flow rate	q	100 l min^{-1}
Feed concentration	C_o	1 mol/l
Feed temperature	T_o	350 K
Inlet coolant temperature	T_C	350 K
CSTR volume	V	100 l
Heat transfer term	hA	$7 \cdot 10^5$ cal min^{-1}K^{-1}
Reaction rate constant	k_o	$7.2 \cdot 10^{10}$ min^{-1}
Activation energy term	E/R	1x 10^4 K
Heat of reaction	ΔH	$-2 \cdot 10^5$ cal/mol
Liquid densities	ρ, ρ_c	$1 \cdot 10^3$ g/l
Specific heats	C_p, C_c	1 cal g^{-1} K^{-1}

Figure 10.3 shows the highly nonlinear open-loop responses of the concentration, when step changes in the coolant temperature are applied to the reactor. As the concentration increases, the reactor behaves as a second-order system. The sampling period of all process measurement was assumed to be 0.2 minutes, this sampling period allowed the faster dynamics to be sampled at about four times the dominant time constant.

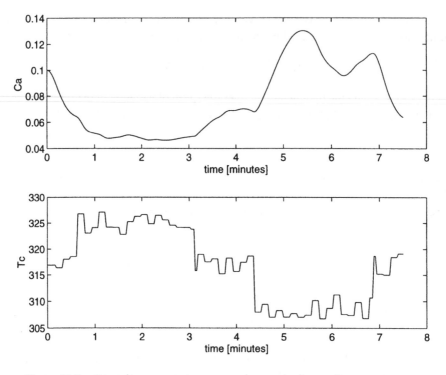

Figure 10.3   Open-loop responses to step changes in the cooling temperature.

A Laguerre linear model described by a seventh-order model with a time scale of 6.66 was identified using the first 25 minutes of data, and then validated with the the remaining data. Figure 10.4 shows that the linear model is unable to model the whole process accurately.

The nonlinear model has six Laguerre local models, describing the estimated concentration, $\hat{C}_A$, as

$$\hat{C}_A(t+1) = \sum_{j=1}^{6} \sum_{i=1}^{7} c_{ij}\phi_j(\alpha)l_i(t) + c_{0j}\phi_j(\alpha), \tag{10.15}$$

where the basis function $\phi(\alpha)$ is defined by

$$\phi(\alpha) = \exp(-0.5(\alpha - m_j)^2/\sigma_j^2)/\phi_T,$$

the variable ϕ_T is defined as $\phi_T = \sum_{j=1}^{6} \phi_j(\alpha)$, and the variable α was chosen as $C_A(t)$.

Figure 10.5 shows that the residual error is smaller than the one obtained with just the linear model, as a result the errors obtained in the validation phase are also smaller.

10.5   PREDICTIVE CONTROL

Nonlinear predictive control has emerged as a promising technique for the control of highly nonlinear processes. At the heart of this strategy lies a model of the process to be controlled.

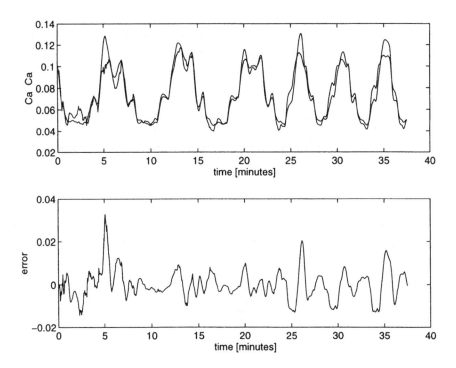

Figure 10.4 Identification and validation of a linear model.

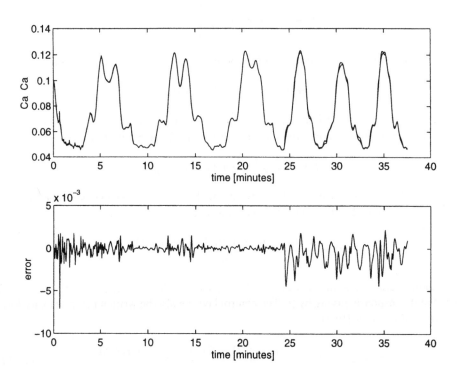

Figure 10.5 Identification and validation of a nonlinear model.

If there is no available *a priori* information about the physical laws describing the system, then a connectionist representation based on a general NARMAX model can be used to implement a predictive controller (Sbarbaro and Hunt 1991).

The decomposition of the control problems into a set of smaller subproblems corresponding to a fixed context can simplify the search for a good solution (Yeung and Bekey 1993). In addition, this approach makes use of modular local models, which deal better with the problem of temporal crosstalk, can be structured more easily, and some prior knowledge can be more easily incorporated than in fully nonlinear models (Johansen 1994).

The design of the controller is based on each local model (Zervos and Dumont 1988). Let us consider again the state representation of equation (10.11),

$$
\begin{aligned}
\mathbf{l}(t+1) &= A\mathbf{l}(t) + Bu \\
y(t) &= C(\alpha(t))^T \mathbf{l}(t) + \mathbf{c}_0(\alpha(t)),
\end{aligned}
$$

where $C(\alpha(t)) = [c_1(\alpha), \dots, c_n(\alpha)]$. The d-step ahead output for an input defined such that $u(t) = \dots = u(t + d - 1)$ is

$$
\begin{aligned}
\mathbf{l}(t+d) &= A^d \mathbf{l}(t) + \sum_{i=0}^{d-1} A^i Bu(t), \\
y(t+d) &= y(t) + C(\alpha(t+d))^T (A^d - I)\mathbf{l}(t) + \beta u(t),
\end{aligned}
\tag{10.16}
$$

where β is defined as

$$
\beta = C(\alpha(t+d))^T \sum_{i=0}^{d-1} A^i B.
$$

The set-point trajectory is defined as a first-order system based on the equation

$$
y_r(t+1) = \gamma y(t) + (1 - \gamma) y_{sp},
$$

where γ represents the desired dominant closed-loop time constant, $0 < \gamma < 1$, and y_{sp} the desired set-point. By recursive substitution $y_r(t + d)$ can be written as

$$
y_r(t+d) = \gamma^d y(t) + (1 - \gamma^d) y_{sp}.
$$

Assuming that α is constant over the relevant time horizon or at least slowly time-varying, then from equation (10.16), $u(t)$ can be calculated as

$$
u(t) = \frac{y_r(t) - y(t) - C(\alpha)^T (A^d - I)\mathbf{l}(t)}{\beta}.
$$

The tuning parameters, in this case, are the horizon of the controller, d, and the desired speed of the response, giving by γ. The control law can also be written in the velocity form (Zervos and Dumont 1988) as

$$
u(t) = u(t-1) + \frac{y_r(t) - y(t) - C(\alpha)^T \sum_{i=0}^{d-1} A^i BA\Delta\mathbf{l}}{\beta},
\tag{10.17}
$$

where $\Delta\mathbf{l} = \mathbf{l}(t) - \mathbf{l}(t-1)$.

10.5.1 The control of a CSTR

The nonlinear predictive controller with six local models, as they were described in section 4, was used to control the process. The initial parameters were set to zero, and the three initial minutes were used to initialise the estimator, during this period the process was run open loop with step changes in its input. The diagonal elements of the covariance matrix were set to 10^4. The forgetting factor was chosen to be 0.98. The controller was implemented using its velocity form, i.e. equation (10.17). Figure 10.6 shows the response of the system to set-point changes when a small noise was added to the output of the process.

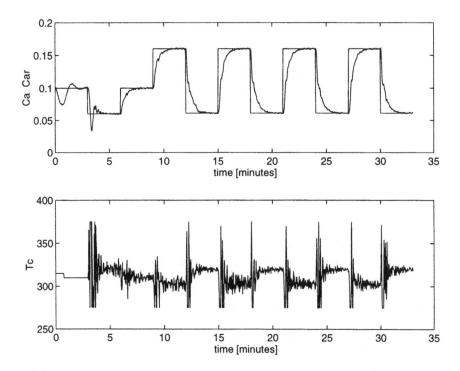

Figure 10.6 Control of a CSTR: concentration set point tracking.

10.6 CONCLUSIONS

This chapter has shown the versatility of local Laguerre models for modelling continuous dynamical systems. The use of local Laguerre models avoids the estimation of order and time delay of the system. Laguerre models not only can cope with noisy data, because of their low pass characteristic, but they also scale very well with higher-dimensional systems, due to the fact that their approximation capabilities do not rely on the exact order of the system. Even though, it is necessary to select the time constant and the number of filters of the Laguerre network, the parameterisation is less senstive to these parameters. The use of a forgetting factor avoids the bias in the estimates due to a lack of reliable estimates of the system initial conditions.

ACKNOWLEDGEMENTS

This research was supported by CONICYT project 1941002 and University of Concepcion project DI # 94.92.29-1. The author would like to thank the editors, Dr. R. Murray-Smith, Dr. T.A. Johansen and also Dr. J. Kalkkuhl, for their comments and suggestions to improve this chapter.

REFERENCES

CHEN, S., S. BILLINGS AND P. GRANT (1990) Nonlinear system identification using neural networks. *Int. Journal of Control* **51**(6), 1191–1214.

CHEN, S. AND S. BILLINGS (1983) Representation of nonlinear systems: The NARMAX model. *Int. J. Control* **49**(3), 1013–1032.

HABER, R. AND L. KEVICZKY (1985) Identification of 'linear' systems having signal-dependent parameters. *Int. Journal of System Sciences* **16**(7), 869–884.

HE, X. AND H. ASADA (1993) A new method for identifying orders of input-output models for nonlinear dynamics systems. In: *Proceedings of The American Control Conference.* pp. 2520–2523.

ISERMANN, R. (1981) *Digital Control Systems.* Springer-Verlag, Berlin.

JACOBS, R.A. AND M.I. JORDAN (1993) Learning piecewise control strategies in a modular neural network architecture. *IEEE Transactions System, Man, and Cybernetics* **23**(2), 337–345.

JOHANSEN, T.A. (1994) *Operating regime based process modeling and control.* PhD Thesis, The Norwegian Institute of Technology, Trondheim.

JOHANSEN, T.A. AND B.A. FOSS (1993) Constructing NARMAX models using ARMAX models. *International Journal of Control* **58**(5), 1125–1153.

KALKKUHL, J.C. AND M.R. KATEBI (1993) Nonlinear control design using gate function approach. In 'Proc. 3rd. IEEE Conference on Control Applications, Glasgow, Scotland'.

MORNINGRED, J.D., B.E. PADEN, D.E. SEBORG AND D. A. MELLICAMP (1992) An adaptive nonlinear predictive controller. *Chemical Engineering Science* **47**(4), 755–762.

NARENDRA, K.S. AND K. PARTHASARATHY (1990) Identification and control of dynamical systems using neural networks. *IEEE Transactions on Neural Networks* **1**(1), 4–27.

SBARBARO, D. AND K.J. HUNT (1991) A nonlinear receding horizon controller based on connectionist models. In 'Proc. IEEE Conference on Decision and Control, Brighton, England'.

SBARBARO, D. (1995) Context Sensitive Networks for Modelling Nonlinear Dynamic Systems. In 'Proc. ECC95, Rome, Italy' (4), pp. 2420–2425.

WAHLBERG, B. (1991) System identification using Laguerre models. *IEEE Trans. on Automatic Control* **36**(5), 551–562.

YEUNG, D.-Y. AND G. A. BEKEY (1993) On reducing learning time in context-dependent mappings. *IEEE Trans. on Neural Networks* **4**(1), 31–42.

ZERVOS, C.C. AND G.A. DUMONT (1988) Deterministic adaptive control based on Laguerre series representation. *Int. Journal of Control* **48**(6), 2333–2356.

CHAPTER ELEVEN

Multiple Model Adaptive Control

KEVIN D. SCHOTT and B. WAYNE BEQUETTE

Multiple model adaptive control (MMAC) is a model-based control strategy which incorporates a set of model/controller pairs rather than relying on a single model and controller to handle all possible operating conditions. A bank of of models is used with a weighting function which chooses the single model or combination of models that best represents the current plant input–output behaviour. A weighted-sum of the controller outputs is then used to supply the control action to the plant.

Although MMAC is model-based, it relaxes the requirement for a single, precise, nonlinear model. Often such a nonlinear model is not available either due to lack of process knowledge or insufficient "non-production" time for development of detailed models. Many initial MMAC applications involved aircraft control where models for different flight conditions were used in the model bank. More recently, MMAC has been used in drug infusion systems where plant (i.e. patient) parameter variability is the largest concern.

In this chapter we review MMAC theory and applications. We present a specific single-input–single-output example based on a drug infusion application to highlight some of the strengths and weaknesses of the MMAC method.

11.1 INTRODUCTION

Process control system design has traditionally involved the use of linear models and linear controllers; however, linearity is an idealisation that is sometimes not a good approximation of actual behaviour. For systems with mild nonlinearities, or which operate continuously at a single operating point, linear control system design (particularly when tuned for robustness) has generally been successful. Even with systems that are known to have significant nonlinearities, linear controllers are typically used because they are well-understood. In the chemical process industries in particular, the proportional-integral-derivative (PID) controller is the standard which is implemented in most control loops. It is easy to tune and has some ability to handle nonlinearities through features such as "anti-reset windup" that turn-off the integral mode when constraints are encountered. Also, parameter-scheduling options often exist where the tuning parameters may be adjusted as a function of the process output.

There are several reasons that nonlinear controllers have not been widely implemented. Due to the existence of an overwhelming number of non-general, nonlinear control techniques, few control engineers are able to maintain an understanding of all the various

approaches available. The lack of time or process understanding to develop detailed nonlinear models to describe system behaviour has also limited the growth of nonlinear control
techniques. Our goal with this chapter is to present an approach for control system design
and implementation that we feel has many of the advantages of traditional linear controller
design without many of the disadvantages of some nonlinear control system design approaches.

11.1.1   Basic idea

Multiple model adaptive control (MMAC) is a model-based control strategy which incorporates a set of model/controller pairs as shown in Figure 11.1. The tuning parameters for
each controller are based on the controller's corresponding model. A weighting function
chooses which model or combination of models from within the model bank best represents
the current plant input–output behaviour. A weighted-sum of the outputs from controllers
in the controller bank is then used to supply the control action.

The blending and switching of models is a unique feature of the MMAC method:
although a single, precise plant model may not be available for all operating regions, model-
based control techniques may still be used in operating regions which are well-known. Continuous systems may operate around one particular setpoint for much of the time and can be
described by just one of the models in the model bank. For this case, a single controller is
chosen (via convergence of the weighting fractions to a single model/controller pair) which
by itself can satisfactorily control the plant. Operating regions where the plant is not well-
known may be adequately described and controlled by a combination of model/controller
pairs, or the control system designer may supply a number of hypothesis models in the
model bank and allow the controller to select the one which most closely represents the
plant.

Although the parameters in all of the model/controller pairs are fixed, MMAC is adaptive
since different combinations of models are used to represent the plant as the plant parameters change. Because all of the models are always available, another MMAC feature is
zero start-up time for identification. Any type of model (linear or nonlinear, state-space or

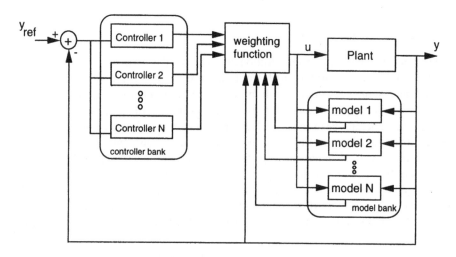

Figure 11.1   Basic multiple-model adaptive control (MMAC) strategy.

input–output, continuous or discrete) and any control method (PID, LQR, MPC, etc.) can be used within the MMAC framework. Indeed, the approach is general enough that a mixture of model and controller types can be used.

There are several important issues in developing an MMAC strategy:

Model bank selection:
        (i) type of models;
        (ii) parameters for each model.
Controller design:
        (i) number of models;
        (ii) type of controllers;
        (iii) tuning of each controller.
Probability weighting:
        (i) method;
        (ii) initialisation;
        (iii) tuning.

Before we discuss many of the important issues in the development of an MMAC strategy, we review previous work on the topic.

11.1.2 Theory

Most of the fundamental background for MMAC is based in optimal control, specifically solving a problem with Kalman filtering and LQG (Linear model, Quadratic cost function, Gaussian noise) control with the assumption of perfect knowledge of the model parameters. A general assumption of MMAC has been that the uncertain constant parameters can take on discrete values. A typical block diagram for MMAC based on KF/LQG is shown in Figure 11.2. Theoretical bases are established by Saridis and Dao (1972), Deshpande *et al.* (1973), Lainiotis (1976a,b), and some basic ideas are presented in Stengel (1986).

Greene and Willsky (1980) study a simple two-state system with two models in the model bank; neither model contains the actual plant. They perform a detailed stability analysis and find that not all control gains which stabilise the individual models result in a stable closed loop for the actual system. This result is very important as it shows that proper design of an MMAC system requires more than simply designing for good nominal performance based on each model.

Maybeck (1989) notes that a basic problem with the MMAC approach is the potentially large number of models required. For example, if there are two uncertain parameters that can each assume 10 values, then 100 separate filters and controllers must be implemented. Maybeck reviews an approach known as Moving-Bank MMAC that he and collaborators present in a number of applications papers. The basic idea is to include only the models closest to the current operating point in the model bank. He presents five different methods for changing the size or moving the active model bank. His paper also contains a concise review of the standard state-space MMAC approach.

11.1.3 Applications

Applications of MMAC have been primarily focused in three different areas: (i) aircraft and space-structure, (ii) drug delivery and (iii) chemical process control. Here we discuss important characteristics of problems in each area.

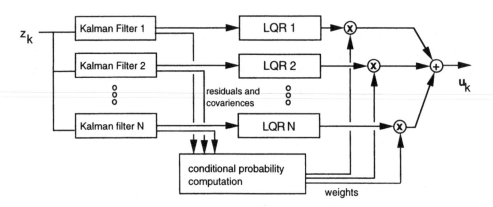

Figure 11.2   Standard state-space multiple-model adaptive controller.

Aircraft and space-structure control

A major challenge in the control of aircraft is to be able to operate under a number of different flight conditions. Although the dynamic behaviour of an aircraft is known to be nonlinear, specifications for control system design are often based on linear models at a number of equilibrium flight conditions. Athans *et al.* (1977) present an MMAC design for the F-8C aircraft. The linear models are characterised by nine states, two manipulated inputs and six measured outputs. At several equilibrium conditions a steady-state, discrete, constant-gain Kalman filter is designed. An LQG compensator is designed based on each Kalman filter, as shown in Figure 11.2. Simulation results are presented using a nonlinear model at a single flight condition. Four models (based on four flight conditions) are used in the model bank. They note that the robustness and sensitivity of the MMAC algorithm appears to rely on careful tuning of the Kalman filters.

Maybeck and coworkers at Wright-Patterson Air Force Base have presented a large number of simulation examples of applications of MMAC to flexible space structure control. Gustafson and Maybeck (1994) apply Moving-Bank MMAC (Maybeck 1989) to a 352-state model of a SPace Integrated Control Experiment (SPICE) structure. There are 18 manipulated variables and 18 measurements (an additional 36 measurements are used for comparison purposes). They reduce the order of the model to obtain an order tractable for implementation. They find that their moving bank approach provides almost instantaneous tracking of parameters and stabilising control over the full-range of parameter variations.

Drug infusion control

An area which has had many experimental applications of MMAC is drug infusion control. He *et al.* (1986) apply MMAC via simulation and animal experiments to the control of blood pressure in dogs by controlling the infusion rate of nitroprusside. Their model bank consists of eight transfer function models and the controller bank consists of eight PI controllers, each based on one of the models. The models all have the same dynamic characteristics with different gains. The model gains are selected so that the corresponding controller would satisfy phase margin requirements for each model interval. Martin *et al.* (1987) combine pole-placement, a Smith predictor and PI control into an MMAC framework for blood pressure control using nitroprusside. The control strategy is applied to a nonlinear simulation model. MMAC-PI is used by Yu *et al.* (1987) to control arterial oxygen by adjusting the inspired oxygen fraction in mechanically ventilated dogs.

Yu *et al.* (1992) use a model predictive controller (MPC) in the MMAC framework to regulate arterial pressure and cardiac output (blood flowrate) by manipulating the infusion rate of two drugs. Thirty-six models are used to span the entire space of expected responses in dogs exhibiting symptoms of cardiac heart failure. To reduce the computational load, only the six highest probability models are used for the control calculation at each time step. Major advantages to using the MPC approach are that constraints are easily taken care of through solution of a quadratic programme and multivariable systems are handled efficiently. A nice review of MPC is provided by Garcia *et al.* (1989).

MMAC is particularly advantageous in drug delivery applications for a number of reasons. Patient to patient sensitivity to drugs can vary significantly and the response of a single patient to a drug can be very time-dependent. Model parameters then vary greatly from patient to patient and in the same patient at different times. There is a major advantage to MMAC during the initial stages of administering a drug to a patient because no initialisation time is required for parameter identification. MMAC begins by controlling with a certain pre-defined weight distribution among the models (all models weighted equally, for example), then new control calculations begin as soon as measurements are available, whereas most adaptive control procedures require an initialisation time to determine the model parameters. For a review of adaptive control see Seborg *et al.* (1986).

It should be noted that a number of special functions are typically used in the feedback loop to achieve better control of drug delivery. He *et al.* (1986) and Martin *et al.* (1987) use a nonlinear unit to freeze the infusion rate of the drug nitroprusside if the patient's blood pressure drops too low; another nonlinear unit places minimum and maximum limits on the allowable infusion rate.

Chemical process control

Nonlinear dynamic behaviour has long been considered a common, challenging chemical process control characteristic. As plant-wide control continues to advance it is more common to find on-line optimisation strategies that change setpoints (and therefore operating regimes) of many of the process variables. Because of nonlinear dynamic behaviour the lower-level controllers designed for one operating regime may not have satisfactory performance at another operating point. Although nonlinear model-based strategies have been developed to be able to control processes over a wide range of operating conditions, these strategies have not had a wide impact because of model development time and somewhat of a lack of understanding of the nonlinear control techniques. MMAC is an ideal candidate for chemical process control because a set of linear models which spans the expected operating space can be developed very quickly and the controllers associated with each model are easily designed and tuned.

Chemical reactors are unit operations that exhibit interesting static and dynamic nonlinear behaviour. In Schott and Bequette (1995), we apply MMAC to simulations of two different chemical reactors; (i) the Van de Vusse reactor and (ii) a classic, exothermic, continuous stirred tank reactor (CSTR). Banerjee *et al.* (1995) present a multiple-model "self-scheduling" approach for the control of the classic CSTR; their multiple-model controller scheduling approach is detailed in Chapter 12.

11.1.4   Introduction summary and chapter outline

In the introduction section we have provided a concise survey of MMAC applications. Note that most of the actual experimental applications have been in drug infusion control, probably due to a number of reasons: (1) developing a nonlinear model valid for a single patient, much less one that can be used for a wide variety of patients, is virtually impossible; (2) there is no clear scheduling variable for gain scheduling; (3) there is extreme patient to patient parameter variability; and (4) it is desirable to bound parameters and control actions.

Multiple model adaptive control is an intuitively pleasing idea and viewing the general diagram shown in Figure 11.1 immediately leads to a number of implementation issues that must be addressed. In section 11.2 we discuss model selection. In section 11.3 we discuss issues associated with determining the probability of each model describing the plant behaviour and how to use this probability at each time step to find the control action to be implemented at the next time step. Construction of the controller bank is addressed in section 11.5.2. Obviously, the issues presented in sections 11.2 – 11.5.2 (model, probability weighting, control calculation) are not independent and must be covered simultaneously or iteratively in any MMAC design and implementation. In section 11.5 we present an example based on a drug infusion control problem. In section 11.6 we summarise the strengths and weaknesses of the MMAC approach.

11.2   MODEL BANK

Controller design is substantially less difficult if an accurate plant model is available. MMAC attempts to provide a good model, and therefore a good controller, over a wide range of operating conditions. MMAC works optimally when a single model, which exactly matches the plant, is identified and the counterpart optimal controller can be applied. When an exact match is not found, MMAC may provide a combination of models, or choose the single closest model available. Such operation is sub-optimal, yet perhaps acceptable for a prescribed performance criteria.

The MMAC approach is quite general and can involve the use of any type of model (including a mixture of models) in the model bank. Linear state-space models have generally been used in aircraft and space-structure control, while input–output models have generally been used in drug delivery control. Both approaches have been used in chemical process control. The distinction between state-space and input–output is largely irrelevant from a model prediction perspective since input–output models are easily cast in state-space form. Either continuous or discrete models may be used, though discrete models are most common; although models are often presented in continuous form, they are normally discretised at a constant sample time when implemented in the feedback loop.

Since MMAC has its roots in optimal control theory, state-space models have generally been implemented in conjunction with LQG controllers and used with Kalman filters for state estimation. The number of models selected is usually related to the number of operating conditions (i.e. different sets of plant parameters) over which the control system is expected to operate. Model parameters are either determined by linearising a nonlinear model and assuming equilibrium at each operating condition, or, if the plant operates in distinct, well-characterised regions, models may be developed from plant data in those operating regimes.

Input–output models have generally been selected when less is known about the system and a fundamental nonlinear model is not readily available or easily developed. For example, in drug delivery control, often only lower and upper bounds on expected patient sensitivities (gains) are known. It is usually assumed that the dynamic behaviour (time constants, time-delays) does not vary widely, or can be approximated by changes in the process gain only. However, this assumption is not a limitation, and in fact one of the biggest assets of the MMAC approach is that structural changes in the plant may be taken into account in a rather straightforward fashion.

The most ill-defined part of designing an MMAC controller is determining the number of models required to span an operating region. This task must be done in conjunction with the individual controller designs since a larger number of models allows more accurate plant identification and hence the controller based on each model may be more tightly tuned. Using fewer models in the model bank forces the controller design based on each model to be tuned more robustly.

11.3 PROBABILITY WEIGHTING AND TUNING

The MMAC method requires some form of model discrimination via a hypothesis or *a posteriori* probability computation. Several methods have been proposed for determining which model or models best represent the plant. The most common method for finding the likelihood that a model fits the plant is with a probability estimate based on Bayes' rule. Other approaches include maximum *a posteriori* probability and horizon-based error tracking, such as minimum mean square error.

11.3.1 Bayes' rule

The conditional probability of A_i given B_i can be stated with Bayes' rule

$$Pr(A_i|B_i) = \frac{Pr(B_i|A_i)\,Pr(A_i)}{Pr(B_i)}, \tag{11.1}$$

where i is one particular outcome out of all possible outcomes. This probability is used in the weighting function of an MMAC system in the following manner: given a measured output and a set of estimated outputs, we would like to generate a probability that a particular model represents the plant. Here we give a very brief summary of how Bayes' rule is implemented within the MMAC framework. Rigorous derivations may be found in a number of stochastic control texts, including those by Stengel (1986) and Maybeck (1982).

From equation (11.1), a discrete, recursive form of Bayes' rule can be developed which gives the conditional probability that the ith model (i.e. the model containing parameters a_i), given the most recent measured outputs z at time-step k, represents the plant

$$Pr(a_i|z_k) = \frac{pr(z_k|a_i)\,Pr(a_i|z_{k-1})}{\sum_{j=1}^{N} pr(z_k|a_j)\,Pr(a_j|z_{k-1})}, \tag{11.2}$$

where $pr(z_k|a_i)$ is a conditional probability density function.

Using a Gaussian probability density function, applying equation (11.2) to a discrete, constant-coefficient system in state-space form, and developing state estimates with an nth order Kalman filter, the probability that the ith model output represents the plant at time-step

k is given by

$$p_{i,k} = \frac{\frac{1}{(2\pi)^{\frac{q}{2}}|S_{i,k}|^{\frac{1}{2}}} \exp(-\frac{1}{2}r_{i,k}^T S^{-1}{}_{i,k}r_{i,k})\, p_{i,k-1}}{\sum_{j=1}^{N}\left[\frac{1}{(2\pi)^{\frac{q}{2}}|S_{j,k}|^{\frac{1}{2}}} \exp(-\frac{1}{2}r_{i,k}^T S^{-1}{}_{i,k}r_{i,k})\, p_{j,k-1}\right]}. \tag{11.3}$$

The measurement residuals $r_{k,i}$ (measured outputs at time-step k minus estimated outputs from i^{th} model at time-step k) are already available from the Kalman filters and the residual covariance matrices $S_{i,k}$ are generated using the covariance matrices, also available from the Kalman filters. Though only the most recent outputs are required, identification is not based solely on that single time-sample of information. Past information on the residuals is contained within former probabilities $p_{i,k-1}$, $p_{i,k-2}$, etc.

An important addition to equation (11.3) is a lower bound preventing $p_{i,k}$ from becoming zero

$$\begin{aligned} P_{i,k} &= p_{i,k} & \textit{for } p_{i,k} > \delta \\ P_{i,k} &= \delta & p_{i,k} \leq \delta \end{aligned} \tag{11.4}$$

Since equation (11.3) is recursive, once a probability reaches zero it will remain zero thereafter. Theoretically the probability can never reach zero, however practically, very small $p_{i,k}$ are rounded to zero due to machine precision. This bounding also limits the number of past observations contained in the current probability estimate. Large values of δ yield faster model switching because non-contributing model probabilities are kept artificially high. Low δ values require the probabilities in equation (11.3) to go through more iterations before ith controller contributes a relevant portion of the overall control signal.

Before calculating a final control action, the probabilities are renormalised, removing the contributions of the models which had their probabilities bounded by δ

$$\begin{aligned} W_{i,k} &= \frac{P_{i,k}}{\sum_{j=1}^{N} P_{j,k}} & \textit{for } P_{i,k} > \delta \\ W_{i,k} &= 0 & P_{i,k} = \delta. \end{aligned} \tag{11.5}$$

The final control action supplied to the plant is a probability-weighted average

$$u_k = \sum_{i=1}^{N} W_{i,k}\, u_{i,k}. \tag{11.6}$$

Note that the inclusion of δ affects the probabilities, no longer assuring that the correct mixture of plants will be generated when operating in regions other than precisely at one of the models. For this reason, MMAC very often excels at finding a single model which represents the plant, yet may not always blend them well. In fact, nearly all published Bayesian-MMAC results show convergence to a single model rather than a combination of more than one model. In many instances, a more accurate description of the MMAC approach may be that it blends or softens *the transition* between individual controllers, not necessarily the controllers.

Since equation (11.3) was developed from an optimal stochastic control approach, it is only truly valid within that framework. However, a form of equation (11.3) can be used successfully outside of the Kalman/LQR implementation by replacing the residual covariance matrix with a "convergence factor" K (He *et al.* 1986)

$$p_{i,k} = \frac{\exp(-\epsilon_{i,k}^T K \epsilon_{i,k}) \, p_{i,k-1}}{\sum_{j=1}^{N} [\exp(-\epsilon_{j,k}^T K \epsilon_{j,k}) \, p_{j,k-1}]}, \tag{11.7}$$

where, to maintain relative distances between models for setpoints of any magnitude, the residual $\epsilon_{i,k}$ (measured outputs minus the ith model outputs, both at time-step k) is normalised with respect to the reference signal z_{ref}

$$\epsilon_{i,k} = \frac{z_{meas} - z_{i,k}}{z_{ref}}. \tag{11.8}$$

Statistical relevance can be lost when using the convergence factor approach since the substitution of K fixes the covariances. Nevertheless, equation (11.7) provides a convenient filter which yields some measure of model closeness and is bounded by one and zero. Parameter K is then a tuning parameter that controls the rate at which the "probability" converges. Large K values cause the probabilities to quickly match the closest model output. Excessively fast convergence, however, can cause rapid switching and incorrect tracking during the dynamic portion of a system response. Small K values give slow convergence, smoothing the probabilities and eventually moving toward a model based on steady-state differences. Unfortunately, current literature does not have any guidelines for selecting K or δ. In addition, since equations (11.3) and (11.7) are recursive, K and δ are functions of sampling rate.

11.3.2 Other weighting techniques

Other weighting schemes can be used in place of the Bayesian-based estimation described above. Since the first part of the control problem is one of identification, other model-based estimations can be implemented. For example, if an appropriate likelihood function can be chosen, then maximum likelihood techniques such as a maximum *a posteriori* probability (MAP) approach can be used. However, such an approach requires that a single model/controller pair be chosen, and hence a large model bank is is required to provide good plant/model agreement. (The use of a single model/controller based on the highest probability from equation (11.3) is also a common alternative.)

Another option is to look at a history of past errors between the plant and each model. Low error by various criteria (ISE, IAE, RMS) over a past-horizon gives some indication of model closeness, yet does not give probability information. When a past-horizon approach is used, generally only the single model with the lowest error is used to control the plant and the blending aspects of MMAC are again lost. This approach may also be considered a form of scheduling (Banerjee *et al.* 1995). A benefit of this approach though may be improved initial tuning since you may be able to choose the length of the past-horizon based on the dynamics of the plant.

11.4 CONTROLLER DESIGN

The first question which must be addressed when designing a multiple-model controller is that of the number of models to use in the model bank. If the plant parameters can only assume certain specific values, then one approach is to use as many models as there are discrete parameter values. This approach can yield an extremely large model bank. Such systems are used successfully though, and measures have been developed to handle large model banks through windowing techniques (Maybeck 1989).

Model number determination becomes much more difficult when plant parameters may assume an infinite number of values. Unfortunately, an algorithm for choosing the number of models which guarantees achievement of a pre-specified performance criteria does not yet exist. Typically, the design procedure must be approached by first finding a minimum number of models that are required so that performance criteria are satisfied for each model region, then simulating the system and including more models until performance criteria are met over the entire operating conditions of the plant. An MMAC controller's performance can be poor and may even become unstable if the plant lies outside the range of models provided in the model bank, so thorough testing is essential.

Controller selection is based primarily on the type of model used. State-space approaches usually employ Kalman filter/LQR pairs, whereas input–output models are generally paired with PID controllers. Nonlinear models and associated model-based controllers may also be used. The tuning for each controller is based on the individual parameter space which the model/controller pair is supposed to cover. Tuning criteria may be based on satisfying gain-margin/phase-margin requirements, or developed through the application of robust performance techniques.

11.5 DRUG INFUSION CONTROL EXAMPLE

As noted in previous sections of this chapter, drug infusion control has been one of the primary application areas for MMAC. In this example we present a portion of a drug infusion control problem. We assume that the plant can be characterised by a first-order plus time-delay model, with uncertainty in all three parameters. Many systems encountered in the chemical process industry can also be modelled in the same fashion. For example, heat exchangers modelled as first-order plus dead-time processes have three parameter variation with changes in process fluid flowrate or the process fluid itself. Also, because of difficulties in data collection or incomplete process knowledge, higher-order dynamics may be neglected during preliminary studies, as is often the case with distillation columns where individual stages are modelled as first-order plus time-delay systems.

There are many issues that must be addressed in the application to a real physiological system which is highly nonlinear. The control strategy must satisfy constraints on the manipulated variable (drug infusion rate) and must have built-in safety features such as setting the drug infusion rate to zero when the patient's blood pressure falls below a certain value. In order to not confuse these special issues with the basic control system design and implementation procedure (and problems), we focus on the time-varying linear problem. Details on these other application-specific issues can be found in He *et al.* (1986), Martin *et al.* (1987) and Yu *et al.* (1992).

Our goal in presenting this example is to provide a straightforward problem which the reader can quickly and easily reproduce so as to get a feel for the complications encountered when implementing an MMAC strategy. Certainly more complicated systems have been successfully controlled using MMAC, however their additional complexity can cloud the fundamental MMAC implementation issues. In section 11.5.1 we define the plant characteristics and the performance criteria which we are trying to meet. Section 11.5.2 addresses the choice of models and design of the controller, and in section 11.5.3 we present simulation results.

11.5.1 Process model

For this example our process model for a human patient requiring the infusion of heart medication is

$$G(s) = \frac{k_p \exp(-\theta_p s)}{\tau_p s + 1}. \tag{11.9}$$

It is known (Slate *et al.* 1979) that patient sensitivities can vary by a factor of 36. The parameters for this model vary over the ranges listed in Table 11.1 and have been normalised for convenience. This problem requires an adaptive approach as a single, fixed-parameter controller will not work over the entire operating range of the plant parameters. If a fixed-parameter controller is used, tuning the controller for the high plant-gain region gives unacceptably sluggish response when the actual plant gain is small, and tuning for the low plant-gain makes the system unstable when the actual plant-gain is high.

The desired performance criteria are: (i) a settling time (time required for the output to remain within a band $\pm 10\%$ of the setpoint change) of 15, and (ii) a maximum overshoot of 20%. The setpoint profile we will use for most of the simulations is

$$z_{ref} = \begin{cases} 0 & \text{for} \quad 0 \le t < 5 \\ 1 & 5 \le t < 30 \\ 0.2 & 30 \le t < 55 \\ 1 & 55 \le t < 80 \\ 0 & 80 \le t < 105 \\ 1 & 105 \le t < 130. \end{cases}$$

Table 11.1 Plant parameter variation.

Parameter range	Nominal
$0.25 \le k_p \le 9.00$	$k_p = 1.00$
$0.75 \le \tau_p \le 1.50$	$\tau_p = 1.00$
$0.50 \le \theta_p \le 1.50$	$\theta_p = 1.125$

11.5.2 Controller design

The objective is to design a controller that can satisfy the performance specifications for all possible plants within the parameter space. Our approach is to first assume perfect, immediate, single-plant identification, then choose the minimum number of models which will individually satisfy the performance criteria, and finally test the controller via simulations to see if it does indeed perform as required. Since we desire the same response regardless of the plant parameters, the controllers are tuned to yield identical responses in all operating regions. Note however that MMAC does not require each controller be tuned identically and this feature can be exploited in other applications.

The first step is to decide on the structure of the model and controller banks. In this example we are using the Internal Model Control (IMC) procedure detailed in (Morari and Zafiriou, 1989) in which we only have a single tuning parameter, the IMC filter-factor λ. We selected IMC because of its straightforward handling of dead-time. The block diagram corresponding to this approach is shown in Figure 11.3.

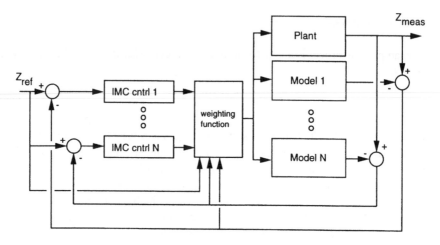

Figure 11.3   IMC MMAC implementation.

Many process systems exhibit behaviour that is well-approximated as a static nonlinearity with linear dynamics. For these systems it may be adequate to develop a model bank consisting of models with identical dynamics but varying gains. Based on dead-time and desired response considerations, we chose a filter factor of $\lambda = 2.5$. Then, through trial-and-error, we found the dynamically "best" and "worst" plants which satisfy the overshoot and settling-time requirements, assuring that the performance requirements are met for the worst-case high-gain and low-gain plants in each region (Figure 11.4 ; Table 11.2).

Since we desire the same response characteristics across the entire plant parameter region, establishing tuning in one region defines the uncertainty bounds for all of the models and a series of models for the model bank can be easily constructed. Consider the admissible gains bounded by

$$k_{p,min} \leq k_p \leq k_{p,max} \tag{11.10}$$

and let the region for some model i span the range

$$k_{i,min} \leq k_i \leq k_{i,max} \tag{11.11}$$

Let the relationship between the minimum and maximum gains for each region be

$$k_{i,max} = A\, k_{i,min} \tag{11.12}$$

Table 11.2   Range of parameters around nominal plant which yield acceptable performance for $\lambda = 2.5$.

Gain	Time Constant	Dead-time	Response
0.58	0.75	0.500	slowest
1.00	1.00	1.125	nominal
1.42	1.50	1.500	fastest

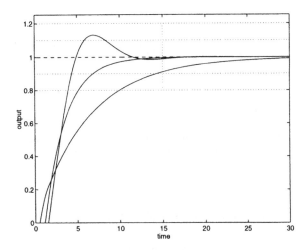

Figure 11.4 Closed-loop response for nominal, slowest, and fastest plants in each region.

Assuming that the model gain for each region is located at the centre of the uncertainty region

$$k_i = \frac{1}{2}(k_{i,min} + k_{i,max}),$$ (11.13)

it is easy to show that

$$A = \left\{ \frac{k_{N,maxN}}{k_{1,min}} \right\}^{\frac{1}{N}}$$ (11.14)

and the ith model gain is

$$k_i = \frac{1}{2}k_{p,min} (A + 1) A^{(i-1)}.$$ (11.15)

Again, this method gives only a minimum number of models. Due to imperfect, less than immediate identification, more models will most likely be needed to meet the performance requirements. Once a single plant-gain range is established (Table 11.2) we find that $A = 2.45$ and then from equation (11.14) the minimum number of plants which have the potential to meet the performance requirements is $N = 4$ (any non-integer N must be increased to the next highest integer). The nominal model gains and the plant-gain ranges they cover are given in Table 11.3. All models have time constants of 1.0 and dead-times of 1.125.

For weighting we use the Bayesian-based estimator given by (11.7). Convergence factor K and bounding factor δ were found by trial-and-error via computer simulations. Due to space limitations, plots of the manipulated-variable are not shown. All manipulated variable responses were stable. No constraints of any type were imposed on the system. All simulations were carried out in *Simulink*(TM) with a sampling time of 0.10. All plots displaying weighting fractions follow this legend: the weighting-fraction for model 1 is displayed as a solid line, model 2 by a dashed line, model 3 by a dash-dot line, and model 4 by a dotted line.

11.5.3 Simulation results

Figures 11.5 and 11.6 show an acceptable MMAC plant identification and subsequent control. The four models defined in Table 11.3 were used with equal initial weighting for each model ($W_i = \frac{1}{4}$). Convergence factor K is set at 0.1 and bounding parameter $\delta = 0.01$. The actual plant parameters are $k_p = 0.55$, $\tau_p = 1.5$, and $\theta_p = 1.5$. These plant dynamics represent the "worst" in terms of aggressive control action leading to overshoot. Figure 11.6 shows the evolution of the weighting factors for each model. The adaptive controller correctly chose model/controller pair 1 as the appropriate one to use.

Identification

Figures 11.5 and 11.6 also illustrate two potential shortcomings with this MMAC implementation. The first problem is encountered when attempting to operate at a setpoint of zero ($t = 80$ to 105). Upon requesting a setpoint of zero the scaling of equation (11.8) becomes ∞ and immediately drives the weighting to $\frac{1}{4}$ via the bounding in equation (11.4). This equal weighting then forces re-identification on the next setpoint change away from zero. If the setpoint is not scaled as in (11.8), then the problem is slightly different. Instead of settling at equal controller weighting, the identification stops. Although all of the models will eventually reach an output of zero, machine precision constraints and bounding during the weighting calculations halt identification before equal weighting is achieved. The last identified weighting remains in effect until a nonzero setpoint is applied. This same "feature" could be applied when using setpoint scaling by disabling the weighting calculation upon receiving a setpoint of zero. In either case though, note the difficulty when attempting to use models in deviation-variable form.

The second shortcoming is closely related to the problem described above. The response due to the initial change in setpoint will be different than subsequent, nonzero setpoint changes. This difference is a result of the initial model weighting. Since the plant and all of the models initially have an output of zero, there is no driving force to change the weighting until there is a difference between the plant and model outputs. The control effort defined by the initial weighting continues through the system dead-time until plant measurements are available to modify the weights. Figures 11.7 and 11.8 show this problem in more detail ($k_p = 1.4$, $\tau_p = 1.5$, and $\theta_p = 1.50$). Although the setpoint changes at $t = 5$, the weighting factors do not begin to change until the first differences between plant and model outputs are seen ($t = 6.2$). As with all adaptive schemes, this problem is only prominent in systems with dead-time since model identification, and hence weighting calculations, cannot take place during the time-delay.

This problem can be eased, but cannot always solved, by using a different initial weighting profile. One may have *a priori* information about which model region the plant

Table 11.3 Model bank.

Model	Gain	Plant Gain Range
1	0.431	0.250 to 0.612
2	1.056	0.612 to 1.500
3	2.587	1.500 to 3.674
4	6.337	3.674 to 9.000

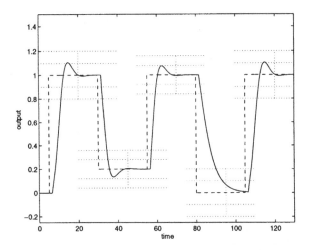

Figure 11.5 Acceptable MMAC response.

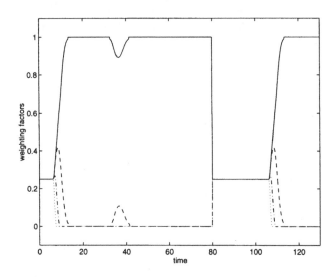

Figure 11.6 Initial weighting is $\frac{1}{4}$ for all controllers (**mod**$_1$ = solid; **mod**$_2$ = dash; **mod**$_3$ = dash-dot; **mod**$_4$ = dot).

will initially exist, or, to be conservative, one may initially weight the more aggressive controllers lower at the start (e.g. Figures 11.9 and 11.10). The problem may also be reduced by using a single, conservatively tuned controller initially, then switching to MMAC (Martin *et al.* 1987) once different weighting factors become available.

Tuning

If plant identification seems too slow, as perhaps in the previous case, one is tempted to increase the convergence factor so that the correct plant is identified more quickly. However, Figure 11.10 shows that when the convergence is "too" high, the wrong model is identified

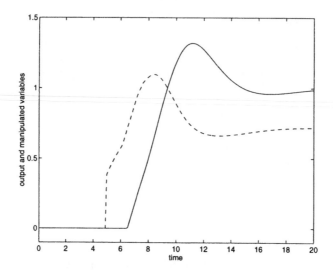

Figure 11.7 Inclusion of dead-time in the model allows control action to start before a change in measured output is felt   (output = solid; manipulated variable = dash).

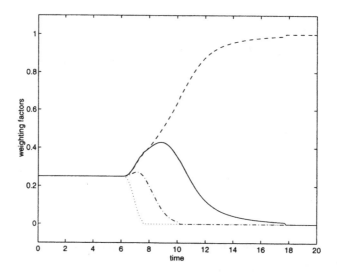

Figure 11.8 Controller weighting remains equal until plant/model output differences are seen ($t = 6.2$)   (**mod**$_1$ = solid; **mod**$_2$ = dash; **mod**$_3$ = dash-dot; **mod**$_4$ = dot).

during the dynamic portion of the response (here $K = 10$, $\delta = 0.01$; plant: $k_p = 1.4$, $\tau_p = 1.5$, $\theta_p = 1.5$; same plant as figures 11.7 and 11.8). With the convergence set "high", the weighting quickly goes to the closest model. In this case, when making a step reduction in setpoint, the plant output is closest to model 3 for a time, making the response sluggish since controller 3 is de-tuned in relation to the plant. The reason for this misidentification is clearly evident when examining the plant and model outputs (Figure 11.11). As stated earlier, there are presently no guidelines available for specifying K; a trial-and-error approach must be taken through simulations.

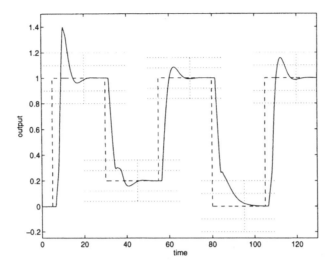

Figure 11.9 Even with more aggressive controllers weighted lower at the start of the simulation, overshoot limit is still violated initially.

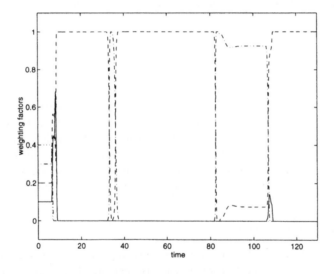

Figure 11.10 Plant misidentification during setpoint changes. Note here that the setpoint scaling was turned off at $t = 75$ to show how identification "locks-up" without it ($t = 85$ to 105) (**mod$_1$** = solid; **mod$_2$** = dash; **mod$_3$** = dash-dot; **mod$_4$** = dot).

Most misidentification problems are due to dynamic differences between the plant and models. The MMAC approach works very well when there are only gain differences. De-tuning can average-out the differences in dynamics and base identification primarily on the gain. To improve performance when there are significant variations in the plant dynamics, a better plant/model match is needed and so the model bank must be increased. Expanding the model bank can be done either by initially specifying a tighter performance requirement than is actually required, or by ad hoc model addition if more is known about the system

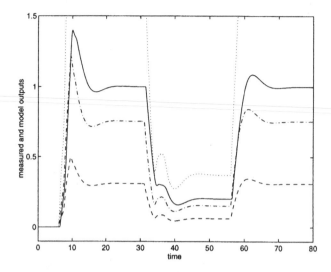

Figure 11.11 Plant and model outputs. During steady state, the plant is closer to model 2, however the uncertainty in dead-time between the plant and models causes misidentification during reductions in setpoint. Plant output = solid; (**mod**$_1$ = dash; **mod**$_2$ = dash-dot; **mod**$_3$ = dot; **mod**$_4$ not shown).

(e.g. the dynamics only vary in a certain region of plant operation). In any case, since selecting the number of models to use with respect to performance is still an open research issue, simulations must be run to verify performance across the entire parameter region. Martin *et al.* (1987) got satisfactory results for this problem using seven models, whereas He *et al.* (1986) used eight. However, both did take special measures to address the initial weighting problems noted earlier. Another way that this problem can be addressed is through the use of model-predictive controllers which explicitly handle constraints (e.g. QDMC).

Time-varying plant-parameters

The MMAC controller is well suited for situations where plant parameters clearly jump from one known set to another. MMAC can also operate effectively when parameters change continuously; however, the identification aspects of the controller become much more important. As with all adaptive controllers, the rate of parameter change has a big effect on the controller's performance. Figures 11.12 and 11.13 show the results of ramping the plant gain over time (k_p=0.50 → 8.25 from t=40 → 180). The controller has difficulty when the plant is rapidly moving through a model region. High bounding (δ) may improve performance here, but more important is the fact each controller is specifically detuned to some degree to be robust to plant/model mismatch. Once again, more models are needed so that each one may be tuned more tightly. Figures 11.14(a) and 11.14(b) again show the effects of ramping the plant gain (k_p = 0.80 → 4.75 from t=50 → 120), however here the plant is changing more slowly with respect to the region covered by each model and the system performs better.

The acceptable responses to step-changes at the beginning and end of these simulations suggest another advantage of the MMAC approach, namely that all the controllers within

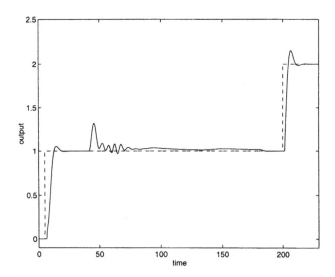

Figure 11.12 System output in response to changing plant gain.

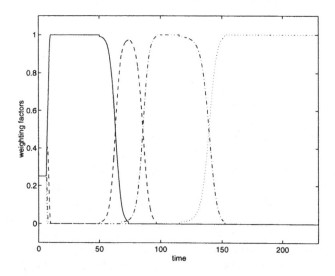

Figure 11.13 Evolution of weighting factors as plant gain changes. Here the gain passes through model region 2 so quickly that model 2 is never identified singly, though at some point the plant and model 2 matched exactly   (**mod**$_1$ = solid; **mod**$_2$ = dash; **mod**$_3$ = dash-dot; **mod**$_4$ = dot).

the controller bank need not be tuned for the same performance. For example, the plant-parameters may vary such that the plant will "start" in a certain parameter region and "end" in another (e.g. batch heating/cooling of a polymer solution). In such a case, the model/controller pairs covering the "ends" of the parameter region may be tuned for good servo response (start-up and shut-down) whereas the "middle" plants may be more tightly tuned for good regulator performance (disturbance rejection).

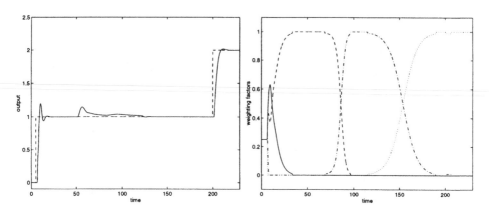

(a) System performance improves when the plant-gain changes more through the models regions.

(b) More model "blending" takes place with slowly changing plant-gain   (mod_1 = solid; mod_2 = dash; mod_3 = dash-dot; mod_4 = dot).

Figure 11.14   Effects of ramping plant gain.

Noise

Due to the averaging effect of the weighting function, MMAC is fairly insensitive to measurement noise unless the models are closely spaced (i.e. have very similar outputs). The model outputs themselves are only affected by the noise through the control signal unless the models specifically include some mechanism for disturbance compensation. Since the comparison between the plant and models occurs over a period of time, zero mean noise does not have a serious impact on the weighting calculations unless the magnitude of the noise is large enough that it encompasses more than one model output. This consequence must be considered in applications which require the use of pseudonoise. In general, large model banks which have many outputs close together are more susceptible to plant misidentification because of noise effects. Figures 11.15(a) and 11.15(b) show the effect of adding band-limited white noise to the plant output (plant: $k_p = 1.1$, $\tau_p = 1.5$, $\theta_p = 1.5$; tuning: $K = 1.0$, $\delta = 0.05$; noise: **peak − to − peak** $= \pm0.35$, $\sigma = 0.1$). Weighting is only seriously affected while maintaining the plant at an output of 0.2. The additional noise on the output is large enough to overlap the outputs of nearby models 1 and 3.

Input disturbance

Disturbances on the manipulated variable can seriously affect an MMAC controller. Since the system is only allowed to adapt within certain bounds, input disturbances may make the plant unmatchable with the available models, resulting in poor system performance and possibly instability. Figures 11.16(a) and 11.16(b) show the effect of a 0.15 additive step decrease in the input variable at $t = 30$. (plant: $k_p = 1.1$, $\tau_p = 1.5$, $\theta_p = 1.5$; $K = 0.5$, $\delta = 0.05$). The disturbance is not large enough to cause the weighting to shift away from model 2 and controller 2 is able to adequately compensate for the change. However, when

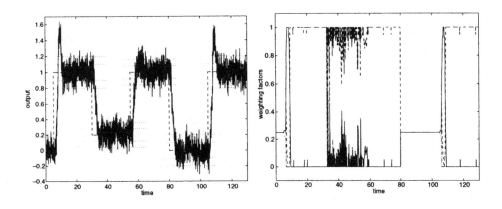

(a) Measurement noise on plant output.

(b) Noise effects on weighting factors.

Figure 11.15 Effect of measurement noise.

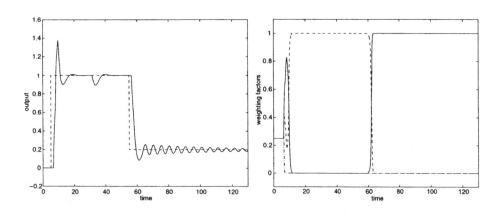

(a) System response to a −0.15 additive step disturbance on the plant input variable.

(b) The apparent reduction in plant gain by the input disturbance causes a lower-gain model to be identified. (**model₁** = solid; **model₂** = dash; **model₃** = dash-dot; **model₄** = dot).

Figure 11.16 System response to an additive step disturbance on the plant input variable.

the setpoint is reduced to 0.2, the disturbance causes the plant to be identified as a lower-gain model (model 1). Since the tuning for controller 1 is based on a lower plant-gain, the resulting control action is unacceptably oscillatory. A similar misidentification occurs on a positive input disturbance, though the system response is then too sluggish since a controller based on a higher plant-gain is applied.

11.6 SUMMARY

We have presented a concise review of MMAC theory and applications. The major steps in MMAC development are determining the type and number of models, the type of controllers and the weighting function. The selection of models, controllers, and weighting functions are all interdependent. A key point is that MMAC is a methodology or framework and not a specific algorithm, since many types of models and controllers can be used.

The MMAC approach has advantages and disadvantages. While an MMAC system designer has many "degrees of freedom" with which to craft a solution to a particular nonlinear control problem, few general guidelines, and even fewer rigorous, theoretically based results, are available to assist the designer in synthesising the controller. There is, however, a basic trade-off between the number of model/controller pairs and closed-loop performance. If available computing resources can tolerate large model/controller banks, then tighter performance specifications can be achieved. Furthermore, the inherent parallel structure of the MMAC framework allows easy implementation of parallel-computing techniques to increase the speed of control computations.

A time-variant SISO system similar to a drug infusion control problem was used to illustrate many of the issues involved with the design of a multiple-model adaptive controller. After specifying an IMC-based strategy and performance requirements, we determined the smallest model bank which has the potential for satisfying this performance criterion. A major advantage to the MMAC approach is that the individual model/controller pairs are easy to tune for desired nominal performance. For our example it was easy to meet specified settling time and overshoot criteria for the region covered by each model. Determining the proper number of models and controller tuning required to *guarantee* performance over the *entire* operating space is a greater challenge though, and only becomes more difficult with multivariable systems.

ACKNOWLEDGEMENTS

A portion of this chapter was written while the second author was on sabbatical at Merck and Co., Rahway, NJ. Financial support from the Whitaker Foundation, Merck and Co., and the National Science Foundation is gratefully acknowledged.

REFERENCES

ATHANS, M., D. CASTANOU, K.-P. DUNN, C.S. GREENE, W.H. LEE, N.R. SANDELL AND A.S WILLSKY (1977) 'The Stochastic Control of the F-8C Aircraft Using a Multiple Model Adaptive Control (MMAC) Method – Part 1: Equilibrium Flight'. *IEEE Trans. Auto. Cont.* AC-22(5), 768–780.

BANERJEE, A., Y. ARKUN, B. OGUNNAIKE AND R. PEARSON (1995) 'Multiple model based control of nonlinear systems'. *Proceedings of the 1995 AIChE.*

DESHPANDE, J., T.N. UPADHYAY AND D.G. LAINIOTIS (1973) 'Adaptive control of linear stochastic systems'. *Automatica* (9), 107–115.

GARCIA, C., D.M. PRETT AND M. MORARI (1989) 'Model predictive control: theory and practice – a survey'. *Automatica* 23(3), 335–248.

GREENE, C. AND A.S. WILLSKY (1980) 'An Analysis of the Multiple Model Adaptive Control Algorithm'. *Proceedings Conf. Decision Control* pp. 1142–1145.

GUSTAFSON, J. AND P.S. MAYBECK (1994) 'Flexible spacestructure control via moving-bank multiple model algorithms'. *IEEE Trans. Aero. Elec. Sys.* **30**(3), 750–757.

HE, W., H. KAUFMAN AND R. ROY (1986) 'Multiple model adaptive control procedure for blood pressure control'. *IEEE Trans. Biomed. Eng.* **BME-33**(1), 10–19.

LAINIOTIS, D. (1976a) 'Partitioning: a unifying framework for adaptive systems. I: Estimation'. *Proc. IEEE* **64**, 1126–1143.

LAINIOTIS, D. (1976b) 'Partitioning: a unifying framework for adaptive systems. II: Control'. *Proc. IEEE* **64**, 1144–1161.

MARTIN, J., A.M. SCHNEIDER AND N.T. SMITH (1987) 'Multiple-model adaptive control of blood pressure using sodium nitroprusside'. *IEEE Trans. Biomed. Eng.* **BME-34**(8), 603–611.

MAYBECK, P. (1982) *Stochastic Models, Estimation, and Control.* Vol. 2. Academic Press. New York, NY.

MAYBECK, P. (1989) *Moving-Bank Multiple Model Adaptive Estimation and Control Algorithms: An Evaluation.* Vol. 31 of *Control and Dynamic Systems.* Academic Press. New York, NY. pp. 1–31.

MORARI, M. AND E. ZAFIRIOU (1989) *Robust Process Control.* Prentice Hall. Englewood Cliffs, NJ.

SARIDIS, G. AND T.K. DAO (1972) 'Learning approach to the parameter-adaptive self-organizing control problem'. *Automatica* **8**, 589–597.

SCHOTT, K. AND B.W. BEQUETTE (1995) Control of Chemical Reactors Using Multiple-Model Adaptive Control (MMAC). In '4th IFAC Symposium on Dynamics and Control of Chemical Reactors, Distillation Columns and Batch Reactors'. preprints of DYCORD+ 95. Helsingor, Denmark. pp. 345–350.

SEBORG, D., T.F. EDGAR AND S.L. SHAH (1986) 'Adaptive control strategies for process control: A Survey'. *AIChE J.* **32**, 881–913.

SLATE, J., L.C. SHEPPARD, V.C. RIDEOUT AND E.H. BLACKSTONE (1979) A Model for Design of a Blood Pressure Controller for Hypertensive Patients. In '5th IFAC Symposium on Identification and System Parameter Estimation'. Darmstadt, Germany.

STENGEL, R. (1986) *Stochastic Optimal Control: Theory and Application.* John Wiley. New York, NY.

YU, C., R.J. ROY, H. KAUFMAN AND B.W. BEQUETTE (1992) 'Multiple-model adaptive predictive control of mean arterial pressure and cardiac output'. *IEEE Trans. Biomed. Eng.* **39**(8), 765–778.

YU, C., W.G. HE, J.M. SO, R. ROY, H. KAUFMAN AND J.C. NEWELL (1987) 'Improvement in arterial oxygen control using multiple-model adaptive control procedures'. *IEEE Trans. Biomed. Eng.* **BME-34**(8), 567–574.

H_∞ Control of Nonlinear Processes Using Multiple Linear Models

A. BANERJEE, Y. ARKUN, R. PEARSON and B. OGUNNAIKE

This chapter addresses the problem of controlling a nonlinear process when linear models have been identified at different operating points. It is motivated by the larger problem of transition control, where a controller has to be designed for a plant that operates in multiple regimes and makes transitions between them.

The multiple local models are combined into a single linear parameter varying (LPV) global model. The parameters of the global model are chosen to be model probabilities estimated on-line using a Bayesian approach. A controller is then designed for this LPV structure that is robust in the $\mathcal{H}_\infty$ sense.

12.1 INTRODUCTION

In this work we address the control of nonlinear systems subject to multiple operating regimes. Different operating conditions are usually initiated by external factors such as changes in product specifications (e.g. polymer product and grade changeovers) or persistent plant disturbances (e.g. variations in feed conditions). Poor transition control during changes in operating conditions leads to long periods of transient operation, usually accompanied by the production of off-specification material and significant economic loss. The objective of this work is to develop a controller design which can successfully regulate the plant not only at particular operating points, but also in regions of transition. Such a controller will be called the transition controller.

Therefore the underlying premise of this chapter is that a controller has to be designed for a nonlinear system that operates in several significantly different modes. This makes it necessary to have a nonlinear model that accurately matches the plant behaviour in all operating regimes. The first principles models are usually difficult to develop if they have to cover a wide range of conditions. Even if such a global model can be obtained it may not be appropriate for controller design. The alternative is to identify an empirical model from plant input–output data. However unmodelled dynamics which are negligible at one operating point may be dominant at another. Therefore it may not be easy to select a model structure that works well in all regimes. Furthermore in order to uncover all the necessary plant dynamics, the inputs required by the identification algorithm may not be practically implementable due to their large amplitude and/or large frequency. Therefore this chapter presents an alternative approach wherein multiple local linear models are identified at the different regions of operation, and controller design is carried out using these models.

12.2 COMBINING MULTIPLE LINEAR MODELS

Henceforth it will be assumed that multiple linear models have been identified to explain plant behaviour at different operating points and that all information about the plant is contained in the local models. Let there be N local linear models with the state-space representation:

$$
\begin{aligned}
\dot{x} &= A_i x + B_i u \\
y &= C_i x + D_i u \\
&\quad (i = 1, \ldots, N).
\end{aligned}
$$

(12.1)

These local models may have been obtained either through identification, or by linearising a first principles model if one is available.

Model validity functions are estimates of the validity, or trustworthiness, of each of the local models. These functions make up a vector $p(t) = [p_1(t), \ldots, p_N(t)]^T \in R^N$, and may be defined such that $p_i(t)$ maps plant input-output data into a range of validity for the ith model:

$$
p_i(t) : (y, u) \rightarrow [0, 1],
$$

where

$$
\begin{aligned}
p_i(t) &\rightarrow 1 \quad \textit{when the ith model is valid,} \\
p_i(t) &\rightarrow 0 \quad \textit{otherwise}
\end{aligned}
$$

(12.2)

and

$$
\sum_i^N p_i(t) = 1.
$$

(12.3)

Equation (12.3) implies that as the plant moves into a region where one of the models becomes more trustworthy than the others, the other models lose their validity. The idea of using similar functions to compare models has also been used by (Johansen and Foss 1992, Johansen and Foss 1995).

The ith local linear model is described by the state-space matrices:

$$
[A_i, B_i, C_i, D_i].
$$

These are then combined with the model validity functions to construct a time-varying global model for the plant which will be denoted by:

$$
M(p_i, A_i, B_i, C_i, D_i).
$$

One way of doing this is to use a linear parameter varying (LPV) system as the global model. LPV systems are fixed affine functions of time-varying parameter vectors $\theta(t)$:

$$
\begin{aligned}
\dot{x} &= A(\theta(t))x + B(\theta(t))u \\
y &= C(\theta(t))x + D(\theta(t))u.
\end{aligned}
\tag{12.4}
$$

Traditionally such systems have been interpreted as LTI systems with time-varying parameters, or as linearisations of a nonlinear system along the trajectory of the parameter θ. However in this work we interpret $\theta(t)$ to be the time-varying model validity function vector $p(t)$. This implies that the nonlinearities of the true plant are captured by the functional dependence of the system state-space matrices on $p(t)$. Therefore the nominal global model is given by a map of the form

$$
M(A(p), B(p), C(p), D(p))
$$

with a specific form yet to be determined.

If such a formulation is used, an implicit restriction on the global model is that if $p_i = 1$, the LPV system should reduce to the ith linear state space model:

$$
\left.
\begin{aligned}
A(p) = A_i \quad B(p) = B_i \\
C(p) = C_i \quad D(p) = D_i
\end{aligned}
\right|_{p_i = 1.}
$$

One way of satisfying these conditions is to use the following state-space representation for the global model:

$$
\begin{aligned}
\dot{x} &= \left[\sum_i^N p_i(t) A_i \right] x + \left[\sum_i^N p_i(t) B_i \right] u \\
y &= \left[\sum_i^N p_i(t) C_i \right] x + \left[\sum_i^N p_i(t) D_i \right] u.
\end{aligned}
\tag{12.5}
$$

This is similar to the global model proposed in (Johansen and Foss 1992), but with a very important difference. In our work we propose to estimate the validity functions p_i on-line, whereas in (Johansen and Foss 1992, Johansen and Foss 1995) the validity functions are obtained off-line, and their dependence on the outputs is therefore fixed *a priori*. Hence we believe that our approach will have better performance in the presence of disturbances.

The main reason for combining the local models into a single time-varying global model is to design a controller for the global model which is parameterised by the validity functions. However, the subsequent controller design requires that all the local models satisfy the following conditions in order to meet robustness requirements for all possible trajectories of the model validity functions.

1   The models are strictly proper.
2   The models are stabilisable and detectable.
3   All models have the same B_i, C_i and D_i matrices. This is not a serious restriction as shown in (Apkarian *et al.* 1994). The condition will be met if the models include sensor and actuator dynamics. If not, then this condition can be satisfied by filtering the inputs and outputs. This will not significantly change the problem if the filter bandwidths are chosen to be sufficiently high.

Therefore the structure that we have chosen to implement for the nominal global model, $M(p)$, is:

$$\dot{x} = \left[\sum_i^N p_i(t) A_i \right] x + Bu$$
$$y = Cx + Du. \tag{12.6}$$

The advantages of using this particular LPV structure as the global model are:

1   The global model satisfies the linearisation property, i.e. each local model is recovered exactly from the global model at the operating point at which it was obtained, provided the $p_i(t)$ are estimated properly.
2   Recent work in (Apkarian *et al.* 1994) has extended robust control theory to cover plants of this structure.

This nominal global model is coupled with uncertainty description $\Delta(t)$ to represent the true unknown plant as shown by Figure 12.1. The uncertainty $\Delta(t)$ incorporates the effects of:

1   the errors and uncertainties in the local models – these represent the plant–model mismatches around the points where each of the models was identified,
2   the errors and uncertainties in the global nominal model – these represent the plant–model mismatches during transition between the regimes.

The controller must be designed with a certain degree of robustness against these uncertainties. Furthermore the uncertainty in the global model, $\Delta(t)$ which may be large during transitions, reduces to the uncertainty in the ith local model when the plant is within the domain of the ith model. This explains the time-varying nature of the uncertainty, which must be accounted for during controller design.

12.3   MODEL VALIDITY FUNCTIONS

There are a number of ways of interpreting the model validity functions, and this is reflected in the different methods that may be used to assign them on-line. Some possible approaches are:

1   Fuzzy logic: Here the model validity functions are interpreted as set membership functions, and are estimated off-line. This is the approach taken in (Kuipers and Åström 1994, Zhao *et al.* 1995), and is similar to that of (Johansen and Foss 1995).
2   Bayesian estimation: Given the conditions (12.2) and (12.3) on the model validity functions, one way of interpreting them is as model probabilities, i.e. $p_i(t)$ is the probability of the ith model being valid. The values of these may then be estimated on-line from plant measurements using Bayes theorem.

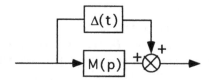

Figure 12.1   Nominal model with uncertainty.

3 Simultaneous state and parameter estimation: The validity functions can be treated as parameters of the global model, and estimated on-line using a moving horizon based estimator. This approach has been demonstrated in (Banerjee *et al.* 1995).

The Bayesian approach will be taken in the remainder of this work. Let the plant measurements be denoted by y_i, and the measurement history by $Y_i = [y_i, y_{i-1} \ldots]^T$. Let $p(j|Y_i)$ denote the probability that model j best describes the plant given the measurement history till time t_i. Then applying Bayes theorem,

$$
\begin{aligned}
p(j|Y_i) &= p(j|y(t_i), Y_{i-1}) \\
&= \frac{f(y_i|j, Y_{i-1}) \, p(j|Y_{i-1})}{p(y_i)} \\
&= \frac{f(y_i|j, Y_{i-1}) \, p(j|Y_{i-1})}{\sum_k f(y_i|k, Y_{i-1}) \, p(k|Y_{i-1})},
\end{aligned}
\tag{12.7}
$$

where $f(y_i|j, Y_{i-1})$ is the probability distribution function (pdf) of the outputs of the jth model at time t_i given the measurement history Y_{i-1}.

Equation (12.7) describes how an incoming plant measurement changes the belief about model validity, i.e. how the measurement relates the *a posteriori* probability to the *a priori* probability. The approach followed by Lainiotis and co-workers (Deshpande *et al.* 1973, Hilborn and Lainiotis 1969, Lainiotis 1971, Lainiotis 1976) is to design a Kalman filter for each model. If the ith model exactly matches the plant, the model residuals, ε_i will be zero-mean, and their covariance will be given by $\Omega_i = C_i P_i C_i^T + R_i$, where P_i is the state error covariance of the ith Kalman filter, based on the ith model, and R_i is the covariance of the measurement noise in the ith regime. Then assuming stationarity,

$$
\begin{aligned}
f(y_i|j, Y_{i-1}) &= f(y_i|j) \\
&= f(\varepsilon_i|j) \\
&= \frac{\exp\left\{-\frac{1}{2}\varepsilon_i(k)\, \Omega_i^{-1}(k)\, \varepsilon_i(k)^T\right\}}{\left[(2\pi)^N \det(\Omega_i(k))\right]^{\frac{1}{2}}}.
\end{aligned}
\tag{12.8}
$$

Therefore equation (12.8) can be substituted into (12.7) to obtain an algorithm for estimating model validity. The algorithm must be initialised by the designer with an estimate for the starting probabilities.

The approach of Lainiotis as discussed above is based on certain assumptions which may not be correct. For example it is assumed that the true plant and all the postulated models are linear-Gaussian, which implies that if a model matches the plant, the filter residuals will be zero-mean, white, and distributed with some pre-specified covariance. However it has been shown in (Maybeck 1979, Willsky *et al.* 1980) that despite any inaccuracies in these assumptions, and the sensitivities of the filters to disturbances, the method will work as long as the residual characteristics of the different models are sufficiently different.

Another shortcoming is the underlying assumption that the true plant does not switch from one model to another. Therefore the probability estimation algorithm can get locked onto one model, so that its probability converges to one and that of the other models converges to zero. Therefore if the plant subsequently changes from one regime to another, it may take a very long time for the probabilities to reflect this change. This is a wind-up problem, and one 'fix', e.g. (Willsky *et al.* 1980), is to specify a lower bound on the probabilities.

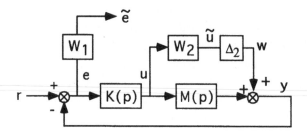

Figure 12.2 Feedback loop.

12.4 CONTROLLER DESIGN

Controller design is performed by assuming the true plant to lie in the family given by $G = M(p) + \Delta(t)$, where $M(p)$ is the LPV model given in equation (12.6) and $\Delta(t)$ is the additive uncertainty.

There are two approaches to robust controller design for such systems.

1 Traditional robust control. Here the p_i are treated as uncertainties, and a single LTI controller for all regimes is designed whose performance and stability is robust to all possible variations in p_i. Such a controller will be overly conservative.
2 Self-scheduled control. Such a controller adjusts to variations in plant dynamics by estimating values of $p_i(t)$ on-line and using them in the control law to adjust to variations in plant dynamics, so as to maintain stability and good performance during transition. This is the approach followed in this work.

As the control law has to include $p(t)$, a logical choice for the controller is an LPV one. Let this be denoted by $K(p)$, with the following structure:

$$\dot{x}_c = \left(\sum_i^N p_i(t) A_{K,i} \right) x_c + \left(\sum_i^N p_i(t) B_{K,i} \right) e \qquad (12.9)$$

$$u = \left(\sum_i^N p_i(t) C_{K,i} \right) x_c + \left(\sum_i^N p_i(t) D_{K,i} \right) e, \qquad (12.10)$$

where e is the tracking error as shown in Figure 12.2. With this controller structure, if the plant goes to the ith operating point, i.e.

$$p_i \to 1, \ p_j \to 0 \ \forall j \neq i,$$

then the LPV controller reduces to a local controller given by the state-space system matrices $[A_{K,i}, B_{K,i}, C_{K,i}, D_{K,i}]$.

The feedback loop is next cast into the $M - \Delta$ form in order to use the $\mathcal{H}_\infty$ framework to obtain robust stability and performance for all possible values of the vector $p(t)$. These requirements can be translated into a condition on performance,

$$\|W_1 S(p)\|_\infty \leq \beta \qquad (12.11)$$

and a condition that the closed loop remain stable for all uncertainties

$$\Delta = \Delta_2 W_2,$$

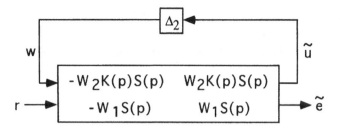

Figure 12.3   Equivalent closed-loop diagram.

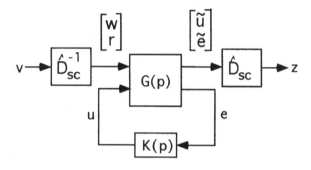

Figure 12.4   Representation for controller design.

where

$$\|\Delta_2\|_\infty \leq \beta. \tag{12.12}$$

$S(p) = (I + M(p) K(p))^{-1}$ is the sensitivity operator mapping r to e; W_1 and W_2 are suitable frequency-dependent weighting matrices, and $\|\cdot\|_\infty$ is the $\mathcal{L}_2$ induced norm. The block diagram of Figure 12.2 can be rearranged to that of Figure 12.3, and so a controller $K(p)$ that satisfies the conditions (12.11) and (12.12) must satisfy :

$$\left\| \begin{pmatrix} -W_2 K(p) S(p) & W_2 K(p) S(p) \\ -W_1 S(p) & W_1 S(p) \end{pmatrix} \right\|_\infty \leq \gamma, \tag{12.13}$$

where

$$0 < \gamma \leq \frac{1}{\beta}. \tag{12.14}$$

However, the uncertainty region is time-varying, and therefore Δ_2 is a time-varying full complex matrix. To take this into account, the robust performance condition of equation (12.13) is modified to (e.g. (Doyle *et al.* 1989)):

$$\inf_{D_{sc}} \left\| \begin{pmatrix} D_{sc} & 0 \\ 0 & I \end{pmatrix} \begin{pmatrix} -W_2 K(p) S(p) & W_2 K(p) S(p) \\ -W_1 S(p) & W_1 S(p) \end{pmatrix} \begin{pmatrix} D_{sc}^{-1} & 0 \\ 0 & I \end{pmatrix} \right\|_\infty \leq \gamma, \tag{12.15}$$

where D_{sc} is a scalar-times-identity scaling matrix. For notational convenience $\hat{D}_{sc}$ is defined to be:

$$\hat{D}_{sc} = \begin{pmatrix} D_{sc} & 0 \\ 0 & I \end{pmatrix}$$

In order to design a controller that satisfies this condition, a state-space representation of the system must be obtained. If the filters W_1 and W_2 are given by the state-space matrices $[\bar{A}_1, \bar{B}_1, \bar{C}_1, \bar{D}_1]$ and $[\bar{A}_2, \bar{B}_2, \bar{C}_2, \bar{D}_2]$ respectively, then the system of Figure 12.3 may be transformed into that of Figure 12.4,

$$v = \hat{D}_{sc} \begin{bmatrix} w \\ r \end{bmatrix} \qquad z = \hat{D}_{sc} \begin{bmatrix} \tilde{u} \\ \tilde{e} \end{bmatrix}$$

and the block $G(p)$ represents the system:

$$\dot{x} = \left(\sum_i p_i \hat{A}_i \right) x + \hat{B}_1 \begin{bmatrix} w \\ r \end{bmatrix} + \hat{B}_2 u \qquad (12.16)$$

$$\begin{bmatrix} \tilde{u} \\ \tilde{e} \end{bmatrix} = \hat{C}_1 x + \hat{D}_{11} \begin{bmatrix} w \\ r \end{bmatrix} + \hat{D}_{12} u$$

$$e = \hat{C}_2 x + \hat{D}_{21} \begin{bmatrix} w \\ r \end{bmatrix} + \hat{D}_{22} u,$$

where

$$\hat{A}_i = \begin{bmatrix} A_i & 0 & 0 \\ \bar{B}_1 C & \bar{A}_1 & 0 \\ 0 & 0 & \bar{A}_2 \end{bmatrix}$$

$$\hat{B}_1 = \begin{bmatrix} 0 & 0 \\ \bar{B}_1 & -\bar{B}_1 \\ 0 & 0 \end{bmatrix} \qquad \hat{B}_2 = \begin{bmatrix} B \\ \bar{B}_1 D \\ \bar{B}_2 \end{bmatrix}$$

$$\hat{C}_1 = \begin{bmatrix} -\bar{D}_1 C & -\bar{C}_1 & 0 \\ 0 & 0 & \bar{C}_2 \end{bmatrix} \qquad \hat{C}_2 = \begin{bmatrix} -C & 0 & 0 \end{bmatrix}$$

$$\hat{D}_{11} = \begin{bmatrix} -\bar{D}_1 & \bar{D}_1 \\ 0 & 0 \end{bmatrix} \qquad \hat{D}_{12} = \begin{bmatrix} -\bar{D}_1 D \\ \bar{D}_2 \end{bmatrix}$$

$$\hat{D}_{21} = \begin{bmatrix} -I & I \end{bmatrix} \qquad \hat{D}_{22} = -D.$$

A D-K iterative scheme (e.g. (Doyle 1985)) has been employed to design a controller that satisfies the condition of equation (12.15) using the state-space representation of equation (12.16), the algorithm for which is given below.

1 Set $D_{sc} = I$.
2 With this value of D_{sc} as constant, find the smallest γ such that there exists a $K(p)$ that satisfies the condition of equation (12.15).
   The trajectory of the model validity function vector $p(t)$ always lies in a polytope whose vertices are:

$$\begin{bmatrix} 1 & 0 & \cdots & 0 \end{bmatrix}^T \cdots \begin{bmatrix} 0 & \cdots & 0 & 1 \end{bmatrix}^T. \qquad (12.17)$$

It can also be shown that the lower fractional transformation of $G(p)$ and $K(p)$, $F_l(G(p), K(p))$ is also an LPV system whose state-space matrices may be given by

$$[A_{cl}(p), B_{cl}(p), C_{cl}(p), D_{cl}(p)]. \qquad (12.18)$$

These matrices evolve in a polytope of matrices whose vertices are:

$$\begin{bmatrix} A_{cl,i}, B_{cl,i}, C_{cl,i}, D_{cl,i} \end{bmatrix}$$

and which are obtained by substituting (12.17) into (12.18). If the closed-loop system satisfies equation (12.15), then the closed-loop system is said to have quadratic $\mathcal{H}_\infty$ performance if a single quadratic Lyapunov function can be found that establishes global stability. Applying the results of (Apkarian *et al.* 1994) to the system in Figure 12.4, the closed loop will have quadratic $\mathcal{H}_\infty$ performance for all possible trajectories of the vector $p(t)$ if and only if there exists a single matrix X such that

$$\begin{pmatrix} A_{cl,i}^T X + X A_{cl,i} & X B_{cl,i} \hat{D}_{sc}^{-1} & C_{cl,i}^T \hat{D}_{sc}^T \\ \hat{D}_{sc}^{-T} B_{cl,i}^T X & -\gamma I & \hat{D}_{sc}^{-T} D_{cl,i}^T \hat{D}_{sc}^T \\ \hat{D}_{sc} C_{cl,i} & \hat{D}_{sc} D_{cl,i} \hat{D}_{sc}^{-1} & -\gamma I \end{pmatrix} < 0 \ (i = 1, \dots, N). \tag{12.19}$$

The Lyapunov function is given by

$$V(x) = x^T X x$$

and it is shown in (Apkarian *et al.* 1994) that designing a controller that satisfies this condition is equivalent to solving a system of $2N + 1$ coupled linear matrix inequalities (LMIs) based on the state-space matrices of equation (12.16). The controller is LPV, and of the form of equation (12.9).

3   With this value of $K(p)$, find the smallest γ for which there exists a D_{sc} that satisfies the left-hand side of equation (12.15).

This is equivalent to finding a single D_{sc} and a single X that satisfy the system of matrix inequalities in (12.19), using the $K(p)$ determined in step 2. However these inequalities are not linear in D_{sc}, and cannot be solved by an LMI solver as such. To get around this problem, the substitution $P = \hat{D}_{sc}^T \hat{D}_{sc}$ must be made. This reduces the inequalities in (12.19) to the standard form of the bounded real lemma (Apkarian *et al.* 1993, Boyd *et al.* 1994) which is linear in P, and which can then be solved.

4   Stop if γ can no longer be reduced or the condition of equations (12.14) and (12.15) have been met, else go to step 2.

12.5 CSTR EXAMPLE

The theory developed above was applied to a first-order exothermic reaction in a continuously stirred tank reactor (CSTR). The dynamic behaviour is described by the following state equations, taken from (Uppal *et al.* 1974).

$$\begin{aligned} \frac{dx_1}{dt} &= -x_1 + D_a (1 - x_1) \exp\left(\frac{x_2}{1 + x_2/\gamma} \right) \\ \frac{dx_2}{dt} &= -x_2 + B D_a (1 - x_1) \exp\left(\frac{x_2}{1 + x_2/\gamma} \right) + \beta (u - x_2) \\ y &= x_1. \end{aligned}$$

The constants are $D_a = 0.072$, $\gamma = 20$, $B = 8$ and $\beta = 0.3$. The system is interesting because it exhibits output multiplicity, as can be seen from the steady-state curve given in Figure 12.5. Furthermore only the upper and lower branches of the curve are stable, while the central branch is unstable.

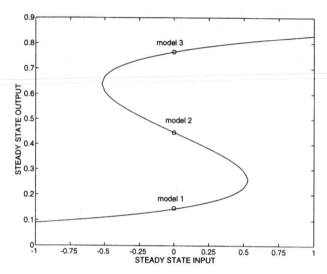

Figure 12.5   Steady-state curve for CSTR.

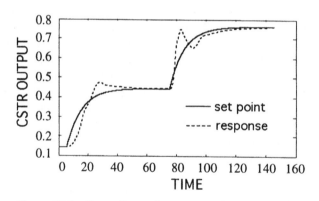

Figure 12.6   Controller performance with three models.

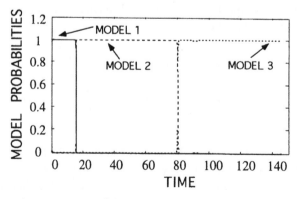

Figure 12.7   Model probabilities with three models.

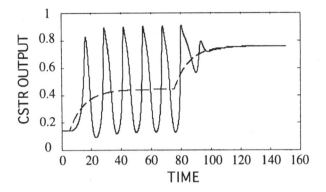

Figure 12.8 Controller performance with two models.

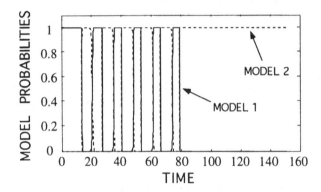

Figure 12.9 Model probabilities with two models.

Three linear models of the process are identified around the three steady-state points corresponding to $u = 0$ on the three branches of the steady-state curve. As the middle branch is unstable, so therefore is the second model. Furthermore the eigenvalues of the model identified on the upper branch have imaginary components. Therefore the dynamics of the system change significantly depending upon the operating point.

Multiple model $\mathcal{H}_\infty$ controller design is carried out as described in the previous section using the weighting functions:

$$W_1 = \frac{(s + 6)^2}{100\,(s + 0.0001)\,(s + 0.6)} \quad W_2 = \frac{10\sqrt{2}s}{s + 35}$$

The Bayesian approach described in section 12.3 is used to calculate model validity functions, but with a lower bound of 0.05 on the model probabilities to prevent wind-up problems. Figure 12.6 shows the performance of the controller when a setpoint trajectory through all three regimes was provided. Figure 12.7 shows how the model validity functions varied with time.

To check whether all the three models are needed, the simulation was repeated but with only the two linear models identified at the upper and lower branches. From Figure 12.8 it is clear that the controller is no longer able to track a setpoint trajectory through the unstable region around the middle branch. Figure 12.9 shows that the model validity functions oscillate between the two models, as the probability estimator is unable to find a model that accurately matches the plant. Thus just the first and third models do not contain enough information about the process to successfully implement a multiple model controller.

12.6 CONCLUSIONS

Many plants are required to operate in multiple regimes and make transitions between them. For such systems, it is often difficult to identify a single global nonlinear model for controller design. This paper presents an alternative approach where different local linear models have been identified at the operating regions. It has been shown that a robust $\mathcal{H}_\infty$ controller can be designed for the nonlinear plant based on these local models. This controller is parameterised by time-varying model validity functions that are estimated on-line.

ACKNOWLEDGEMENT

The first two authors gratefully acknowledge the financial support of E. I. DuPont de Nemours & Co., Inc, and the National Science Foundation through the grant CTS-9522564.

REFERENCES

APKARIAN, P., J. P. CHRETIEN, P. GAHINET AND J. M. BIANNIC (1993) μ synthesis by D-K iterations with constant scaling. In 'Proceedings of the American Control Conference'. pp. 3192–3196.

APKARIAN, P., P. GAHINET AND G. BECKER (1994) Self-scheduled H_∞ control of linear parameter-varying systems. In 'Proceedings of the American Control Conference'. pp. 856–860.

BANERJEE, A., Y. ARKUN, B. OGUNNAIKE AND R. PEARSON (1995) 'Multiple model based estimation of nonlinear systems'. Paper# 183f, presented at the AIChE annual meeting at Miami, FL.

BOYD, S., L. EL. GHAOUI, E. FERON AND V. BALAKRISHNAN (1994) *Linear Matrix Inequalities in System and Control Theory*. SIAM. Philadelphia.

DESHPANDE, J. G., T. N. UPADHYAY AND D. G. LAINIOTIS (1973) 'Adaptive control of linear stochastic systems'. *Automatica* **9**, 107–115.

DOYLE, F. J., A. K. PACKARD AND M. MORARI (1989) 'Robust controller design for a nonlinear CSTR'. *Chemical Engineering Science* **44**(9), 1929–1947.

DOYLE, J. (1985) Structured uncertainty in control system design. In 'Proceedings of the IEEE Conference on Decision and Control'. pp. 260–265.

HILBORN, C. G. AND D. G. LAINIOTIS (1969) 'Optimal estimation in the presence of unknown parameters'. *IEEE Trans. Man. & Cyb.*

JOHANSEN, T. A. AND B. A. FOSS (1992) Representing and learning unmodeled dynamics with neural network memories. In 'Proceedings of the American Control Conference'. pp. 3037–3043.

JOHANSEN, T. A. AND B. A. FOSS (1995) 'Identification of non-linear system structure and parameters using regime decomposition'. *Automatica* **31**(2), 321–326.

KUIPERS, B. AND K. ÅSTRÖM (1994) 'The composition and validation of heterogeneous control laws'. *Automatica* **30**(2), 233–249.

LAINIOTIS, D. G. (1971) 'Optimal adaptive estimation: Structure and parameter adaptation'. *IEEE Trans. Automatic Control* **AC-16**(2), 160–170.

LAINIOTIS, D. G. (1976) 'Partitioning : A unifying framework for adaptive systems. I. Estimation'. *Proc. IEEE* **64**, 1126.

MAYBECK, P. S. (1979) *Stochastic Models, Estimation and Control*. Academic Press. London.

UPPAL, A., W. H. RAY AND A. B. POORE (1974) 'On the dynamic behavior of continuous stirred tank reactors'. *Chemical Engineering Science*.

WILLSKY, A. S., E. Y. CHOW, S. B. GERSHWIN, C. S. GREENE, P. K. HOUPT AND A. L. KURKJIAN (1980) 'Dynamic model-based techniques for the detection of incidents on freeways'. *IEEE Transactions on Automatic Control* **AC-25**(3), 347–359.

ZHAO, J., V. WERTZ AND R. GOREZ (1995) Design a stabilizing fuzzy and/or non-fuzzy state-feedback controller using LMI method. In 'Proceedings of the 3rd European Control Conference, Rome, Italy'. pp. 1201–1206.

CHAPTER THIRTEEN

Synthesis of Fuzzy Control Systems Based on Linear Takagi–Sugeno Fuzzy Models

J. ZHAO, R. GOREZ and V. WERTZ

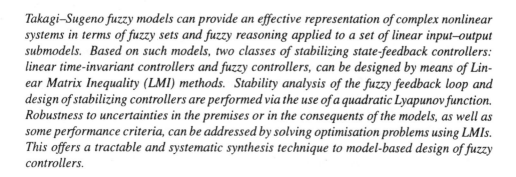

Takagi–Sugeno fuzzy models can provide an effective representation of complex nonlinear systems in terms of fuzzy sets and fuzzy reasoning applied to a set of linear input–output submodels. Based on such models, two classes of stabilizing state-feedback controllers: linear time-invariant controllers and fuzzy controllers, can be designed by means of Linear Matrix Inequality (LMI) methods. Stability analysis of the fuzzy feedback loop and design of stabilizing controllers are performed via the use of a quadratic Lyapunov function. Robustness to uncertainties in the premises or in the consequents of the models, as well as some performance criteria, can be addressed by solving optimisation problems using LMIs. This offers a tractable and systematic synthesis technique to model-based design of fuzzy controllers.

13.1 INTRODUCTION

Over the last 20 years, fuzzy control has found successful applications not only in consumer products but also in industrial processes. However, this does not mean that there exist tractable systematic methods for designing a fuzzy controller governed by a set of fuzzy control rules. The point which is probably most often raised in discussions on the synthesis of fuzzy controllers is that they are usually designed in an *ad hoc* manner, whereas rigorous synthesis methodologies, which would include stability, robustness and other performance requirements, do not exist (Passino and Yurkovich 1996). This peculiarity of fuzzy control may limit its applications and arouse relevant suspicion about its reliability from a theoretical viewpoint.

Model-free approaches to fuzzy controller synthesis need no explicit model of the process to be controlled. Fuzzy control rules are derived by extracting expertise from engineers and operators, by modelling the operator's control actions or by a learning mechanism embedded in the fuzzy control system (Lee 1990). In a certain sense, derivation and tuning of fuzzy control rules are performed as a trial-and-error procedure. Such a model-free approach has dominated the first generation of fuzzy controllers. One of its drawbacks is that often there is not enough information on process operation to generate fuzzy controllers; and even when a fuzzy controller can be obtained, stability, robustness and performance issues cannot be addressed with undoubted answers.

An alternative to model-free approaches is a model-based approach, which can extract a fuzzy control policy from a representative knowledge set of the process to be controlled. Typically, if an input–output description of a process to be controlled is available in the form of a fuzzy model, obtained by identification or other means, it should be reasonable and strongly attractive to generate fuzzy control rules based on this model. Here, the fuzzy controller consists of a set of fuzzy control rules each of which is derived from the corresponding rule of the process model. In this sense, this type of fuzzy controller is operating in a "region by region" fashion since each *IF-THEN* rule can be viewed as a "local" model valid in a particular operating region. Therefore, such a fuzzy controller provides a clearer linguistic interpretation. For example, in (Tanaka and Sano 1994) fuzzy controllers were constructed from sub-controllers, derived by solving Riccati equations with respect to the corresponding fuzzy *IF-THEN* rules of a fuzzy model of the controlled process; Wang proposed fuzzy model-based adaptive control algorithms, see (Wang 1993); Johansen considered fuzzy model-based controllers obtained via feedback linearisation, see (Johansen 1994).

A model-based approach obviously requires fuzzy models. Through a fuzzy modelling procedure, the system to be controlled can be represented by a set of linguistic statements, i.e. a set of *IF-THEN* rules. Thus, designing a fuzzy controller amounts to translating the linguistic description of the process into a set of linguistic statements for the control policy, considering design requirements such as stability, robustness and performance. Translation mechanisms from fuzzy models to fuzzy controllers form the central issue dealt with in this chapter. Thanks to the fact that linear Takagi–Sugeno fuzzy models (Takagi and Sugeno 1985) can be embedded in a general class of "discrete" polytopic linear differential inclusions, fuzzy controller synthesis based on such fuzzy models can be expressed in terms of linear matrix inequalities (Boyd *et al.* 1994), if quadratic Lyapunov functions are employed. Solving problems involving linear matrix inequalities results in fuzzy controllers which can meet the design requirements. Such a translation mechanism is systematic, tractable and reliable like conventional controller synthesis techniques. Virtually this synthesis technique combines the powerful description capability of fuzzy models with well-developed linear control theory and methods, in such a way that it provides a new approach to handling nonlinear control problems.

This chapter is organised as follows. Section 2 recalls some fundamentals of linear Takagi–Sugeno fuzzy models. Section 3 gives a sufficient stability condition for linear Takagi–Sugeno fuzzy models and its expression in terms of linear matrix inequalities. Section 4 is devoted to the design of stabilising fuzzy controllers. Section 5 discusses the synthesis of robust fuzzy controllers, with respect to premise or consequent uncertainties. Finally, section 6 ends this chapter with some conclusions and comments.

13.2 LINEAR TAKAGI–SUGENO FUZZY MODELS

13.2.1 Takagi–Sugeno fuzzy models for dynamic systems

A general single-input–single-output discrete-time dynamic system can be represented by the following model:

$$y(k + 1) = f(y(k), ..., y(k - n + 1); u(k), ..., u(k - m + 1)), \tag{13.1}$$

where u and y are the input and output variables of the system, k denotes the current time, and n, m are integers whose largest one determines the order of the model. A Takagi–Sugeno (TS) fuzzy model can be used to represent such a system, including a set of N *IF-THEN* rules such as

$$
\begin{aligned}
R^i \quad : \quad & IF \quad y(k) \; is \; \mathbf{A}_1^i \; and \; ... \; and \; y(k - n + 1) \; is \; \mathbf{A}_n^i \; and \\
& u(k) \; is \; \mathbf{B}_1^i \; and \; ... \; and \; u(k - m + 1) \; is \; \mathbf{B}_m^i \\
& THEN \quad y^i(k + 1) = a_1^i y(k) + ... + a_n^i y(k - n + 1) \\
& \qquad\qquad\qquad + b_1^i u(k) + ... + b_m^i u(k - m + 1), \tag{13.2}
\end{aligned}
$$

where $\mathbf{A}_j^i$ and $\mathbf{B}_j^i$ are fuzzy sets defined on the corresponding model premise variables; y^i is the output of the ith fuzzy rule; and a_j^i, b_j^i are coefficients of linear combinations of current and past values of the input and output variables. Note that such a fuzzy model does not involve constant terms p_0^i in its consequents, as opposed to the original form proposed in (Takagi and Sugeno 1985). To point out the difference, in the sequel, the above fuzzy model will be referred to as a Linear Takagi–Sugeno (LTS) fuzzy model. In the literature, as mentioned in Chapter 2, various names are used, for example, composite linear models, operating regime-based models, polytopic models, etc.

The output of a LTS fuzzy model can then be written as:

$$y(k + 1) = \sum_{i=1}^{N} \lambda_i(k) y^i(k + 1), \tag{13.3}$$

where $\lambda_i(k)$s are defined as fuzzy weights and calculated from the firing strengths $\omega_i(k)$ of various rules by

$$\lambda_i(k) = \frac{\omega_i(k)}{\sum_{j=1}^{N} \omega_j(k)}, \tag{13.4}$$

hence $\lambda_i(k) \in [0, 1]$ and $\sum_{i=1}^{N} \lambda_i(k) = 1$. It should be stressed that even if the rules in a LTS fuzzy model involve only linear combinations of the model inputs, the dynamic model (13.3) is truly nonlinear since the coefficients of the model depend nonlinearly on the system input and output via the fuzzy weights. The word "linear" simply means that the output is a linear combination of the model inputs. It will be seen later on that such a LTS fuzzy model is convenient for model-based design of fuzzy controllers.

13.2.2 Identification of linear Takagi–Sugeno fuzzy models

Linear Takagi–Sugeno fuzzy models of a given nonlinear system can be obtained by identification techniques. An integrated identification procedure based on fuzzy clustering and compatibility analysis of fuzzy clusters, hence able to integrate not only premise and consequent identification but also structure and parameter identification, has been proposed in (Zhao 1995). It is slightly different from similar techniques described in Chapter 2, first by the emphasis put on LTS fuzzy models, second by the use of distinct compatibility conditions. Consequently, this procedure is presented briefly and illustrated by an industrial application.

Description of an identification procedure

The proposed identification procedure for LTS fuzzy models includes the following steps (Zhao *et al.* 1994, Zhao 1995):

- fuzzy clustering of the input/output data samples in the input–output product space;
- performing compatibility analysis of clusters and merging compatible clusters;
- extracting fuzzy rules from fuzzy clustering results in order to achieve a fuzzy partition of the input product space and obtain consequents of fuzzy *IF-THEN* rules;
- simplifying and approximating fuzzy *IF-THEN* rules to obtain the final fuzzy model.

Fuzzy clustering has been introduced by Bezdek for pattern recognition purposes (Bezdek 1981). Linear consequents in TS fuzzy models ask for fuzzy clustering providing hyperplanar clusters. Various fuzzy clustering algorithms have been applied for identification of TS fuzzy models, e.g. (Yoshinari *et al.* 1993) using Fuzzy C-Elliptotype (FCE) algorithm, (Zhao *et al.* 1994) using Gustafson–Kessel algorithm, etc.; most likely the resulting hyperplanes will not contain the origin of the feature vector space. For identifying LTS fuzzy models, constraints have to be put on the prototypes used in fuzzy clustering, so that linear relations without constant terms can be obtained through extraction from the fuzzy clusters. Such a fuzzy clustering algorithm has been implemented by modifying the original Fuzzy C-Elliptotype (FCE) algorithm (Bezdek *et al.* 1981); it is called Modified FCE (MFCE) algorithm (Zhao 1995) and its convergence can be analysed using an approach similar to that in (Bezdek *et al.* 1981).

If there is no *a priori* information about the number of clusters, a compatibility analysis based on Compatible Cluster Merging (CCM) conditions can be used to detect geometrical features of clusters, in order to merge compatible clusters, hence reduce the number of clusters. This must be seen as one of the basic ingredients of the identification procedure and was introduced originally in (Zhao *et al.* 1994). With the MFCE algorithm, two CCM conditions must be met simultaneously to allow merging of clusters: first hyperplanes must be approximately parallel, second they must be located sufficiently close to each other.

The complete procedure is illustrated by the example of identifying a LTS fuzzy model for an industrial glass furnace.

Example – identification of a glass furnace

The dynamics of the temperature in a glass-melting furnace can be described by two input and one output variables: *crown temperature* $u(k)$, *glass flow pull* $p(k)$ and *bottom temperature* $y(k)$. Two data sets obtained over two different periods, with a sampling interval equal to two hours, were used for identification and model validation, see Figure 13.1.

Based on some physical interpretation, first-order models: $y(k + 1) = a \cdot y(k) + b \cdot u(k) + c \cdot p(k)$ were selected beforehand for the consequents of the fuzzy rules. Starting with an initial number of clusters $N_{max} = 5$, repetitive use of CCM conditions eventually leads to two clusters. Consequently, there are only two *IF-THEN* rules in the fuzzy model.

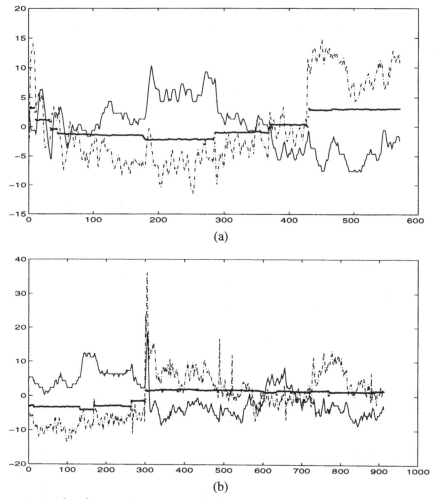

Figure 13.1 Identification data set (a) and validation data set (b) with sampling interval = 2 hours: *bottom temperature* (full lines), *crown temperature* (dashed lines) and *pull* (bold lines); plotted data are deviations of actual values from mean values.

Extracting fuzzy rules from the results of fuzzy clustering and smoothing the fuzzy sets, by fitting membership functions into trapezoidal shapes and by adjusting the parameters of *IF-THEN* rule consequents, result in the LTS fuzzy model shown in Figure 13.2.

The outputs predicted by the LTS fuzzy model for the identification data set and for the validation data set are shown in Figure 13.3. Agreement with the actual process outputs is very good.

13.2.3   Linear Takagi–Sugeno fuzzy models and PLDI models

Consider a LTS fuzzy model of a single-input–single-output system, with N *IF-THEN* rules such as (13.2) where it is supposed that all coefficients b_j^i, $j > 1$, are null. Defining a n-dimensional state vector $\mathbf{x}(k) = [y(k), y(k-1)..., y(k-n+1)]^T$, one can write down

Final membership functions for the variables $y(k)$, $u(k)$ and $p(k)$:

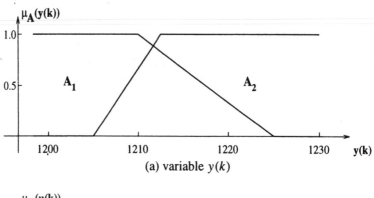

(a) variable $y(k)$

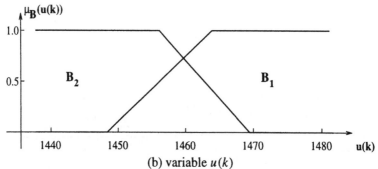

(b) variable $u(k)$

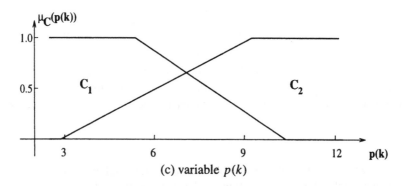

(c) variable $p(k)$

Final fuzzy rules:

R^1:   *IF $y(k)$ is $\mathbf{A}_1$ and $u(k)$ is $\mathbf{B}_1$ and $p(k)$ is $\mathbf{C}_1$*
       THEN $y^1(k+1) = 0.9961y(k) + 0.0045u(k) - 0.1975p(k)$;
R^2:   *IF $y(k)$ is $\mathbf{A}_2$ and $u(k)$ is $\mathbf{B}_2$ and $p(k)$ is $\mathbf{C}_2$*
       THEN $y^2(k+1) = 0.9048y(k) + 0.0818u(k) - 0.6541p(k)$.

Figure 13.2   LTS fuzzy model of an industrial glass furnace.

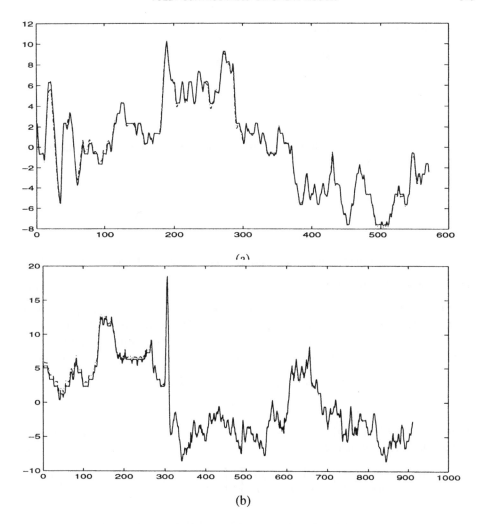

Figure 13.3 Outputs of the actual process (full lines) and one-step-ahead prediction of the fuzzy model (dashed lines) on the identification data set (a) and on the validation data set (b); plotted data are deviations from the mean bottom temperature. Sampling interval is two hours.

the consequent part of the i-th *IF-THEN* rule as (Tanaka and Sugeno 1992):

$$
\begin{aligned}
x_1^i(k+1) &= a_1^i x_1(k) + a_2^i x_2(k) + \ldots + a_n^i x_n(k) + b_1^i u(k) \\
x_2^i(k+1) &= x_1(k) \\
&\ldots \\
x_n^i(k+1) &= x_{n-1}(k),
\end{aligned}
$$

hence in the following state-space description:

$$
\mathbf{x}^i(k+1) = A_i \mathbf{x}(k) + B_i u(k), \tag{13.5}
$$

with matrices A_i and B_i in the controllability canonical form:[1]

$$A_i = \begin{pmatrix} a_1^i & a_2^i & \cdots & a_{n-1}^i & a_n^i \\ 1 & 0 & \cdots & 0 & 0 \\ \vdots & \vdots & \ddots & \vdots & \vdots \\ 0 & 0 & \cdots & 1 & 0 \end{pmatrix}, \qquad B_i = \begin{pmatrix} b_1^i \\ 0 \\ \vdots \\ 0 \end{pmatrix}. \qquad (13.6)$$

The above controllable pair describes a "partial" behaviour of a system, and the complete fuzzy model can be written as:

$$\begin{aligned} \mathbf{x}(k+1) &= \left[\sum_{i=1}^{N} \lambda_i(k) A_i \right] \mathbf{x}(k) + \left[\sum_{i=1}^{N} \lambda_i(k) B_i \right] u(k) \qquad (13.7) \\ y(k) &= (1, 0, ..., 0) \mathbf{x}(k), \end{aligned}$$

where $\lambda_i(k)$s are the fuzzy weights defined in (13.4). Since the latter take their values in the interval [0, 1], with $\sum_{i=1}^{N} \lambda_i(k) = 1$, the problem of designing a controller for a LTS fuzzy model in the form (13.7) can be embedded in the problem of designing a controller for the whole class of systems described by the discrete-time version of Polytopic Linear Differential Inclusions (PLDI) introduced in (Boyd et al. 1994). Typically a PLDI model representing the dynamics of a discrete-time linear time-varying system is in the form:

$$\begin{aligned} \mathbf{x}(k+1) &= A(k)\mathbf{x}(k) + B(k)\mathbf{u}(k) \qquad (13.8) \\ \mathbf{y}(k) &= C(k)\mathbf{x}(k) + D(k)\mathbf{u}(k), \end{aligned}$$

where the four system matrices may be time-varying and belong to the convex hull of a set of N given 4-tuples of matrices:

$$\begin{bmatrix} A(k) & B(k) \\ C(k) & D(k) \end{bmatrix} \in Co \left\{ \begin{bmatrix} A_1 & B_1 \\ C_1 & D_1 \end{bmatrix}, \begin{bmatrix} A_2 & B_2 \\ C_2 & D_2 \end{bmatrix}, ..., \begin{bmatrix} A_N & B_N \\ C_N & D_N \end{bmatrix} \right\}. \qquad (13.9)$$

Specifically, in the LTS fuzzy model (13.7), $A(k)$ and $B(k)$ are sparse matrices with many zero entries, $C(k) = (1, 0, ..., 0)$ and $D(k) = 0$. The design techniques presented in the sequel have been developed for such a model.

13.3 STABILITY OF LINEAR TS FUZZY MODELS

13.3.1 A sufficient stability condition

Assume that the LTS fuzzy model (13.7) may have zero input. For this free system, the state $\mathbf{x} = 0$ is an equilibrium state. It is asymptotically stable, in other words, all the trajectories of the free system converge to zero as $k \to \infty$, if there exists a quadratic Lyapunov function $V(\mathbf{x}) = \mathbf{x}^T P \mathbf{x}$, $P > 0$, which decreases along every nonzero trajectory of the fuzzy model. A free LTS fuzzy model is then said to be quadratically stable also. A sufficient stability condition has been given in (Tanaka and Sugeno 1992); it is recalled in the following theorem.

---

[1]A similar approach can be used with a general state-space representation, provided the premises of the fuzzy rules are expressed in terms of measurable variables including state variables, and controllability is ensured.

Theorem 1 *The equilibrium* $\mathbf{x} = 0$ *of a fuzzy free system:*

$$\mathbf{x}(k+1) = \sum_{i=1}^{N} \lambda_i(k) A_i \mathbf{x}(k) \tag{13.10}$$

is asymptotically stable if there exists a common positive definite matrix P such that

$$A_i^T P A_i - P < 0, \tag{13.11}$$

for $i = 1, 2, ..., N$.[2]

Proof: See (Tanaka and Sugeno 1992).

The key point is the fact that stability of all the submodels (or A_is) is proven by means of a common matrix P. If different P_is had to be used for proving stability of different A_is, there is no guarantee that the entire fuzzy model is stable; it may be unstable as shown by an example later. An equivalent form of the above theorem can be obtained through the transformation represented below.

Lemma 1 *If* $P = P^T > 0$ *and* $Q = P^{-1}$, *then*

$$P - A^T P A > 0 \iff \begin{pmatrix} P & PA \\ A^T P & P \end{pmatrix} > 0, \tag{13.12}$$

$$Q - AQA^T > 0 \iff \begin{pmatrix} Q & AQ \\ QA^T & Q \end{pmatrix} > 0. \tag{13.13}$$

Proof: See (Zhao *et al.* 1995*a*).

13.3.2 Linear matrix inequality

The matrix inequalities (13.12) and (13.13) turn out to be linear matrix inequalities. Typically, a Linear Matrix Inequality (LMI) is of the form:

$$F(z) = F_0 + \sum_{i=1}^{m} z_i F_i > 0, \tag{13.14}$$

where $z = (z_1, z_2, ..., z_m)$ is a m-tuple of variables or parameters, $F_i = F_i^T, i = 0, 1, ..., m$, are given $n \times n$ real symmetric matrices. Inequality (13.14) means that matrix $F(z)$ is positive definite. A set of multiple LMIs: $F^{(1)}(z) > 0$, $F^{(2)}(z) > 0$, ..., $F^{(N)}(z) > 0$, is equivalent to a single LMI: $diag\{F^{(1)}(z), F^{(2)}(z), ..., F^{(N)}(z)\} > 0$. In (Boyd *et al.* 1993, Boyd *et al.* 1994) it is shown that some problems involved in systems and control can reduce to LMI problems or to convex optimisation problems with constraints expressed as LMIs.

When a matrix variable, e.g. a positive definite $n \times n$ symmetric matrix P, is encountered in an inequality such as (13.12), denoting by $P_1, P_2, ..., P_m, m = n(n+1)/2$, the usual basis of elementary matrices[3] in the space of $n \times n$ symmetric matrices allows transformation of the matrix inequality into an LMI

$$\begin{pmatrix} P & PA \\ A^T P & P \end{pmatrix} = \sum_{i=1}^{m} z_i \begin{pmatrix} P_i & P_i A \\ A^T P_i & P_i \end{pmatrix} > 0, \tag{13.15}$$

---

[2]Strictly speaking, one should only consider fuzzy rules which are fired for $u(k) = 0$.

[3]Elementary matrices are matrices whose entries are zero, except one diagonal entry or two symmetric off-diagonal entries equal to 1.

where z_i are the entries of P. Therefore an inequality like (13.12) is an implicit LMI.

Some standard problems with respect to LMIs have been listed in (Boyd *et al.* 1994), for example, LMI Problems and Generalized EigenValue Problems transcribed as follows for the convenience of later expression.

LMI problems: Given a LMI: $F(z) > 0$, the corresponding LMI Problem (LMIP) is to find z^{feas} such that $F(z^{feas}) > 0$, or determine that the LMI is infeasible (has no solution).

Generalised eigenvalue problems: A Generalised EigenValue Problem (GEVP) is to minimise the maximum generalised eigenvalue of a pair of matrices which depend affinely on a variable z, subject to a LMI constraint, or to determine that the constraint is infeasible, i.e.

$$\underset{z}{\text{minimise}} \quad \lambda \tag{13.16}$$

$$\text{subject to} \quad \lambda B(z) - A(z) > 0, \quad B(z) > 0, \quad C(z) > 0,$$

where matrices $A(z)$, $B(z)$ and $C(z)$ are symmetric matrices depending affinely on the optimisation variable z.

The above standard problems with LMIs can be solved via well developed algorithms, by which one is able to determine whether or not the problem is feasible, and if it is, to obtain feasible points and the corresponding optimal objective values with a prespecified accuracy (Boyd *et al.* 1994). In this work, all computations involving LMIs are performed by means of the software *LMI-Lab* (Gahinet and Nemirovskii 1993).

13.4 MODEL-BASED SYNTHESIS OF STABLE FUZZY CONTROL SYSTEMS

13.4.1 Stabilising state-feedback controllers

For LTS fuzzy models (13.7), the use of LMI methods allows the design of various stabilising state-feedback controllers (Zhao *et al.* 1995a). Here two such controllers are considered. The first one is a conventional linear time-invariant state-feedback controller: $u(k) = K\mathbf{x}(k)$. The second one can be viewed as a linear time-varying state-feedback controller: $u(k) = K(k)\mathbf{x}(k)$, but as a matter of fact, it is a truly nonlinear controller as its gain matrix $K(k)$ is obtained from fuzzy implications. Terminologically, these two controllers will be referred to respectively as non-fuzzy and fuzzy controllers.

Non-fuzzy state-feedback controllers

Using a state-feedback controller $u(k) = K\mathbf{x}(k)$ in a LTS fuzzy model such as (13.7) leads to a feedback control loop described by:

$$\mathbf{x}(k + 1) = \sum_{i=1}^{N} \lambda_i(k)\{A_i + B_i K\}\mathbf{x}(k). \tag{13.17}$$

A sufficient condition for stability of the feedback loop is given by the following theorem.

Theorem 2 *The equilibrium* $\mathbf{x} = 0$ *of the fuzzy control system (13.17) is asymptotically stable if there exist a positive definite matrix Q and a matrix Y such that for all $i \in [1, N]$:*

$$\begin{pmatrix} Q & A_i Q + B_i Y \\ (A_i Q + B_i Y)^T & Q \end{pmatrix} > 0. \tag{13.18}$$

The state-feedback gain matrix is given by $K = Y Q^{-1}$.

Proof: According to Theorem 1, the fuzzy control system (13.17) is asymptotically stable if there is a positive definite matrix P and a matrix K such that

$$(A_i + B_i K)^T P (A_i + B_i K) - P < 0,$$

and then use of Lemma 1 yields

$$\begin{pmatrix} Q & (A_i + B_i K)Q \\ Q(A_i + B_i K)^T & Q \end{pmatrix} > 0,$$

where $Q = P^{-1}$. By denoting $Y = KQ$, one obtains (13.18).

$\square$

Solving the matrix inequality (13.18) with matrix variables Q and Y is a standard LMIP. Any feasible solution for this LMIP provides a linear time-invariant state-feedback controller, $u(k) = YQ^{-1}\mathbf{x}(k)$, stabilising the given fuzzy model.

Fuzzy state-feedback controllers

Based on the LTS fuzzy model of the system to be controlled, one can search for a stabilising fuzzy state-feedback controller, represented by a set of N *IF-THEN* rules having the same premises as the fuzzy model. Logical consistency then requires that the term $u(k)$ is not involved in premises of any fuzzy rule. This restriction will be removed further by introducing a delay in the control loop. Thus the fuzzy rules of the controller can be written as:

$$\begin{aligned} R_u^i \quad : \quad & IF \ y(k) \ is \ \mathbf{A}_1^i \ and \ ... \ and \ y(k - n + 1) \ is \ \mathbf{A}_n^i \ and \\ & u(k - 1) \ is \ \mathbf{B}_2^i \ and \ ... \ and \ u(k - m + 1) \ is \ \mathbf{B}_m^i \\ & THEN \ u^i(k) = K_i \mathbf{x}(k). \end{aligned} \tag{13.19}$$

Each of the *IF-THEN* rules can be viewed as describing a "local" state-feedback controller with the gain matrix K_i associated with the corresponding "local" submodel of the system to be controlled (they have the same premise). However, as a consequence of Theorem 1, such a "local" controller cannot be designed on the basis of the corresponding "local" submodel independently of the others.

The real output of the fuzzy controller is then computed by

$$u(k) = \left[\sum_{i=1}^N \lambda_i(k) K_i \right] \mathbf{x}(k) = K(k)\mathbf{x}(k), \tag{13.20}$$

where $\lambda_i(k)$s are the fuzzy weights obtained in the fuzzy model of the controlled system. According to their definition, $\lambda_i(k)$s involve nonlinearly the premise variables of the fuzzy models, so the controller (13.20) is indeed nonlinear. Such a controller is also referred to as a "gain scheduling controller" (Boyd *et al.* 1994). As a result, the fuzzy feedback control loop can be represented by:

$$\mathbf{x}(k + 1) = \sum_{i=1}^N \lambda_i(k) \left\{ A_i + B_i \sum_{j=1}^N \lambda_j(k) K_j \right\} \mathbf{x}(k), \tag{13.21}$$

which is obviously far more complicated than the corresponding description (13.17) in the previous case of a time-invariant controller. However, using a similar approach, a sufficient condition for the stability of the feedback loop can be expressed in the following theorem.

Theorem 3 *The equilibrium* $\mathbf{x} = 0$ *of the fuzzy control system (13.21) is asymptotically stable if there exist a positive definite matrix* Q *and a set of matrices* Y_i *such that for all* $i \in [1, N]$ *and* $j \in (i, N)$:

$$\begin{pmatrix} Q & A_i Q + B_i Y_i \\ (A_i Q + B_i Y_i)^T & Q \end{pmatrix} > 0, \tag{13.22}$$

$$\begin{pmatrix} Q & \dfrac{(A_i Q + B_i Y_j + A_j Q + B_j Y_i)}{2} \\ \dfrac{(A_i Q + B_i Y_j + A_j Q + B_j Y_i)^T}{2} & Q \end{pmatrix} > 0. \tag{13.23}$$

The fuzzy state-feedback gain matrices are given by $K_i = Y_i Q^{-1}$, $i = 1, 2, ..., N$.

Proof: Following the same lines as in (Tanaka and Sano 1994) and in the previous constant controller case, and denoting $Y_i = K_i Q$, $i = 1, 2, ..., N$, the sufficient stability conditions (13.22), (13.23), can be derived from Theorem 1.

$\square$

As in the previous case, solving LMIP (13.22) and (13.23) yields a common positive definite matrix $P = Q^{-1}$ and all the "local" state-feedback gain matrices $K_i = Y_i Q^{-1}$, $i = 1, 2, ..., N$, which guarantee the stability of the feedback loop associated with the fuzzy controller (13.20).

Remark 1: The dimensions of the matrix inequalities (13.23) depend on the number of rules. However, in some cases, for example, with triangular or trapezoidal membership functions, the fact that some rules can never be fired simultaneously due to their particular premise definitions leads to possible reduction in the dimension of the problem, by dropping matrix inequalities corresponding to "mutually independent" rules from the LMI constraints.

Fuzzy control systems with a reset action

The LMI approach presented above can be extended to the design of a servo control system including a fuzzy state-feedback controller and an integrator for achieving tracking and disturbance rejection with zero steady-state error. A block-diagram of this feedback loop is shown in Figure 13.4, where $\mathbf{x}(k)$, $y(k)$, $u(k)$ and $r(k)$ are respectively the state vector, output, control and reference input of the control system, the controlled plant being described in the form of a LTS fuzzy model (13.7).

Introducing an additional state variable $x_{n+1}(k)$, corresponding to accumulation of successive tracking errors,

$$\begin{aligned} x_{n+1}(k+1) &= x_{n+1}(k) + e(k) \\ &= x_{n+1}(k) + r(k) - C\mathbf{x}(k), \end{aligned} \tag{13.24}$$

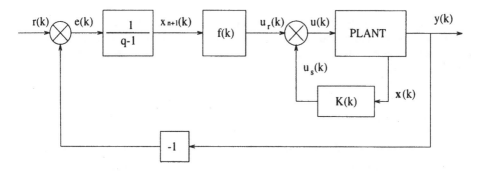

Figure 13.4 Block diagram of a fuzzy servo control system.

leads to the following augmented state-space description:

$$
\begin{aligned}
\tilde{\mathbf{x}}(k+1) &= \begin{pmatrix} \mathbf{x}(k+1) \\ x_{n+1}(k+1) \end{pmatrix} \\
&= \sum_{i=1}^{N} \lambda_i(k) \{ \begin{pmatrix} A_i & \mathbf{0} \\ -C & 1 \end{pmatrix} \begin{pmatrix} \mathbf{x}(k) \\ x_{n+1}(k) \end{pmatrix} + \begin{pmatrix} B_i \\ 0 \end{pmatrix} u(k) + \begin{pmatrix} \mathbf{0} \\ 1 \end{pmatrix} r(k) \}.
\end{aligned}
$$
(13.25)

Now the controller is composed of two parts: the first one is a fuzzy state-feedback controller based on the state variables of the controlled system:

$$
u_s(k) = \left[\sum_{i=1}^{N} \lambda_i(k) K_i \right] \mathbf{x}(k);
$$
(13.26)

the other is an incremental output feedback fuzzy controller, based on the additional state:

$$
u_r(k) = \left[\sum_{i=1}^{N} \lambda_i(k) f_i \right] x_{n+1}(k),
$$
(13.27)

where f_is are additional gains. The output of the augmented state-feedback controller is then given as

$$
\begin{aligned}
u(k) &= u_s(k) + u_r(k) \\
&= \left[\sum_{i=1}^{N} \lambda_i(k)(K_i, f_i) \right] \begin{pmatrix} \mathbf{x}(k) \\ x_{n+1}(k) \end{pmatrix},
\end{aligned}
$$
(13.28)

with augmented state-feedback gain matrices: $\tilde{K}_i = (K_i, f_i)$.

Defining the augmented system matrices:

$$
\tilde{A}(k) = \sum_{i=1}^{N} \lambda_i(k) \begin{pmatrix} A_i & \mathbf{0} \\ -C & 1 \end{pmatrix} = \sum_{i=1}^{N} \lambda_i(k) \tilde{A}_i,
$$
(13.29)

$$
\tilde{B}(k) = \sum_{i=1}^{N} \lambda_i(k) \begin{pmatrix} B_i \\ 0 \end{pmatrix} = \sum_{i=1}^{N} \lambda_i(k) \tilde{B}_i,
$$
(13.30)

transforms the design problem for a control system with reset action into a standard problem of seeking an augmented state-feedback controller (13.28). This is quite similar to the previous case with fuzzy state-feedback controllers, except that now the augmented state matrices (13.29) and (13.30) are used instead of the matrices of the original controlled system.

13.4.2   Example

In order to illustrate the effectiveness of this design technique for fuzzy control systems, take the second-order system studied in (Tanaka and Sugeno 1992); it is described by a LTS fuzzy model including two *IF-THEN* rules:

$$R^1 \quad : \quad IF \ \ y(k-1) \ is \ \mathbf{A}_1$$
$$THEN \ \ \mathbf{x}^1(k+1) = A_1\mathbf{x}(k) + B_1u(k);$$
$$R^2 \quad : \quad IF \ \ y(k-1) \ is \ \mathbf{A}_2$$
$$THEN \ \ \mathbf{x}^2(k+1) = A_2\mathbf{x}(k) + B_2u(k);$$

where state $\mathbf{x}(k) = [y(k), y(k-1)]^T$, the membership functions of the fuzzy sets $\mathbf{A}_1$ and $\mathbf{A}_2$ are defined in Figure 13.5, and the system matrices are

$$A_1 = \begin{pmatrix} 1.0 & -0.5 \\ 1.0 & 0 \end{pmatrix}, \qquad B_1 = \begin{pmatrix} 0.2 \\ 0 \end{pmatrix},$$

$$A_2 = \begin{pmatrix} -1.0 & -0.5 \\ 1.0 & 0 \end{pmatrix}, \qquad B_2 = \begin{pmatrix} 0.4 \\ 0 \end{pmatrix}.$$

Although the two submodels are stable, the fuzzy model with zero input is proven to be unstable (Tanaka and Sugeno 1992).

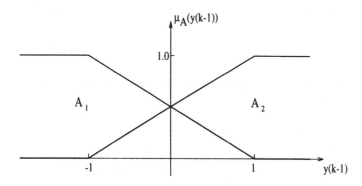

Figure 13.5   Membership functions of the fuzzy sets $\mathbf{A}_1$ and $\mathbf{A}_2$.

Solving the associated LMIP (13.18) yields feasible solutions for constant state-feedback controllers. For example, depending on the choice of different parameter settings used in the *LMI-lab* software, various controllers can be obtained, see Table 13.1.

Here again, solving the associated LMIP (13.22) and (13.23) with different parameter settings leads to different feasible solutions for fuzzy controllers, see Table 13.2.

Table 13.1   Constant controllers.

	Controller I	Controller II	Controller III
K	$\begin{pmatrix} 0.8479 & 1.5567 \end{pmatrix}$	$\begin{pmatrix} 0.7011 & 1.6279 \end{pmatrix}$	$\begin{pmatrix} 0.6241 & 1.6582 \end{pmatrix}$

Table 13.2   Fuzzy controllers.

	Controller I	Controller II	Controller III
K_1	$\begin{pmatrix} -1.5434 & 0.8042 \end{pmatrix}$	$\begin{pmatrix} -1.8515 & 0.6534 \end{pmatrix}$	$\begin{pmatrix} -3.0702 & 2.2017 \end{pmatrix}$
K_2	$\begin{pmatrix} 1.5885 & 0.6397 \end{pmatrix}$	$\begin{pmatrix} 2.0893 & 0.3586 \end{pmatrix}$	$\begin{pmatrix} 2.8616 & 1.2015 \end{pmatrix}$

Lastly, based on the same fuzzy model of the controlled system, a fuzzy state-feedback controller with additional reset action has been designed. A feasible solution of the LMIP associated with the augmented system leads to the state-feedback gains K_i and the gains f_i:

$$\tilde{K}_1 = (K_1, f_1) = \begin{pmatrix} -4.1302 & 1.1914 & 0.7249 \end{pmatrix},$$
$$\tilde{K}_2 = (K_2, f_2) = \begin{pmatrix} 2.5337 & 0.6548 & 0.3581 \end{pmatrix}.$$

Figures 13.6 and 13.7 show the responses of the fuzzy control system for two sets of initial conditions: $[0.9, -0.7]$ and $[-6, -4]$. It can be seen that fuzzy controllers give performances slightly better than linear controllers in the first case but dramatically better in the second case. This may be due to the fact that, in each case, we have selected controllers within a complete family of stabilising controllers and the selected linear controllers were probably not the best ones. Nevertheless, this is again an illustration of the instability of the system to be controlled, in spite of the stability of the two submodels. The selected linear controllers are able to stabilise both submodels considered individually, but they have difficulties to cope with the instability of the global model; on the contrary, fuzzy controllers seem to be more appropriate to take it into account. Figure 13.8 displays the tracking responses of the fuzzy control system with reset action, first for steps in the reference variable, second for a ramp signal. In the first case the steady-state error is zero thanks to the reset action; in the second case there is a "velocity error", the value of which can change slightly according to the operating region of the nonlinear system.

13.4.3   Additional requirements in the synthesis of fuzzy control systems

Performance issues in fuzzy control systems

In the synthesis of control systems, meeting some design requirements and achieving desired performances should be considered in addition to stability. If fuzzy control systems are analysed or synthesised on the basis of quadratic Lyapunov functions, one can represent certain performance specifications and design requirements in the form of LMIs. Here it is shown that solving the optimisation problems related to these LMIs leads to a state-feedback controller such that the resulting fuzzy control system has the desired performance or satisfies the design requirements.

Due to space limitation, only the convergence rate to equilibrium is considered as an example of performance specifications, and only fuzzy state-feedback controllers will be dealt

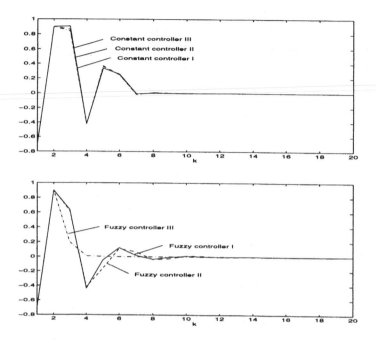

Figure 13.6   Responses of fuzzy control systems for initial conditions $[0.9, -0.7]$.

with. Similar but simpler results can be obtained for constant state-feedback controllers. In (Zhao 1995), one can find results for other performance specifications or design requirements, such as rejection of external disturbances, limitations on the control variables, etc.

A discrete-time system with state vector $\mathbf{x}(k)$ is said to be *globally exponentially stable* if there exist positive constants α and $0 < \rho < 1$ such that

$$\| \mathbf{x}(k) \| \le \alpha \rho^k \| \mathbf{x}(0) \|, \tag{13.31}$$

for all initial states $\mathbf{x}(0)$. The number ρ characterises the rate of convergence of the system states to the equilibrium $\mathbf{x} = 0$, smaller ρ providing faster convergence. Thus, ρ will be called *decay rate*.

If a quadratic Lyapunov function $V(\mathbf{x}(k)) = \mathbf{x}^T(k)P\mathbf{x}(k)$, $P > 0$, satisfies the inequality:

$$\Delta V(\mathbf{x}(k)) = V(\mathbf{x}(k+1)) - V(\mathbf{x}(k)) \le (\rho^2 - 1)V(\mathbf{x}(k)), \tag{13.32}$$

instead of the simple stability requirement: $\Delta V(\mathbf{x}(k)) < 0$, one can write:

$$\begin{aligned} V(\mathbf{x}(k+1)) &\le \rho^2 V(\mathbf{x}(k)) & (13.33) \\ &\le \rho^{2(k+1)} V(\mathbf{x}(0)), & (13.34) \end{aligned}$$

i.e.

$$\mathbf{x}^T(k+1)P\mathbf{x}(k+1) \le \rho^{2(k+1)}\mathbf{x}^T(0)P\mathbf{x}(0). \tag{13.35}$$

Therefore, as a consequence of standard inequalities of linear algebra, the condition (13.35) leads to

$$\| \mathbf{x}(k) \| \le \rho^k \kappa(P)^{1/2} \| \mathbf{x}(0) \|, \tag{13.36}$$

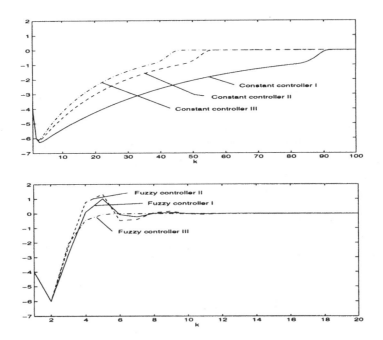

Figure 13.7 Responses of fuzzy control systems for initial conditions $[-6, -4]$.

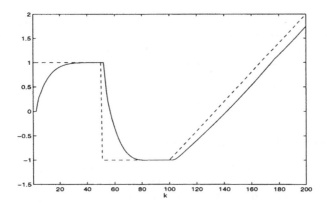

Figure 13.8 Responses (full line) of the fuzzy servo control system for different reference inputs (dashed line).

where $\| \, . \, \|$ is the Euclidean norm, $\kappa(P)$ is the ratio of the maximum and minimum eigenvalues of the positive definite matrix P. Hence the condition (13.31) holds with $\alpha = \kappa(P)$.

The above considerations allow the designer of fuzzy control systems to introduce a performance specification in the form of an optimal decay rate. For a fuzzy controller with the feedback gain matrix $K(k) = \sum_{j=1}^{N} \lambda_j(k) K_j$, such a design requirement can be described as a matrix inequality derived from (13.32):

$$
\begin{pmatrix}
P & P\{ \sum_{i=1}^{N} \lambda_i(k)[A_i + B_i K(k)]\} \\
\{ \sum_{i=1}^{N} \lambda_i(k)[A_i + B_i K(k)]\}^T P & \rho^2 P
\end{pmatrix} \geq 0.
\tag{13.37}
$$

Now, with the same notation as in section 12.4.1, and setting: $\tau = \rho^2$, the synthesis problem reduces to the solution of a Generalised EigenValue Problem (GEVP), hence the following theorem.

Theorem 4 *There is a fuzzy state-feedback controller optimising the decay rate of the fuzzy control system (13.21) if the optimisation problem:*

$$
\underset{Q, Y_1, Y_2, \ldots, Y_N}{minimise} \quad \tau
$$

$$
subject\ to \quad
\begin{pmatrix}
Q & A_i Q + B_i Y_i \\
(A_i Q + B_i Y_i)^T & \tau Q
\end{pmatrix} \geq 0,
\tag{13.38}
$$

$$
\begin{pmatrix}
Q & \frac{(A_i Q + B_i Y_j + A_j Q + B_j Y_i)}{2} \\
\frac{(A_i Q + B_i Y_j + A_j Q + B_j Y_i)^T}{2} & \tau Q
\end{pmatrix} \geq 0,
\tag{13.39}
$$

$$
i \in [1, N], \quad j \in (i, N].
$$

has a solution for a positive definite matrix Q and matrices Y_i. The fuzzy state-feedback gain matrices are then given by $K_i = Y_i Q^{-1}$, $i = 1, 2\ldots, N$, and the optimal decay rate by $\rho = \tau^{1/2}$.

Proof: As for the proof of Theorem 3, there exists a fuzzy state-feedback controller, which leads to a stable fuzzy control system with a decay rate ρ, if the LMIs (13.38) and (13.39) have feasible solutions. In order to obtain an optimal decay rate, the optimisation problem: minimise τ subject to LMIs (13.38) and (13.39), has to be solved. It is easily proven that such an optimisation problem reduces to a GEVP.

□

Logical consistency of fuzzy control systems

Previously, one has considered constant or fuzzy state-feedback controllers for LTS fuzzy models which have no term $u(k)$ in premises of the *IF-THEN* rules. This was motivated by the desire to avoid a difficulty resulting from logical inconsistency. If a LTS fuzzy model with $u(k)$ in premises of some rules has to be used in the synthesis of a fuzzy control system, the preceding approach can still be used for finding a fuzzy state-feedback controller, but such a fuzzy controller could not be implemented since the fuzzy controller output $u(k)$ should be present in the premises of rules in the model and also in the controller. Specifically, the determination of the fuzzy controller output $u(k)$ is dependent on fuzzy weights $\lambda_i(k)$, which in turn are determined by the value of $u(k)$ in the premises of rules. In other words, $u(k)$ is determined by itself. This kind of dependence between the premises and consequents is contradictory in terms of logic. Such a phenomenon is referred to as *logical inconsistency*. To allow a fuzzy control system to run causally, logical consistency should be guaranteed, in such a way that fuzzy inference in fuzzy controllers can be performed effectively.

To handle the problem of logical consistency in the synthesis of a fuzzy control system, the controller should put dependences of $u(k)$ and $\lambda_i(k)$s in the right order to overcome confusions in the computation of $u(k)$. The simplest solution would consist in delaying the control action or using incremental control, i.e., deliberately computing the control variable $u(k + 1)$ which should be applied at the next sampling time. In both cases, the system to be controlled is augmented by an additional state variable, which is the actual control variable: $x_{n+1}(k) = u(k)$. The new control variable to be computed by the controller is either $u(k+1)$ or the increment $\Delta u(k + 1) = u(k + 1) - u(k)$, which is not in the premises of the fuzzy rules. Therefore, this satisfies the requirement of logical consistency and the LMI method proposed in this chapter can be used. However, as shown in (Zhao *et al.* 1996), the existence of a fuzzy state-feedback controller for this augmented system is determined by that of a constant state-feedback controller for the original system. Nevertheless, it should be noted that if one cannot find such a constant state-feedback controller through the approach proposed here, it does not exclude the possibility of finding a fuzzy state-feedback controller by another way. This is an open challenge.

13.5 DESIGN OF ROBUST FUZZY CONTROL SYSTEMS

Since LTS fuzzy models are usually used to describe complex nonlinear systems, there may be uncertainties resulting not only from the identification procedure (modelling uncertainty), but also from inherent uncertainties in the real system (system uncertainty). Consequently, we have to consider a class of models including the nominal model which has been constructed or identified and perturbed models differing from the nominal one by some type of perturbations. Obviously, when designing a stabilising controller, robust stability should be ensured for the nominal model, in order to cope with all uncertainties (Zhao *et al.* 1995*b*).

13.5.1 Uncertainties on LTS fuzzy models

Assume that uncertainties are included in *IF-THEN* rules of the fuzzy model. Specifically, fuzzy models may contain *premise uncertainties*, namely uncertainties derived from membership functions describing the fuzzy partitions of the universe of discourse, and *consequent uncertainties*, which are in the system matrices A_i and B_i. In fact, this is consistent with the representation of reasoning knowledge of human beings, in which usually there exists a gap between a real phenomenon and the antecedents and conclusions in the linguistic description of this phenomenon by human beings.

Uncertainties on premises of LTS fuzzy models

The existence of uncertainties in premises of *IF-THEN* rules of LTS fuzzy models is a distinct problem in comparison with conventional models. The premise uncertainties stem from the uncertainties of membership functions. This leads to uncertainties on the firing strengths of *IF-THEN* rules. With respect to the premise uncertainties, in (Tanaka and Sano 1994) one has considered the robust stabilisation problem of fuzzy control systems for simple models.

The equivalent model (13.7) of the system to be controlled and the resulting fuzzy controller (13.20) make explicit use of fuzzy weights not of firing strengths. Therefore, analysing the influence of premise uncertainties on fuzzy control systems must be tackled through fuzzy weights rather than through firing strengths. Denoting by $\lambda_i(k)$ the nominal value and by $\Delta\lambda_i(k)$ the uncertainty on the ith fuzzy weight, we have

$$0 \le [\lambda_i(k) + \Delta\lambda_i(k)] \le 1, \tag{13.40}$$

$$\sum_{i=1}^{N}[\lambda_i(k) + \Delta\lambda_i(k)] = 1, \tag{13.41}$$

as straightforward consequences of the definition of fuzzy weights.

In fact, the inequality (13.40) or

$$-\lambda_i(k) \le \Delta\lambda_i(k) \le 1 - \lambda_i(k) \tag{13.42}$$

holds for arbitrary changes in the membership functions. Therefore, if a controller is found with respect to such constraints on premise uncertainties, it will be an "absolutely" robust controller. That is to say, the resulting fuzzy control system can bear arbitrary uncertainties from the premises of *IF-THEN* rules. In practice, it is more meaningful to study a non-trivial case where the weight uncertainty $\Delta\lambda_i(k)$ is described by

$$-\gamma_i^{pre}\lambda_i(k) \le \Delta\lambda_i(k) \le \gamma_i^{pre}(1 - \lambda_i(k)), \tag{13.43}$$

where a parameter $0 \le \gamma_i^{pre} \le 1$ measures the size of the admissible range of the uncertainty $\Delta\lambda_i(k)$, $\gamma_i^{pre} = 1$ corresponding to all uncertainties compatible with (13.40). For example, assuming a trapezoidal shape of the fuzzy weight $\lambda_i(k)$ and a given γ_i^{pre}, an admissible region for $\Delta\lambda_i(k)$ is shown in Figure 13.9.

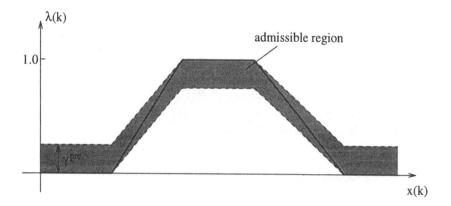

Figure 13.9   Admissible region of the uncertainty on a fuzzy weight.

Even if $\Delta\lambda_i(k)$ corresponds functionally to uncertainties on firing strengths, any unknown uncertainty of $\Delta\lambda_i(k)$ at kth sampling time can be described as a convex combination of the left- and right-hand sides of (13.43):

$$\begin{aligned}
\Delta\lambda_i(k) &= \gamma_i^{pre}[\alpha_i(k)(1 - \lambda_i(k)) + (1 - \alpha_i(k))(-\lambda_i(k))] \\
&= \gamma_i^{pre}(\alpha_i(k) - \lambda_i(k)),
\end{aligned} \tag{13.44}$$

where $0 \leq \alpha_i(k) \leq 1$. Thus, the uncertainties on premises of the fuzzy model can be reduced to that on the matrix expressions of the fuzzy model:

$$
\begin{aligned}
\mathbf{x}(k+1) &= \sum_{i=1}^{N}(\lambda_i(k) + \Delta\lambda_i(k))\{A_i\mathbf{x}(k) + B_i u(k)\} \\
&\triangleq \sum_{i=1}^{N}\lambda_i(k)\{[A_i + \Delta A_i^{pre}(k)]\mathbf{x}(k) + [B_i + \Delta B_i^{pre}(k)]u(k)\}, \quad (13.45)
\end{aligned}
$$

where assuming a common value γ^{pre} for all the rules, matrix uncertainties can be expressed as:

$$
\Delta A_i^{pre}(k) = -\gamma^{pre}A_i + \sum_{j=1}^{N}\alpha_j(k)\gamma^{pre}A_j = \gamma^{pre}\sum_{j=1}^{N}\alpha_j(k)(-A_i + A_j), \quad (13.46)
$$

$$
\Delta B_i^{pre}(k) = -\gamma^{pre}B_i + \sum_{j=1}^{N}\alpha_j(k)\gamma^{pre}B_j = \gamma^{pre}\sum_{j=1}^{N}\alpha_j(k)(-B_i + B_j). \quad (13.47)
$$

Because of the equality $\sum_{j=1}^{N}\alpha_j(k) = 1$ due to the constraint (13.41), the above unknown matrices $\Delta A_i^{pre}(k)$ and $\Delta B_i^{pre}(k)$ are bounded in a polytope:

$$
\begin{aligned}
[\Delta A_i^{pre}(k), \Delta B_i^{pre}(k)] &\in \gamma^{pre}Co\{[\Delta A_{i1}^{pre}, \Delta B_{i1}^{pre}], ..., [\Delta A_{iN}^{pre}, \Delta B_{iN}^{pre}]\} \quad (13.48) \\
&\triangleq \gamma^{pre}\Omega_i^{pre},
\end{aligned}
$$

where the vertices $[\Delta A_{il}^{pre}, \Delta B_{il}^{pre}], l = 1, 2, ..., N$, of polytope Ω_i^{pre} are defined as

$$
\Delta A_{il}^{pre} = -A_i + A_l, \qquad \Delta B_{il}^{pre} = -B_i + B_l. \quad (13.49)
$$

Uncertainties on consequents of LTS fuzzy models

Other uncertainties resulting from the modelling procedure and the true system may appear in the consequent parts of *IF-THEN* rules of a LTS fuzzy model, so they will be called consequent uncertainties. A LTS fuzzy model with consequent uncertainties can be written as

$$
\mathbf{x}(k+1) = \sum_{i=1}^{N}\lambda_i(k)\{[A_i + \Delta A_i^{con}(k)]\mathbf{x}(k) + [B_i + \Delta B_i^{con}(k)]u(k)\}, \quad (13.50)
$$

where $\Delta A_i^{con}(k)$ and $\Delta B_i^{con}(k)$ are the uncertainties on the consequent of the ith *IF-THEN* rule. Such consequent uncertainties imply that all the consequents of *IF-THEN* rules in a LTS fuzzy model can suffer simultaneously from uncertainties independent of each other.

In the current study, it will be assumed that the uncertainties of matrices $\Delta A_i^{con}(k)$ and $\Delta B_i^{con}(k)$ are bounded in a polytope with L vertices $[\gamma^{con}\Delta A_{il}^{con}, \gamma^{con}\Delta B_{il}^{con}], l = 1, 2, ..., L, \gamma^{con} \geq 0$. It should be noted that, because of the special form of matrices A_i and B_i in (13.6) where only entries $(a_1^i, ..., a_n^i, b_1^i)$ may be uncertain, the description of uncertainties on matrices A_i and B_i may be simplified, i.e., the number of vertices can be reduced drastically.

In the same way as for the the parameter γ^{pre} in the previous subsection, a parameter γ^{con} can be used to characterise the size of consequent uncertainties. Hence, one has the following expressions for the consequent uncertainties:

$$[\Delta A_i^{con}(k), \Delta B_i^{con}(k)] \quad \in \quad \gamma^{con} Co\{[\Delta A_{i1}^{con}, \Delta B_{i1}^{con}], ..., [\Delta A_{iL}^{con}, \Delta B_{iL}^{con}]\} \quad (13.51)$$
$$\stackrel{\Delta}{=} \quad \gamma^{con} \Omega_i^{con},$$

in other words,

$$[\Delta A_i^{con}(k), \Delta B_i^{con}(k)] \in \gamma^{con} \sum_{l=1}^{L} \alpha_{il}(k)[\Delta A_{il}^{con}, \Delta B_{il}^{con}], \quad (13.52)$$

where $0 \leq \alpha_{il}(k) \leq 1$ and $\sum_{l=1}^{L} \alpha_{il}(k) = 1$.

13.5.2   Robust state-feedback controllers

So far, all the uncertainties appearing in LTS fuzzy models have been expressed in the form of polytopic uncertainties on system matrices A_i and B_i. The parameter γ^{pre} (or γ^{con}) may be referred to as a *stability margin* for premise uncertainties (or for consequent uncertainties). Now, robust state-feedback controllers based on the nominal LTS fuzzy model will be sought in such a way that the resulting fuzzy control system can be given maximum stability margin either for the premise uncertainties or for the consequent uncertainties. In view of the similar polytopic expressions of premise uncertainties and of consequent uncertainties, for convenience, we will omit the superscript "*pre*" or "*con*" in the following description. The obtained solutions can be directly applied in the two different cases.

Non-fuzzy state-feedback controllers

Now, a conventional linear time-invariant (or non-fuzzy) state-feedback controller: $u(k) = K\mathbf{x}(k)$, is searched for robustly stabilising a nominal LTS fuzzy model, stability being guaranteed for all possible systems resulting from the nominal model and the uncertainties. The feedback control loop can be represented as:

$$\mathbf{x}(k+1) = \sum_{i=1}^{N} \lambda_i(k)\{A_i(k) + B_i(k)K\}\mathbf{x}(k), \quad (13.53)$$

where $A_i(k) \stackrel{\Delta}{=} A_i + \Delta A_i(k)$ and $B_i(k) \stackrel{\Delta}{=} B_i + \Delta B_i(k)$. From Theorem 2, the fuzzy control system (13.53) is robustly stable if there exist a common positive definite matrix Q and a matrix Y such that

$$\begin{pmatrix} Q & A_i(k)Q + B_i(k)Y \\ (A_i(k)Q + B_i(k)Y)^T & Q \end{pmatrix} > 0, \quad (13.54)$$

for all $[\Delta A_i(k), \Delta B_i(k)] \in \gamma^{pre}\Omega_i^{pre}$ (or $\in \gamma^{con}\Omega_i^{con}$), $i = 1, 2, ..., N$.

For the fuzzy control system with polytopic uncertainties (13.48) or (13.51), letting $\eta = 1/\gamma$ and substituting the expression of $[\Delta A_i(k), \Delta B_i(k)]$ in the above matrix inequality yield

$$\sum_{l=1}^{L} \alpha_{il}(k)[\eta V_{il}(Q, Y) + U_{il}(Q, Y)] > 0, \quad (13.55)$$

where

$$U_{il}(Q, Y) \triangleq \begin{pmatrix} 0 & \Delta A_{il}Q + \Delta B_{il}Y \\ (\Delta A_{il}Q + \Delta B_{il}Y)^T & 0 \end{pmatrix},$$

$$V_{il}(Q, Y) \triangleq V_{i0}(Q, Y) = \begin{pmatrix} Q & A_iQ + B_iY \\ (A_iQ + B_iY)^T & Q \end{pmatrix}.$$

Thus, a robust constant controller design problem can be reduced to the following GEVP.

Theorem 5 *The equilibrium of the fuzzy control system (13.53) is asymptotically stable if the optimisation problem:*

$$\begin{aligned} \underset{Q.Y}{minimise} \quad & \eta \\ subject\ to \quad & \eta V_{il}(Q, Y) + U_{il}(Q, Y) > 0, \\ & V_{i0}(Q, Y) > 0, \\ & i \in [1, N], \quad l \in [1, L], \end{aligned} \tag{13.56}$$

has a solution for a positive definite matrix Q and a matrix Y. As a result, the constant state feedback gain matrix is $K = YQ^{-1}$ and the corresponding fuzzy control system is provided with a stability margin $\gamma = \dfrac{1}{\eta}$ with respect to the polytopic uncertainties (13.48) or (13.51).

Remark 2: The LMI constraint conditions $V_{i0}(Q, Y) > 0$, $i = 1, 2, ..., N$, imply that the nominal fuzzy control system is Lyapunov quadratically stable, according to Theorem 1. Obviously, since robust stability is deduced on the basis of quadratic Lyapunov functions, the obtained γ is a quadratic stability margin.

Remark 3: For the premise uncertainties (13.48), it is easy to conclude that, if a constant state-feedback controller satisfying (13.54) exists, the resulting fuzzy control system has a stability margin of $\gamma^{pre} = 1$. However, such a stability margin cannot always be obtained as the constant controller problem may be infeasible.

Fuzzy state-feedback controllers

The fuzzy control system with the fuzzy controller (13.20) can be written as:

$$\mathbf{x}(k+1) = \sum_{i=1}^{N} \lambda_i(k)\{A_i(k) + B_i(k)\sum_{j=1}^{N} \lambda_j(k)K_j\}\mathbf{x}(k). \tag{13.57}$$

According to Theorem 3, the fuzzy control system (13.57) is robustly stable if there exist a common positive definite matrix Q and a set of matrices Y_i, $i = 1, 2, ..., N$, such that

$$\begin{pmatrix} Q & A_i(k)Q + B_i(k)Y_i \\ (A_i(k)Q + B_i(k)Y_i)^T & Q \end{pmatrix} > 0, \tag{13.58}$$

$$\begin{pmatrix} Q & \frac{(A_i(k)Q + B_i(k)Y_j + A_j(k)Q + B_j(k)Y_i)}{2} \\ \frac{(A_i(k)Q + B_i(k)Y_j + A_j(k)Q + B_j(k)Y_i)^T}{2} & Q \end{pmatrix} > 0, \tag{13.59}$$

for all the $[\Delta A_i(k), \Delta B_i(k)] \in \gamma^{pre}\Omega_i^{pre}$ (or $\in \gamma^{con}\Omega_i^{con}$), $i \in [1, N]$ and $j \in (i, N]$.

For convenience, defining

$$U_{il}(Q, Y_j) \triangleq \begin{pmatrix} 0 & \Delta A_{il}Q + \Delta B_{il}Y_j \\ (\Delta A_{il}Q + \Delta B_{il}Y_j)^T & 0 \end{pmatrix},$$

$$V_{il}(Q, Y_j) \triangleq V_{i0}(Q, Y_j) = \begin{pmatrix} Q & A_i Q + B_i Y_j \\ (A_i Q + B_i Y_j)^T & Q \end{pmatrix}.$$

As in the preceding subsection, a robust fuzzy state-feedback controller can be derived by solving the following GEVP.

Theorem 6 *The equilibrium* $\mathbf{x} = 0$ *of the fuzzy control system (13.57) is asymptotically stable if the optimisation problem:*

$$\begin{aligned}
&\underset{Q, Y_1, \ldots, Y_N}{\text{minimise}} && \eta \\
&\text{subject to} && \eta V_{il}(Q, Y_i) + U_{il}(Q, Y_i) > 0, \\
&&& \eta \left[\frac{V_{il_1}(Q, Y_j) + V_{jl_2}(Q, Y_i)}{2} \right] + \frac{U_{il_1}(Q, Y_j) + U_{jl_2}(Q, Y_i)}{2} > 0, \quad (13.60) \\
&&& V_{i0}(Q, Y_i) > 0, \\
&&& \frac{V_{i0}(Q, Y_j) + V_{j0}(Q, Y_i)}{2} > 0, \\
&&& i \in [1, N], \quad j \in (i, N] \quad and \quad l, l_1, l_2 \in [1, L],
\end{aligned}$$

has a solution for a positive definite matrix Q *and a set of matrices* Y_i. *As a result, the fuzzy state feedback controller (13.20) with the gain matrices* $K_i = Y_i Q^{-1}$, $i = 1, 2, \ldots, N$, *is robustly stable and the corresponding fuzzy control system has a stability margin* $\gamma = \dfrac{1}{\eta}$ *with respect either to the premise uncertainties (13.48) or to the consequent uncertainties (13.51).*

Remark 4: As stated previously, the stability requirement for the nominal fuzzy control systems is indicated in the optimisation procedures by the LMI constraints $V_{i0}(Q, Y_i) > 0$ and $\dfrac{V_{i0}(Q, Y_j) + V_{j0}(Q, Y_i)}{2} > 0$, $i \in [1, N]$ and $j \in (i, N]$.

Remark 5: The dimensions of the matrix inequalities involved in the optimisation problems (13.60) depend on the number of vertices of polytopes representing the uncertainties, as well as on the number of rules, hence they may be very large. However, they may be decreased for the same reason as in Remark 1.

13.5.3   Examples

The methods presented here are used to design robust constant and fuzzy state-feedback controllers for LTS fuzzy models with premise or consequent uncertainties. Results show that the latter have better performances, particularly in the case of consequent uncertainties.

Example 1

Consider the same second-order system as in Section 13.4.2. A constant or fuzzy state-feedback controller has to be designed to stabilise the nominal LTS fuzzy model and tolerate maximum uncertainties in the consequents of the fuzzy model. It is assumed that the consequent uncertainties of the fuzzy model are bounded inside two polytopes with eight vertices, constructed from the corresponding nominal system matrices A_i, B_i by combining

$\pm a_1^i$ (or $\pm a_2^i$) and $\pm b_1^i$ and by setting the remaining entries to zero. For example, a vertex in the first polytope is

$$\left[\begin{pmatrix} -1 & 0 \\ 0 & 0 \end{pmatrix}, \begin{pmatrix} 0.2 \\ 0 \end{pmatrix} \right]. \tag{13.61}$$

Solving the GEVP (13.56) for a constant state-feedback controller:

$$u(k) = K\mathbf{x}(k),$$

yields $\eta^{con} = 29.2851$, hence the corresponding maximum quadratic stability margin $\gamma^{con} = \dfrac{1}{\eta^{con}} = 0.0341$ and the constant state feedback gain matrix:

$$K = \begin{pmatrix} 0.5711 & 1.6559 \end{pmatrix}.$$

Now solving the GEVP (13.60) for a fuzzy state-feedback controller:

$$u(k) = [\lambda_1(k)K_1 + \lambda_2(k)K_2]\mathbf{x}(k),$$

yields $\eta^{con} = 3.0605$, hence a quadratic stability margin: $\gamma^{con} = 1/\eta^{con} = 0.3267$, and feedback gain matrices:

$$\begin{aligned} K_1 &= \begin{pmatrix} -2.3560 & 2.2637 \end{pmatrix}, \\ K_2 &= \begin{pmatrix} 2.9673 & 1.1212 \end{pmatrix}. \end{aligned}$$

It can be seen that the stability margin γ^{con} is 10 times higher for a fuzzy controller than for a constant one. This shows that fuzzy control systems with fuzzy controllers can be designed with more robust stability with respect to consequent uncertainties.

The following experiments have been performed in order to verify the robust stability of the above fuzzy control systems. The nominal values of a_1^i, a_2^i and b_1^i are perturbed respectively by Δa_1^i, Δa_2^i and Δb_1^i, considering the following cases shown in Table 13.3.

All these perturbations can be reduced within the considered polytopic uncertainties. Figures 13.10(a) and 13.10(b) show simulation results when applying either the robust fuzzy controller or the robust constant controller to the nominal model and the perturbed models, for the same initial conditions as in Section 13.4.2. Results show the robustness of the fuzzy controller, whereas some perturbations make the fuzzy control system with the constant controller unstable. For comparison, Figure 13.10(c) shows simulation results when applying Fuzzy Controller I obtained in Section 13.4.2 to perturbed systems. It turns out that for some perturbations, the response is oscillatory with very low damping, hence the control system is at least close to the instability boundary.

Table 13.3 Perturbations on the matrices A_i and B_i with $\gamma = 0.3267$.

	I	II	III	IV	V	VI	VII	VIII
ΔA_1	γa_1^1	γa_1^1	$-\gamma a_1^1$	$-\gamma a_1^1$	γa_2^1	γa_2^1	$-\gamma a_2^1$	$-\gamma a_2^1$
ΔB_1	γb_1^1	$-\gamma b_1^1$	$-\gamma b_1^1$	γb_1^1	γb_1^1	$-\gamma b_1^1$	$-\gamma b_1^1$	γb_1^1
ΔA_2	γa_2^2	γa_2^2	$-\gamma a_1^2$	$-\gamma a_2^2$	γa_1^2	$-\gamma a_1^2$	γa_2^2	$-\gamma a_2^2$
ΔB_2	γb_1^2	$-\gamma b_1^2$	γb_1^2	$-\gamma b_1^2$	$-\gamma b_1^2$	$-\gamma b_1^2$	γb_1^2	γb_1^2

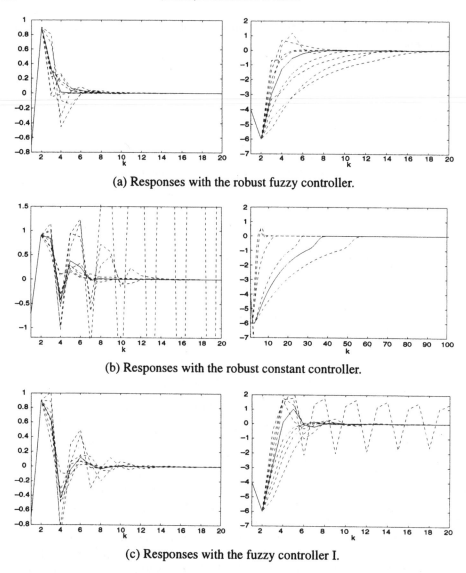

(a) Responses with the robust fuzzy controller.

(b) Responses with the robust constant controller.

(c) Responses with the fuzzy controller I.

Figure 13.10   Responses of fuzzy control systems: nominal model (full lines) and perturbed models (dashed lines).

Example 2

When there are only premise uncertainties, a constant state-feedback controller is no doubt the best choice for the purpose of robust stability if any, (see Remark 3). However, it is not exceptional that the sufficient stability condition used here does not lead to a constant state feedback controller which is able to stabilise simultaneously all the different sub-models involved in the LTS fuzzy model. This is the case when substituting

$$A_1 = \begin{pmatrix} 1.0 & 0.5 \\ 1.0 & 0 \end{pmatrix} \tag{13.62}$$

for the original matrix A_1 in the previous example. Now, a fuzzy state-feedback controller is an option for the achievement of a feasible robust fuzzy control system.

For bounded premise uncertainties such as defined by (13.48), solving the optimisation problem (13.60) yields the fuzzy state-feedback gain matrices:

$$K_1 = (\begin{array}{cc} -7.7485 & -2.8428 \end{array}),$$
$$K_2 = (\begin{array}{cc} 3.8867 & 2.2958 \end{array}),$$

and the maximum stability margin:

$$\gamma^{pre} = \frac{1}{\eta^{pre}} = 0.3146.$$

In order to check the robust stability of the fuzzy control system, the nominal model has been perturbed by premise uncertainties related to the fuzzy partition of the premise variable $y(k-1)$ (see Figure 13.11), bounded by polytopes with $\gamma^{pre} = 0.3146$. The robustness of the fuzzy control system associated with the nominal model and such premise uncertainties is illustrated by simulation results in Figure 13.12.

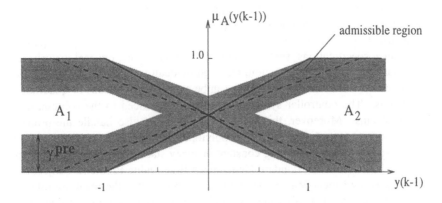

Figure 13.11 Uncertainties in membership function: original membership functions (full lines), perturbed membership functions (dashed lines) and admissible region (shaded part).

Remark 6: Constant and fuzzy state feedback controllers are considered independently for systems with either premise or consequent uncertainties. Experience reveals that the exactness of the membership functions is of less importance for TS fuzzy models (Johansen 1994), so it may be more realistic in practice to design a robust fuzzy controller with respect to the consequent uncertainties.

13.6 CONCLUSIONS

This chapter has presented an approach to the design of state-feedback controllers based on linear Takagi–Sugeno fuzzy models of the system to be controlled. First, a procedure combining a modified Fuzzy C-Elliptotype algorithm and compatibility analysis has

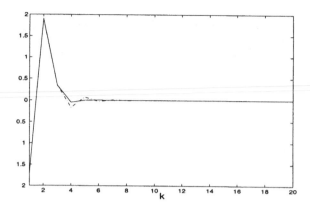

Figure 13.12 Simulation: the nominal fuzzy control system (full line) and the perturbed fuzzy control system (dashed line).

been proposed for the identification of such models. The procedure has been illustrated by identifying a simple fuzzy model of an industrial glass furnace. The model-based control design technique uses Linear Matrix Inequality methods and allows the search for linear time-invariant or fuzzy stabilising controllers. In addition, some design requirements and performance specifications, such as minimum response time, optimal disturbance rejection, limitations on the control action, etc., can be dealt with in the design procedure. The only condition is that the specifications can be expressed by means of quadratic Lyapunov functions. The controller synthesis can then be reduced to the solution of standard LMI problems. Moreover, this synthesis method can also handle the problem of robustness of fuzzy control systems. With polytopic definitions for premise and consequent uncertainties, robust stabilising constant or fuzzy state-feedback controllers can be derived from the solutions of standard Generalised EigenValue Problems subject to constraints expressed by Linear Matrix Inequalities. Consequently, the resulting robust controllers provide maximum stability margins with respect to premise or consequent uncertainties.

The synthesis technique presented here is systematic and effective. It starts with a procedure for determining a common positive matrix P, which guarantees the stability of the resulting fuzzy control system. In approaches such as in (Tanaka and Sugeno 1992, Tanaka and Sano 1994), there is basically no guide on how to construct such a common positive definite matrix, and heavy computations and numerous trials are needed for checking the stability of the control system. Since it is based on a state-space representation of linear Takagi–Sugeno fuzzy models, the design method proposed in this chapter has been applied to simple models whose consequents involve merely the current value of the control variable but not its past values. However, this restriction could be removed by defining an augmented state space representation with auxiliary state variables which are the past values of the control variable. The price to pay is an increase of the computational burden since the particular structure properties of the augmented system and input matrices are not taken into account in the design procedure.

Simulation results presented in this chapter show that a fuzzy state-feedback controller is capable of superior performance compared to a constant controller; even more there may be cases where it is not possible to find a constant controller able to stabilise simultaneously all the different sub-models involved in the fuzzy model, while a fuzzy controller exists. This is due to the fact that the latter, consisting of a set of different "local" state-feedback controllers, can pay more attention to the different "local" behaviours of the controlled model. Eventually, it should be pointed out that since the approach is based on stability analysis and design in terms of quadratic Lyapunov functions and the potential effects of the fuzzy weights on the stability are not taken into account, the solution may be too conservative. Achieving closed-loop stability by means of a common positive definite matrix P is sufficient, but not necessary. It may happen that the loop is stable even if one was not able to find such a matrix P, see e.g. (Tanaka and Sano 1994). Nevertheless, through this approach, the powerful capability of Takagi–Sugeno fuzzy models to represent complex systems is combined with the well-developed linear control theory and its available design methods, which opens a new feasible way for the control of complex nonlinear systems.

REFERENCES

BEZDEK, J., C. CORAY, R. GUNDERSON AND J. WATSON (1981) 'Detection and characterization of cluster substructure: I. Linear structure: fuzzy c-lines, II. Fuzzy c-varieties and convex combinations thereof'. *SIAM J. Appl. Math.* **40**(2), 339–372.

BEZDEK, J. (ED.) (1981) *Pattern Recognition with Fuzzy Objective Function Algorithms*. Plenum Press. New York.

BOYD, S., L. EL GHAOUI, E. FERON AND V. BALAKRISHNAN (1994) *Linear Matrix Inequalities in System and Control Theory*. SIAM. Philadelphia, PA, USA.

BOYD, S., V. BALAKRISHNAN, E. FERON AND L. EL GHAOUI (1993) Control system analysis and synthesis via linear matrix inequalities. In 'Proc. of American Control Conference, ACC'93'. pp. 2147–2154.

GAHINET, P. AND A. NEMIROVSKII (1993) *LMI Lab: A package for manipulating and solving LMI's, Version 2.0*.

JOHANSEN, T. (1994) 'Fuzzy model based control: stability, robustness and performance issues'. *IEEE Trans. on Fuzzy Systems* **2**(3), 221–234.

LEE, C. (1990) 'Fuzzy logic in control systems: fuzzy logic controller, Part I, Part II'. *IEEE Trans. on Systems, Man, and Cybernetics* **20**(2), 404–435.

PASSINO, K. AND S. YURKOVICH (1996) Fuzzy control. To appear in 'Handbook on Control', W. Levine (Ed.), CRC Press, Boca Raton, FL, pp. 1001–1017.

TAKAGI, T. AND M. SUGENO (1985) 'Fuzzy identification of systems and its application to modeling and control'. *IEEE Trans. on Systems, Man, and Cybernetics* **SMC-15**(1), 116–132.

TANAKA, K. AND M. SANO (1994) 'A robust stabilization problem of fuzzy control systems and its application to backing up control of a truck-trailer'. *IEEE Trans. on Fuzzy Systems* **2**(2), 119–134.

TANAKA, K. AND M. SUGENO (1992) 'Stability analysis and design of fuzzy control systems'. *Fuzzy Sets and Systems* **45**, 135–156.

WANG, L. (1993) 'Stable adaptive fuzzy control of nonlinear systems'. *IEEE Trans. on Fuzzy Systems* **1**(2), 146–155.

YOSHINARI, Y., W. PEDRYCZ AND K. HIROTA (1993) 'Construction of fuzzy models through clustering techniques'. *Fuzzy Sets and Systems* **54**, 157–165.

ZHAO, J. (1995) System Modeling, Identification and Control Using Fuzzy Logic. PhD thesis. Université Catholique de Louvain, Belgium.

ZHAO, J., V. WERTZ AND R. GOREZ (1994) A fuzzy clustering method for the identification of fuzzy models for dynamic systems. In 'Proc. of 9th IEEE International Symposium on Intelligent Control, ISIC'94'. Columbus, Ohio, USA. pp. 172–177.

ZHAO, J., V. WERTZ AND R. GOREZ (1995a) Design a stabilizing fuzzy and/or non-fuzzy state-feedback controller using LMI method. In 'Proc. of 3rd European Control Conference, ECC'95'. Rome, Italy. pp. 1201–1206.

ZHAO, J., V. WERTZ AND R. GOREZ (1995b) Linear TS models based robust stabilizing controller design. In 'Proc. of 34th Conference on Decision and Control, CDC'95'. New Orleans, LA, USA. pp. 255–260.

ZHAO, J., V. WERTZ AND R. GOREZ (1996) Dynamic fuzzy state-feedback controller and its limitations. Preprints of the 13th IFAC World Congress, IFAC'96. San Francisco, CA. **F**, pp. 1201–1206.

Index

T - #0049 - 071024 - C0 - 254/178/19 [21] - CB - 9780748405954 - Gloss Lamination